New Trends in Quantum Electrodynamics

New Trends in Quantum Electrodynamics

Special Issue Editor
Roberto Passante

MDPI • Basel • Beijing • Wuhan • Barcelona • Belgrade • Manchester • Tokyo • Cluj • Tianjin

Special Issue Editor
Roberto Passante
Dipartimento di Fisica e Chimica—Emilio Segrè,
Università degli Studi di Palermo
Italy

Editorial Office
MDPI
St. Alban-Anlage 66
4052 Basel, Switzerland

This is a reprint of articles from the Special Issue published online in the open access journal *Symmetry* (ISSN 2073-8994) (available at: https://www.mdpi.com/journal/symmetry/special_issues/New_trends_Quantum_Electrodynamics).

For citation purposes, cite each article independently as indicated on the article page online and as indicated below:

LastName, A.A.; LastName, B.B.; LastName, C.C. Article Title. *Journal Name* **Year**, *Article Number*, Page Range.

ISBN 978-3-03928-524-2 (Hbk)
ISBN 978-3-03928-525-9 (PDF)

© 2020 by the authors. Articles in this book are Open Access and distributed under the Creative Commons Attribution (CC BY) license, which allows users to download, copy and build upon published articles, as long as the author and publisher are properly credited, which ensures maximum dissemination and a wider impact of our publications.

The book as a whole is distributed by MDPI under the terms and conditions of the Creative Commons license CC BY-NC-ND.

Contents

About the Special Issue Editor vii

Preface to "New Trends in Quantum Electrodynamics" ix

Roberto Passante
Dispersion Interactions between Neutral Atoms and the Quantum Electrodynamical Vacuum
Reprinted from: *Symmetry* **2018**, *10*, 735, doi:10.3390/sym10120735 1

David L. Andrews
Symmetries, Conserved Properties, Tensor Representations, and Irreducible Forms in Molecular Quantum Electrodynamics
Reprinted from: *Symmetry* **2018**, *10*, 298, doi:10.3390/sym10070298 35

Alessandro Sergi, Gabriel Hanna, Roberto Grimaudo and Antonino Messina
Quasi-Lie Brackets and the Breaking of Time-Translation Symmetry for Quantum Systems Embedded in Classical Baths
Reprinted from: *Symmetry* **2018**, *10*, 518, doi:10.3390/sym10100518 65

Hidemasa Yamane and Satoshi Tanaka
Ultrafast Dynamics of High-Harmonic Generation in Terms of Complex Floquet Spectral Analysis
Reprinted from: *Symmetry* **2018**, *10*, 313, doi:10.3390/sym10080313 93

Stefan Yoshi Buhmann and A. Salam
Three-Body Dispersion Potentials Involving Electric Octupole Coupling
Reprinted from: *Symmetry* **2018**, *10*, 343, doi:10.3390/sym10080343 111

Wenting Zhou, Roberto Passante and Lucia Rizzuto
Resonance Dipole–Dipole Interaction between Two Accelerated Atoms in the Presence of a Reflecting Plane Boundary
Reprinted from: *Symmetry* **2018**, *10*, 185, doi:10.3390/sym10060185 133

About the Special Issue Editor

Roberto Passante is Professor of Physics at the University of Palermo, Italy. His main research interests are currently in the fields of quantum electrodynamics and quantum optics, in particular Casimir and Casimir-Polder interactions, the Unruh effect, cavity quantum electrodynamics, quantum optomechanics, and atomic radiative processes in static or dynamical external environments and backgrounds.

Preface to "New Trends in Quantum Electrodynamics"

Quantum electrodynamics is one of the most successful physical theories, and its predictions agree with experimental results with exceptional accuracy. Nowadays, after several decades since its introduction, quantum electrodynamics is still a very active research field from both the theoretical and experimental points of view. The aim of this Special Issue is to present recent relevant advances in quantum electrodynamics, both theoretical and experimental, and related aspects in quantum field theory and quantum optics.

Roberto Passante
Special Issue Editor

Review

Dispersion Interactions between Neutral Atoms and the Quantum Electrodynamical Vacuum

Roberto Passante [1,2]

[1] Dipartimento di Fisica e Chimica, Università Degli Studi di Palermo, via Archirafi 36, I-90123 Palermo, Italy; roberto.passante@unipa.it
[2] INFN, Laboratori Nazionali del Sud, I-95123 Catania, Italy

Received: 30 October 2018; Accepted: 28 November 2018; Published: 10 December 2018

Abstract: Dispersion interactions are long-range interactions between neutral ground-state atoms or molecules, or polarizable bodies in general, due to their common interaction with the quantum electromagnetic field. They arise from the exchange of virtual photons between the atoms, and, in the case of three or more atoms, are not additive. In this review, after having introduced the relevant coupling schemes and effective Hamiltonians, as well as properties of the vacuum fluctuations, we outline the main properties of dispersion interactions, both in the nonretarded (van der Waals) and retarded (Casimir–Polder) regime. We then discuss their deep relation with the existence of the vacuum fluctuations of the electromagnetic field and vacuum energy. We describe some transparent physical models of two- and three-body dispersion interactions, based on dressed vacuum field energy densities and spatial field correlations, which stress their deep connection with vacuum fluctuations and vacuum energy. These models give a clear insight of the physical origin of dispersion interactions, and also provide useful computational tools for their evaluation. We show that this aspect is particularly relevant in more complicated situations, for example when macroscopic boundaries are present. We also review recent results on dispersion interactions for atoms moving with noninertial motions and the strict relation with the Unruh effect, and on resonance interactions between entangled identical atoms in uniformly accelerated motion.

Keywords: Casimir–Polder interactions; van der Waals forces; vacuum fluctuations; vacuum energy; many-body dispersion interactions

1. Introduction

Van der Waals and Casimir–Polder dispersion interactions are long-range interactions between two or more neutral atoms or molecules in the vacuum space, arising from their common interaction with the quantum electromagnetic field [1–3]. A related effect is the atom–surface Casimir–Polder force between a neutral atom and a conducting or dielectric surface [4]. A complete description of such interactions requires a quantum description of both matter and radiation. These forces are non additive, that is, the interaction between three on more atoms is not simply the sum of pairwise interactions; non-additive terms are present, involving coordinates of all atoms [5–7]. Nonadditive contributions are usually small for dilute systems, but they can become relevant for dense systems [8,9]. In this paper, we review some properties of two- and three-body dispersion interactions between neutral atoms, both in the nonretarded (van der Waals) and in the retarded (Casimir–Polder) regime, and recent advances in this subject. We will give a particular emphasis on physical models stressing their relation with the zero-point fluctuations of the electromagnetic radiation field and vacuum energy. We will also briefly review some recent results for atoms in an excited state and when a boundary condition such as a conducting plane boundary is present, as well as dispersion and resonance interactions for atoms in noninertial motion. All of these results show that zero-point field fluctuations, and their change consequent to the presence of macroscopic magnetodielectric bodies or to a noninertial motion, can be probed through dispersion forces on atoms or, in general, polarizable bodies. This review mainly deals with two- and three-body dispersion interactions between atoms, in several physical situations, and their physical origin in terms of the zero-point fluctuations of the electromagnetic field; other relevant aspects of dispersion interactions, for example the effect of magnetodielectric bodies on these interactions in the framework of macroscopic quantum electrodynamics, have been reviewed in [10,11].

This review is organized as follows. In Section 2, we introduce the minimal and multipolar atom-field coupling schemes, which will be used to calculate the dispersion interactions. In Section 3, we derive effective Hamiltonians that allow considerable simplification of the calculations, in particular for three- and many-body dispersion interactions, or in the presence of macroscopic boundaries. Section 4 is dedicated to a brief discussion of the vacuum fluctuations, and vacuum energy densities, of the quantum electromagnetic field. Section 5 deals with the two-body van der Waals and Casimir–Polder dispersion interactions, while Section 6 is dedicated to three-body forces. In Section 7, two- and three-body dispersion interactions are discussed in detail, in relation to physical models of dispersion forces based on dressed vacuum field energy densities, and on bare and dressed vacuum field spatial correlations. In Section 8, these results will be extended to the case in which a reflecting boundary is present. Finally, in Section 9, dispersion and resonance interactions between atoms in noninertial motion are discussed, as well as recent proposals to exploit these interactions as an indirect signature of the Unruh effect. Section 10 is devoted to our final remarks.

2. Atom-Field Interaction Hamiltonian: Minimal and Multipolar Coupling

In nonrelativistic quantum electrodynamics, the (transverse) vector potential field operator, using Gauss units, is given by [12,13]

$$\mathbf{A}(\mathbf{r},t) = \sum_{\mathbf{k}\lambda} \sqrt{\frac{2\pi\hbar c^2}{\omega_k V}} \hat{\mathbf{e}}_{\mathbf{k}\lambda} \left(a_{\mathbf{k}\lambda}(t) e^{i\mathbf{k}\cdot\mathbf{r}} + a^\dagger_{\mathbf{k}\lambda}(t) e^{-i\mathbf{k}\cdot\mathbf{r}} \right), \tag{1}$$

where we have used the Coulomb gauge, $\nabla \cdot \mathbf{A}(\mathbf{r},t) = 0$, the polarization unit vectors $\hat{\mathbf{e}}_{\mathbf{k}\lambda}$ ($\lambda = 1,2$) are assumed real, $\omega_k = ck$ in the vacuum space, V is the quantization volume, and $a_{\mathbf{k}\lambda}$ and $a^\dagger_{\mathbf{k}\lambda}$ are respectively annihilation and creation operators satisfying the usual bosonic commutation rules. Due to the Coulomb gauge condition $\nabla \cdot \mathbf{A} = 0$, we have $\mathbf{k} \cdot \hat{\mathbf{e}}_{\mathbf{k}\lambda} = 0$ for all $\mathbf{k}$. The transverse electric field and the magnetic field are given by

$$\mathbf{E}_\perp(\mathbf{r},t) = -\frac{1}{c}\dot{\mathbf{A}}(\mathbf{r},t) = i\sum_{\mathbf{k}\lambda} \sqrt{\frac{2\pi\hbar\omega_k}{V}} \hat{\mathbf{e}}_{\mathbf{k}\lambda} \left(a_{\mathbf{k}\lambda}(t) e^{i\mathbf{k}\cdot\mathbf{r}} - a^\dagger_{\mathbf{k}\lambda}(t) e^{-i\mathbf{k}\cdot\mathbf{r}} \right), \tag{2}$$

$$\mathbf{B}(\mathbf{r},t) = \nabla \times \mathbf{A}(\mathbf{r},t) = i\sum_{\mathbf{k}\lambda} \sqrt{\frac{2\pi\hbar c^2}{\omega_k V}} \mathbf{k} \times \hat{\mathbf{e}}_{\mathbf{k}\lambda} \left(a_{\mathbf{k}\lambda}(t) e^{i\mathbf{k}\cdot\mathbf{r}} - a^\dagger_{\mathbf{k}\lambda}(t) e^{-i\mathbf{k}\cdot\mathbf{r}} \right), \tag{3}$$

where the subscript $\perp$ indicates the transverse part. The vacuum state is defined by $a_{\mathbf{k}\lambda} \mid \{0_{\mathbf{k}\lambda}\} \rangle = 0$, for all $(\mathbf{k}\lambda)$. In the presence of boundaries or macroscopic bodies, appropriate boundary conditions must be set on the field operators, and the dispersion relation can change too (in a photonic crystal, for example). The vacuum state is thus strongly dependent on the presence of microscopic or macroscopic matter.

The Hamiltonian of the free field is

$$H_F = \frac{1}{8\pi} \int_V d^3r \left[\mathbf{E}_\perp^2(\mathbf{r}) + \mathbf{B}^2(\mathbf{r}) \right] = \sum_{\mathbf{k}\lambda} \hbar\omega_k \left(a^\dagger_{\mathbf{k}\lambda} a_{\mathbf{k}\lambda} + \frac{1}{2} \right). \tag{4}$$

In the presence of neutral atoms, the total Hamiltonian is the sum of three terms, H_F, H_{atoms} and H_I, respectively the field, atomic and interaction Hamiltonians. In order to obtain the full Hamiltonian and discuss different matter–radiation coupling schemes, we start with the Lagrangian formalism in the Coulomb gauge. The generalized coordinates are the coordinates $\mathbf{r}_\xi$ of the charged particles (the subscript ξ indicates the particles present), the vector potential $\mathbf{A}(\mathbf{r},t)$ and the scalar potential $\phi(\mathbf{r},t)$. In the Coulomb gauge, only the vector potential is second-quantized, while the scalar potential is described as a classical function. The Helmholtz theorem allows a unique separation, with given boundary conditions at infinity, of the electric field $\mathbf{E}(\mathbf{r},t)$ in a transverse (solenoidal) part $\mathbf{E}_\perp(\mathbf{r},t)$, with $\nabla \cdot \mathbf{E}_\perp(\mathbf{r},t) = 0$, and a longitudinal (irrotational) part $\mathbf{E}_\parallel(\mathbf{r},t)$, with $\nabla \times \mathbf{E}_\parallel(\mathbf{r},t) = 0$,

$$\mathbf{E}(\mathbf{r},t) = \mathbf{E}_\perp(\mathbf{r},t) + \mathbf{E}_\parallel(\mathbf{r},t); \quad \mathbf{E}_\parallel(\mathbf{r},t) = -\nabla \phi(\mathbf{r},t); \quad \mathbf{E}_\perp(\mathbf{r},t) = -\frac{1}{c}\dot{\mathbf{A}}(\mathbf{r},t). \tag{5}$$

The magnetic field is completely transverse, due to the Maxwell's equation $\nabla \cdot \mathbf{B}(\mathbf{r},t) = 0$.

For point particles with electric charge q_ξ, position $\mathbf{r}_\xi$ and mass m_ξ, we define the charge and current densities

$$\rho(\mathbf{r},t) = \sum_\xi q_\xi \delta(\mathbf{r} - \mathbf{r}_\xi); \quad \mathbf{j}(\mathbf{r},t) = \sum_\xi q_\xi \dot{\mathbf{r}}_\xi \delta(\mathbf{r} - \mathbf{r}_\xi). \tag{6}$$

The current density can be separated in its transverse and longitudinal parts, $\mathbf{j}(\mathbf{r},t) = \mathbf{j}_\perp(\mathbf{r},t) + \mathbf{j}_\parallel(\mathbf{r},t)$. The Lagrangian of the coupled system is [13–15]

$$\begin{aligned} L^{\min}(\mathbf{r}_\xi, \mathbf{A}, \phi) &= \int d^3 r \mathcal{L}^{\min}(\mathbf{r}) = \sum_\xi \frac{1}{2} m_\xi \dot{\mathbf{r}}_\xi^2 + \frac{1}{8\pi}\int d^3 r \left(\frac{1}{c^2}\dot{\mathbf{A}}^2(\mathbf{r}) - (\nabla \times \mathbf{A}(\mathbf{r}))^2 \right) \\ &+ \frac{1}{8\pi}\int d^3 r \, |\nabla \phi(\mathbf{r})|^2 - \int d^3 r \, \rho(\mathbf{r})\phi(\mathbf{r}) + \frac{1}{c}\int d^3 r \, \mathbf{j}_\perp(\mathbf{r}) \cdot \mathbf{A}(\mathbf{r}), \end{aligned} \tag{7}$$

where $\mathcal{L}^{\min}(\mathbf{r})$ indicates the Lagrangian density. The last two terms in the second line of Equation (7) give the matter-field interaction. This Lagrangian yields the correct Maxwell-Lorentz equations of motion. It defines the so-called minimal coupling scheme. The momenta conjugate to $\mathbf{r}_\xi$ and $\mathbf{A}(\mathbf{r})$ are

$$\begin{aligned} \mathbf{p}_\xi^{\min} &= \frac{\partial L^{\min}}{\partial \dot{\mathbf{r}}_\xi} = m_\xi \dot{\mathbf{r}}_\xi + \frac{q_\xi}{c}\mathbf{A}(\mathbf{r}_\xi), \\ \mathbf{\Pi}^{\min}(\mathbf{r}) &= \frac{\partial \mathcal{L}^{\min}}{\partial \dot{\mathbf{A}}} = \frac{1}{4\pi c^2}\dot{\mathbf{A}}(\mathbf{r}) = -\frac{1}{4\pi c}\mathbf{E}_\perp(\mathbf{r}), \end{aligned} \tag{8}$$

while the momentum conjugate to the scalar potential $\phi(\mathbf{r},t)$ vanishes.

We can now obtain the Hamiltonian of the interacting system in the Coulomb gauge

$$\begin{aligned} H^{\min}(\mathbf{r}_\xi, \mathbf{A}, \mathbf{p}_\xi^{\min}, \mathbf{\Pi}^{\min}) &= \sum_\xi \mathbf{p}_\xi^{\min} \cdot \dot{\mathbf{r}}_\xi + \int d^3 r \, \mathbf{\Pi}^{\min}(\mathbf{r}) \cdot \dot{\mathbf{A}}(\mathbf{r}) - L^{\min} \\ &= \sum_\xi \frac{1}{2m_\xi}\left(\mathbf{p}_\xi^{\min} - \frac{q_\xi}{c}\mathbf{A}(\mathbf{r}_\xi) \right)^2 + \frac{1}{8\pi}\int d^3 r \left((4\pi c)^2 (\mathbf{\Pi}^{\min}(\mathbf{r}))^2 + (\nabla \times \mathbf{A}(\mathbf{r}))^2 \right) + \frac{1}{8\pi}\int d^3 r \, |\nabla \phi(\mathbf{r})|^2. \end{aligned} \tag{9}$$

The last term includes all electrostatic (longitudinal) interactions: in the case of neutral atoms, they are the interactions between the atomic nucleus and the electrons of the same atom, eventually included in the Hamiltonians of the single atoms, as well as electrostatic (dipole–dipole, for example) interactions between different atoms, yielding nonretarded London–van der Waals forces [15,16]. Thus, in the minimal coupling scheme, electromagnetic interactions between atoms are mediated by both longitudinal and transverse fields.

In the case of two neutral one-electron atoms, A and B, respectively located at the fixed positions $\mathbf{r}_A$ and $\mathbf{r}_B$, and within dipole approximation, Equation (9), using Equation (8), assumes the following form [13,15]:

$$H^{\min} = H_A^{\min} + H_B^{\min} + H_F^{\min} + H_I^{\min} + H_I^{dd}, \tag{10}$$

where H_A^{min} and H_B^{min} are respectively the Hamiltonian of atoms A and B, H_F^{min} is the free transverse field Hamiltonian

$$H_F^{min} = \frac{1}{8\pi} \int d^3r \left(\mathbf{E}_\perp^2(\mathbf{r}) + \mathbf{B}^2(\mathbf{r}) \right), \tag{11}$$

and H_I^{min} and H_I^{dd} are, respectively, the interaction terms of the two atoms with the transverse field and the electrostatic dipole–dipole interaction between the atoms. They are given by

$$H_I^{min} = \sum_{\xi=A,B} \left(\frac{e}{mc} \mathbf{p}_\xi \cdot \mathbf{A}(\mathbf{r}_\xi) + \frac{e^2}{2mc^2} \mathbf{A}^2(\mathbf{r}_\xi) \right), \tag{12}$$

$$H_I^{dd} = \frac{1}{R^3} \left(\boldsymbol{\mu}_A \cdot \boldsymbol{\mu}_B - 3(\boldsymbol{\mu}_A \cdot \hat{\mathbf{R}})(\boldsymbol{\mu}_B \cdot \hat{\mathbf{R}}) \right), \tag{13}$$

where $-e$ is the electron charge, $\boldsymbol{\mu}_A = -e\mathbf{r}_A$ and $\boldsymbol{\mu}_B = -e\mathbf{r}_B$ are respectively the dipole moment operators of atoms A and B, and $\mathbf{R}$ is the distance between the two atoms. Extension to more-electrons atoms, and to the case of three or more atoms, is straightforward.

A different and useful form of the matter–radiation interaction can be obtained by exploiting gauge invariance; it is the so-called multipolar coupling scheme, and it is more convenient than the minimal-coupling one in evaluating radiation-mediated interactions between neutral atoms [14,17–20]. The advantage of the multipolar coupling scheme is that the interaction is expressed in terms of the transverse displacement field, and not the vector potential, and that the electrostatic interaction between the atoms is embedded in the coupling with the transverse fields. This in general allows considerable simplification of the calculations and a clear physical picture. The passage from the minimal to the multipolar coupling scheme can be obtained in two different, and equivalent, ways: by adding a total time-derivative to the Lagrangian or through a unitary transformation on the Hamiltonian [14,17–19,21–23]. In view of the relevance to dispersion interactions, we will now briefly outline this procedure, within dipole approximation.

We first introduce the atomic polarisation operator, in electric dipole approximation, assuming that the (multi-electron) atoms are well separated, i.e., their distance is much larger that the Bohr's radius,

$$\mathbf{P}(\mathbf{r}) = -\sum_\xi e_\xi \mathbf{r}_\xi \delta(\mathbf{r} - \mathbf{r}_\xi) = \sum_\ell \boldsymbol{\mu}_\ell \delta(\mathbf{r} - \mathbf{r}_\ell), \tag{14}$$

where the subscript ξ indicates the atomic electrons, with positions $\mathbf{r}_\xi$ with respect to their atomic nucleus. We have also introduced the atomic dipole moment operator of atom ℓ, $\boldsymbol{\mu}_\ell = -\sum_{\xi_\ell} e_{\xi_\ell} \mathbf{r}_{\xi_\ell}$, with $\mathbf{r}_{\xi_\ell}$ the positions of the electrons of atom ℓ. The longitudinal field of the bound charges in the atoms can be expressed as

$$\mathbf{E}_\parallel(\mathbf{r}) = -\nabla \phi(\mathbf{r}) = -4\pi \mathbf{P}_\parallel(\mathbf{r}), \tag{15}$$

where $\mathbf{P}_\parallel(\mathbf{r})$ is the longitudinal part of the polarization operator, $\mathbf{P}(\mathbf{r}) = \mathbf{P}_\perp(\mathbf{r}) + \mathbf{P}_\parallel(\mathbf{r})$. In addition, in the electric dipole approximation,

$$\mathbf{j}(\mathbf{r}, t) = \dot{\mathbf{P}}(\mathbf{r}, t). \tag{16}$$

The new Lagrangian, where a total time-derivative has been added to (7), is given by

$$\begin{aligned} L^{mult} &= \int d^3 r \mathcal{L}^{mult}(\mathbf{r}) = L^{min} - \frac{1}{c} \frac{d}{dt} \int d^3 r \mathbf{P}(\mathbf{r}) \cdot \mathbf{A}(\mathbf{r}) \\ &= \sum_\xi \frac{1}{2} m_\xi \dot{\mathbf{r}}_\xi^2 + \frac{1}{8\pi} \int d^3 r \left(\frac{1}{c^2} \dot{\mathbf{A}}^2(\mathbf{r}) - (\nabla \times \mathbf{A}(\mathbf{r}))^2 \right) + \frac{1}{8\pi} \int d^3 r \mid \nabla \phi(\mathbf{r}) \mid^2 - \int d^3 r \rho(\mathbf{r}) \phi(\mathbf{r}) \\ &\quad - \frac{1}{c} \int d^3 r \mathbf{P}_\perp(\mathbf{r}) \cdot \dot{\mathbf{A}}(\mathbf{r}), \end{aligned} \tag{17}$$

where $\mathcal{L}^{mult}(\mathbf{r})$ is the new Lagrangian density, and we have used Equations (9) and (15). The subscript ξ identifies the (point-like) atoms located at position $\mathbf{r}_\ell$. The new conjugate momenta are

$$\mathbf{p}_\xi^{\text{mult}} = \frac{\partial L^{\text{mult}}}{\partial \dot{\mathbf{r}}_\xi} = m_\xi \dot{\mathbf{r}}_\xi,$$

$$\mathbf{\Pi}^{\text{mult}}(\mathbf{r}) = \frac{\partial \mathcal{L}^{\text{mult}}}{\partial \dot{\mathbf{A}}} = \frac{1}{4\pi c^2}\dot{\mathbf{A}}(\mathbf{r}) - \frac{1}{c}\mathbf{P}_\perp(\mathbf{r}) = -\frac{1}{4\pi c}\mathbf{D}_\perp(\mathbf{r}), \tag{18}$$

where $\mathbf{D}_\perp(\mathbf{r})$ is the transverse part of the electric displacement field $\mathbf{D}(\mathbf{r}) = \mathbf{E}(\mathbf{r}) + 4\pi\mathbf{P}(\mathbf{r})$. Comparison of (18) with (8) shows that the momenta conjugate to the generalized coordinates, on which canonical commutation relations are imposed in the quantization procedure, are different in the two coupling schemes. In the multipolar coupling, the momentum conjugate to the vector potential has a contribution from the atomic polarization vector.

From the Lagrangian (17), and using (18), we obtain the Hamiltonian in the multipolar scheme within the dipole approximation,

$$\begin{aligned} H^{\text{mult}}(\mathbf{r}_\xi, \mathbf{A}, \mathbf{p}_\xi^{\text{mult}}, \mathbf{\Pi}^{\text{mult}}) &= \sum_\xi \mathbf{p}_\xi^{\text{mult}} \cdot \dot{\mathbf{r}}_\xi + \int d^3 r \mathbf{\Pi}^{\text{mult}}(\mathbf{r}) \cdot \dot{\mathbf{A}}(\mathbf{r}) - L^{\text{mult}} \\ &= \sum_\ell \left(\frac{1}{2m}\sum_{\xi\ell}(\mathbf{p}_{\xi\ell}^{\text{mult}})^2 + V_\ell \right) + \frac{1}{8\pi}\int d^3 r \left(\mathbf{D}_\perp^2(\mathbf{r}) + (\nabla \times \mathbf{A}(\mathbf{r}))^2 \right) \\ &\quad - \sum_\ell \boldsymbol{\mu}_\ell \cdot \mathbf{D}_\perp(\mathbf{r}_\ell) + 2\pi \sum_\ell \int d^3 r \mid \mathbf{P}_{\perp\ell}(\mathbf{r}) \mid^2, \end{aligned} \tag{19}$$

where V_ℓ contains the electrostatic nucleus–electrons interaction of each atom. The term containing the integral of $\mathbf{P}_{\perp\ell}^2(\mathbf{r})$ (the squared transverse polarization vector of atom ℓ) contributes only to single-atom self-interactions, and does not contribute to the interatomic interactions we will deal with in the next sections. Using the Hamiltonian (19), all interatomic interactions are mediated by the transverse field only, without electrostatic interatomic terms.

The passage from the minimal to the multipolar coupling scheme can be equivalently accomplished through a unitary transformation, the so-called Power–Zienau–Woolley transformation, given by [14,15,17–19,21,24]

$$S = \exp\left(\frac{i}{\hbar c}\int d^3 r \mathbf{P}_\perp(\mathbf{r}) \cdot \mathbf{A}(\mathbf{r})\right). \tag{20}$$

The action of this transformation on the transverse electric field operator is

$$S^{-1}\mathbf{E}_\perp(\mathbf{r})S = \mathbf{E}_\perp(\mathbf{r}) + 4\pi\mathbf{P}_\perp(\mathbf{r}) = \mathbf{D}_\perp(\mathbf{r}), \tag{21}$$

where the equal-time commutator between the transverse electric field and the vector potential has been used

$$[A_i(\mathbf{r},t), E_{\perp j}(\mathbf{r}',t)] = -4\pi i c\hbar \delta_{ij}^\perp(\mathbf{r}-\mathbf{r}'), \tag{22}$$

which can be obtained from the expressions (1) and (2) of the field operators, and where $\delta_{ij}^\perp(\mathbf{r})$ is the transverse Dirac delta function. Transverse and longitudinal Dirac delta functions, which allow for extracting the transverse and longitudinal parts of a vector field, are defined as

$$\begin{aligned} \delta_{ij}^\parallel(\mathbf{r}) &= \frac{1}{(2\pi)^3}\int d^3 k \hat{k}_i \hat{k}_j e^{i\mathbf{k}\cdot\mathbf{r}}, \\ \delta_{ij}^\perp(\mathbf{r}) &= \frac{1}{(2\pi)^3}\int d^3 k \left(\delta_{ij} - \hat{k}_i\hat{k}_j\right) e^{i\mathbf{k}\cdot\mathbf{r}}. \end{aligned} \tag{23}$$

In addition,

$$\delta_{ij}\delta(\mathbf{r}) = \delta_{ij}^\parallel(\mathbf{r}) + \delta_{ij}^\perp(\mathbf{r}). \tag{24}$$

When comparing the expressions of the minimal coupling Hamiltonian (9) with the multipolar coupling Hamiltonian (19), some important considerations are in order. As already mentioned, in the multipolar scheme, the momentum field conjugate to the vector potential, the transverse displacement field, is different than that in the minimal coupling scheme (i.e., the transverse electric field). When these Hamiltonians are quantized, the photons in the two schemes are thus different objects. In addition, in the multipolar scheme, the transverse displacement field and the magnetic field appear in the multipolar field Hamiltonian, while the minimal free-field Hamiltonian is expressed in terms of the transverse electric field and the magnetic field. Finally, only the transverse displacement field appears in the interaction Hamiltonian, as shown in Equation (19); no electrostatic interactions are present in the interaction term, contrarily to minimal coupling case (10). When the multipolar Hamiltonian is used, all interactions among atoms are mediated by the quantized transverse fields only, and this makes the multipolar coupling scheme much more convenient for calculating dispersion and resonance interactions between atoms. Another important advantage of the multipolar scheme is that the transverse displacement field $\mathbf{D}_\perp(\mathbf{r})$ is fully retarded, contrarily to the transverse electric field $\mathbf{E}_\perp(\mathbf{r})$, that, as it is well known, contains a non-retarded part [18,25]. This is also evident from the fact that

$$\begin{aligned}\mathbf{D}_\perp(\mathbf{r}) &= \mathbf{E}_\perp(\mathbf{r}) + 4\pi\mathbf{P}_\perp(\mathbf{r}) = \mathbf{E}_\perp(\mathbf{r}) + 4\pi\mathbf{P}(\mathbf{r}) - 4\pi\mathbf{P}_\parallel(\mathbf{r}) = \mathbf{E}_\perp(\mathbf{r}) + \mathbf{E}_\parallel(\mathbf{r}) + 4\pi\mathbf{P}(\mathbf{r}) \\ &= \mathbf{E}(\mathbf{r}) + 4\pi\mathbf{P}(\mathbf{r}),\end{aligned} \quad (25)$$

where we have used Equation (15). Taking into account Equation (14), at points $\mathbf{r}$ different from the positions $\mathbf{r}_\ell$ of the (point-like) atoms, we have

$$\mathbf{D}_\perp(\mathbf{r}) = \mathbf{E}(\mathbf{r}) \quad (\mathbf{r} \neq \mathbf{r}_\ell). \quad (26)$$

In other words, the transverse displacement field in all points of space but the atomic positions coincides with the total (transverse plus longitudinal) electric field, which is a retarded field. This is particularly relevant when issues related to field propagation and causality are considered [25–27].

3. Effective Hamiltonians

Dispersion interactions between two atoms (van der Waals and Casimir–Polder) are effects starting at fourth order in the radiation-matter coupling, and three-body dispersion interactions start at sixth-order [24,28–31]. Application of fourth- or higher-order perturbation theory yields a large number of diagrams, and considerable amount of calculations. It is therefore desirable obtaining an effective Hamiltonian that could simplify calculations, reducing the number of diagrams. In this section, we introduce an effective Hamiltonian, with an interaction term quadratic in the atom-field coupling (i.e., the electric charge), which thus allows to halve the perturbative order necessary to evaluate dispersion interactions. For instance, it allows for evaluating the two-body dispersion potential through a second-order calculation and the non-additive three-body potential through a third-order calculation, in the case of ground-state atoms or molecules. This effective Hamiltonian is expressed in terms of the dynamical polarizability of the atoms.

We start with the multipolar Hamiltonian (19), considering for simplicity only one atom (A), and neglecting the term quadratic in the atomic transverse polarization vector, which does not play any role for the considerations that follow. Our Hamiltonian is thus $H = H_A + H_F - \boldsymbol{\mu}_A \cdot \mathbf{D}_\perp(\mathbf{r}_A)$, where $\mathbf{r}_A$ is the position of the atom. We now perform the following transformation on the Hamiltonian

$$\begin{aligned}H_{\text{new}} &= e^{iQ}He^{-iQ} = H_A + H_F - \boldsymbol{\mu}_A \cdot \mathbf{D}_\perp(\mathbf{r}) + i[Q, H_A + H_F] - \frac{1}{2}[Q, [Q, H_A + H_F]] \\ &\quad + i[Q, -\boldsymbol{\mu}_A \cdot \mathbf{D}_\perp(\mathbf{r}_A)] + \ldots,\end{aligned} \quad (27)$$

where we have used the Baker–Hausdorff relation, and the operator Q is chosen in such a way to eliminate the term linear in the coupling, that is,

$$i[Q, H_A + H_F] = \boldsymbol{\mu}_A \cdot \mathbf{D}_\perp(\mathbf{r}_A). \tag{28}$$

This condition allows for obtaining the generic matrix elements of Q between atom+field states, except on the energy shell where they are undefined. Substitution of (28) into (27) finally yields an effective Hamiltonian that is quadratic in the electric charge because the linear terms have been eliminated by the transformation. For ground-state atoms with random orientations, the effective Hamiltonian has an interaction term of the following form [32,33]

$$H_{\text{int}} = -\frac{1}{2} \sum_{\mathbf{k}\lambda} \alpha_A(k) \mathbf{D}_\perp(\mathbf{k}\lambda, \mathbf{r}_A) \cdot \mathbf{D}_\perp(\mathbf{r}_A), \tag{29}$$

where $\mathbf{D}_\perp(\mathbf{k}\lambda, \mathbf{r})$ are the Fourier components of the transverse displacement field, $\mathbf{D}_\perp(\mathbf{r}) = \sum_{\mathbf{k}\lambda} \mathbf{D}_\perp(\mathbf{k}\lambda, \mathbf{r})$, and

$$\alpha_A(k) = \frac{2}{3} \sum_m \frac{E_{mg} |\boldsymbol{\mu}_A^{mg}|^2}{E_{mg}^2 - \hbar^2 c^2 k^2} \tag{30}$$

is the dynamical polarizability of the isotropic atom. In Equation (30), g indicates the ground state of the atom with energy E_g, m a generic atomic state with energy E_m. In addition, $E_{mg} = E_m - E_g$, and $\boldsymbol{\mu}_A^{mg}$ are matrix elements of the dipole moment operator between states m and g.

The effective Hamiltonian (29) has a clear classical-type interpretation: any $(\mathbf{k}\lambda)$ component of the field induces a dipole moment in the atom with Fourier components determined by its dynamical polarizability $\alpha_A(k)$ at the frequency ck (linear response); this induced dipole moment, then, interacts with the field at the atom's position (the factor $1/2$ in (29) appears because it is an interaction energy between induced, and not permanent, dipoles).

If all relevant field modes have a frequency $\omega_k = ck$ much smaller than the relevant atomic transition frequencies, that is, $ck \ll E_{mg}/\hbar$, in (29), after taking into account the expression (30), we can replace the dynamical polarizability with the static one, $\alpha_A = \alpha_A(0)$, obtaining the so-called Craig–Power effective Hamiltonian [34]

$$H_{\text{int}} = -\frac{1}{2} \alpha_A D_\perp^2(\mathbf{r}_A). \tag{31}$$

The effective Hamiltonian (31) is valid whenever only low-frequency field modes are relevant, for example in the case of dispersion interactions between ground-state atoms in the far zone (retarded Casimir–Polder regime). In the case of two or more atoms, the effective interaction term in the Hamiltonian is simply the sum of those relative to the single atoms [32]. An important point for some following considerations in this review is that Equation (31) shows that, taking also into account (26), the interaction energy with a polarizable body, in the appropriate limits, involves the total electric field. This is true when the effective interaction term can be represented by its static polarizability, that is, when all relevant field modes have frequencies lower than relevant atomic transition frequencies, allowing us to probe the electric field energy density $E^2(\mathbf{r})/8\pi$, through the far-zone retarded Casimir–Polder interaction energy. A similar consideration holds for the magnetic energy density too. This will be an essential point for our following discussion in Section 7.1 on the relation between retarded Casimir–Polder interactions and vacuum field energies and fluctuations.

4. Vacuum Fluctuations

A striking consequence of the quantum theory of the electromagnetic field is the existence of zero-point field fluctuations. In the ground state of the field, where the number of photons is zero, the electric and magnetic field have quantum fluctuations around their average value zero [2]. This means that

$$\langle \{0_{\mathbf{k}\lambda}\} | \mathbf{E}_\perp^2(\mathbf{r}, t) | \{0_{\mathbf{k}\lambda}\} \rangle \neq 0; \quad \langle 0 | \mathbf{B}^2(\mathbf{r}, t) | \{0_{\mathbf{k}\lambda}\} \rangle \neq 0, \tag{32}$$

while $\langle \{0_{\mathbf{k}\lambda}\} \mid \mathbf{E}_\perp(\mathbf{r},t) \mid \{0_{\mathbf{k}\lambda}\}\rangle = \langle \{0_{\mathbf{k}\lambda}\} \mid \mathbf{B}(\mathbf{r},t) \mid \{0_{\mathbf{k}\lambda}\}\rangle = 0$, where $\mid \{0_{\mathbf{k}\lambda}\}\rangle$ indicates the vacuum state of the field. Field fluctuations are directly related to the field energy density, of course. In addition, we have

$$\langle \{0_{\mathbf{k}\lambda}\} \mid \int d^3r \left[\mathbf{E}_\perp^2(\mathbf{r}) + \mathbf{B}^2(\mathbf{r}) \right] \mid \{0_{\mathbf{k}\lambda}\}\rangle = \sum_{\mathbf{k}\lambda} \frac{1}{2}\hbar\omega_k. \tag{33}$$

Electric and magnetic zero-point fluctuations are a direct consequence of the non-commutativity, at equal time, of specific components of the transverse electric and magnetic field operators, given by

$$\left[E_{\perp m}(\mathbf{r},t), B_n(\mathbf{r}',t) \right] = -4\pi i\hbar c \epsilon_{mn\ell} \frac{\partial}{\partial r_\ell} \delta(\mathbf{r}-\mathbf{r}'), \tag{34}$$

where $\epsilon_{mn\ell}$ is the totally antisymmetric symbol, and the Einstein convention of repeated symbols has been used. Equation (34) can be directly obtained from the expressions (2) and (3) of the field operators.

The strength of these fluctuations, and the related field energy densities, depends on the contributions of all field modes, and thus it depends on the presence of boundary conditions, magnetodielectric bodies or matter in general. Zero-point field fluctuations are infinite and, in the unbounded space, spatially uniform. In deriving the specific expressions of (32), a sum over all allowed field modes is involved and, as mentioned, the allowed modes and the dispersion relation depend on the presence of boundaries, for example metallic or magnetodielectric bodies or even single atoms or molecules, or a structured environment. Although vacuum fluctuations diverge, their difference for two different configurations of the boundaries is usually finite, and this leads to the Casimir effect, which is a tiny force between two neutral macroscopic bodies (two parallel conducting plates, in the original formulation of the effect), which has no classical analogue [2,35]. In other words, the zero-point energy is boundary-conditions-dependent, and can be varied by changing the boundary conditions. The presence of atoms or molecules also changes the vacuum electric and magnetic energy density, and we will see in the next sections that this is deeply related to van der Waals and Casimir–Polder forces between neutral atoms, or between atoms and macroscopic objects. This point gives clear insights on the physical origin of such forces of a pure quantum origin.

Another relevant property of zero-point field fluctuations, strictly related to Casimir–Polder interactions, is their spatial correlation. Using expression (2) of the transverse electric field operator, the equal-time spatial correlation function for the $(\mathbf{k}\lambda)$ component of the free transverse electric field is

$$\langle \{0_{\mathbf{k}\lambda}\} \mid E_{\perp i}(\mathbf{k}\lambda, \mathbf{r}', t) E_{\perp j}(\mathbf{k}',\lambda',\mathbf{r}'',t) \mid \{0_{\mathbf{k}\lambda}\}\rangle = \frac{2\pi\hbar c}{V}(\hat{\mathbf{e}}_{\mathbf{k}\lambda})_i(\hat{\mathbf{e}}_{\mathbf{k}\lambda})_j k e^{i\mathbf{k}\cdot\mathbf{r}} \delta_{\mathbf{k}\mathbf{k}'}\delta_{\lambda\lambda'}, \tag{35}$$

where $\mathbf{r} = \mathbf{r}' - \mathbf{r}''$ and the subscripts i,j indicate Cartesian components. The equal-time spatial correlation of the electric field is [26]

$$\langle \{0_{\mathbf{k}\lambda}\} \mid E_{\perp i}(\mathbf{r}',t) E_{\perp j}(\mathbf{r}'',t) \mid \{0_{\mathbf{k}\lambda}\}\rangle = \frac{\hbar c}{4\pi^2} \int_0^\infty dk k^3 \int d\Omega \left(\delta_{ij} - \hat{k}_i\hat{k}_j \right) e^{i\mathbf{k}\cdot\mathbf{r}}$$
$$= -\frac{\hbar c}{4\pi^2} \left(\delta_{ij}\nabla^2 - \nabla_i\nabla_j \right) \frac{i}{r} \int_0^\infty dk \int d\Omega e^{i\mathbf{k}\cdot\mathbf{r}} = -\frac{4\hbar c}{\pi} \left(\delta_{ij} - 2\hat{r}_i\hat{r}_j \right) \frac{1}{r^4}, \tag{36}$$

where the differential operators in (36) act on $\mathbf{r}$, and the sum over polarizations has been carried out by using

$$\sum_\lambda (\hat{\mathbf{e}}_{\mathbf{k}\lambda})_i(\hat{\mathbf{e}}_{\mathbf{k}\lambda})_j = \delta_{ij} - \hat{k}_i\hat{k}_i, \tag{37}$$

and we have also used $\int d\Omega e^{i\mathbf{k}\cdot\mathbf{r}} = 4\pi \sin(kr)/(kr)$. Equation (36) shows that vacuum fluctuations of the electric field have an equal-time spatial correlation scaling with the distance as r^{-4}; as we will discuss in more detail later on, this implies that vacuum fluctuations are able to induce and correlate dipole moments in distant atoms, and this eventually leads to their Casimir–Polder interaction energy [36].

Field fluctuations and vacuum energy densities can be modified by the presence of matter—for example, a metallic or dielectric boundary (they also change in the presence of single atoms or molecules, as we will discuss in Section 7). In such cases, they depend on the position, contrarily to the unbounded-space case, where they are uniform in space.

In the case of perfectly conducting boundaries, appropriate boundary conditions must be set on the field operators at the boundaries' surface, and the electric and magnetic free field operators assume the form [2,37]

$$\mathbf{E}_\perp(\mathbf{r}) = \sum_{\mathbf{k}\lambda}\mathbf{E}_\perp(\mathbf{k}\lambda,\mathbf{r}) = i\sum_{\mathbf{k}\lambda}\sqrt{\frac{2\pi\hbar\omega_k}{V}}\mathbf{f}(\mathbf{k}\lambda,\mathbf{r})\left(a_{\mathbf{k}\lambda}-a^\dagger_{\mathbf{k}\lambda}\right), \tag{38}$$

$$\mathbf{B}(\mathbf{r}) = \sum_{\mathbf{k}\lambda}\mathbf{B}(\mathbf{k}\lambda,\mathbf{r}) = i\sum_{\mathbf{k}\lambda}\sqrt{\frac{2\pi\hbar c^2}{\omega_k V}}\nabla\times\mathbf{f}(\mathbf{k}\lambda,\mathbf{r})\left(a_{\mathbf{k}\lambda}-a^\dagger_{\mathbf{k}\lambda}\right), \tag{39}$$

where $\mathbf{f}(\mathbf{k}\lambda,\mathbf{r})$ are mode functions, assumed real, which take into account the boundaries present. For an infinite perfectly conducting plate located in the plane $z=0$, renormalized electric and magnetic energy densities (that is after subtraction of the densities found in the unbounded space) are

$$\frac{1}{8\pi}\langle\{0_{\mathbf{k}\lambda}\}\mid \mathbf{E}^2(z)\mid\{0_{\mathbf{k}\lambda}\}\rangle_{\text{ren}} = \frac{1}{8\pi}\langle\{0_{\mathbf{k}\lambda}\}\mid \mathbf{E}^2(z)\mid\{0_{\mathbf{k}\lambda}\}\rangle_{\text{plate}} - \frac{1}{8\pi}\langle\{0_{\mathbf{k}\lambda}\}\mid \mathbf{E}^2(z)\mid\{0_{\mathbf{k}\lambda}\}\rangle_{\text{vac}}$$
$$= \frac{3\hbar c}{32\pi^2 z^4}, \tag{40}$$

$$\frac{1}{8\pi}\langle\{0_{\mathbf{k}\lambda}\}\mid \mathbf{B}^2(z)\mid\{0_{\mathbf{k}\lambda}\}\rangle_{\text{ren}} = \frac{1}{8\pi}\langle\{0_{\mathbf{k}\lambda}\}\mid \mathbf{B}^2(z)\mid\{0_{\mathbf{k}\lambda}\}\rangle_{\text{plate}} - \frac{1}{8\pi}\langle\{0_{\mathbf{k}\lambda}\}\mid \mathbf{B}^2(z)\mid\{0_{\mathbf{k}\lambda}\}\rangle_{\text{vac}}$$
$$= -\frac{3\hbar c}{32\pi^2 z^4}, \tag{41}$$

where z is the distance from the plate [38]. Equations (40) and (41) show that the presence of the (perfectly conducting) boundary decreases the electric zero-point energy density in all space, while it increases the magnetic one. In Section 7.1, we will show that something similar occurs in the space around an electric dipolar source of the electromagnetic field. In the expressions above, the renormalized field energy densities diverges at the surface of the conducting plate, $z=0$. This is due to the unrealistic assumption of a perfectly conducting boundary with a fixed position in space; it has been shown that these divergences can be cured by assuming a fluctuating position of the plate [38], or introducing an appropriate cut-off to simulate a non-ideal metal boundary [39,40], thus allowing to investigate the structure of field fluctuations in the very vicinity of the boundary [39]. Field fluctuations near a conducting boundary that is allowed to move, with its translational degrees of freedom treated quantum mechanically and thus, with quantum fluctuations of its position, have been recently considered for both the scalar and the electromagnetic field, also in relation with the surface divergences of energy densities and field fluctuations [41–43].

In the next sections, we will discuss how vacuum energy densities, and their space dependence in the presence of matter are strictly related to van der Waals and Casimir–Polder interactions between atoms and to the atom–surface Casimir–Polder interaction. These considerations will allow us to get a clear physical interpretation of such pure quantum interactions, as well as useful tools to evaluate them in different situations.

5. The Van Der Waals and Casimir–Polder Dispersion Interaction between Two Neutral Ground-State Atoms

Let us consider two ground-state neutral atoms, A and B, located at $\mathbf{r}_A$ and $\mathbf{r}_B$ respectively, and be $\mathbf{r}=\mathbf{r}_B-\mathbf{r}_A$ their distance. We assume that the atoms are in the vacuum space at zero temperature, and at a distance such that there is no overlap between their electronic wavefunctions. We describe their interaction with the quantum electromagnetic field in the multipolar coupling scheme. From Equation (19), the Hamiltonian of the system in the multipolar coupling scheme is

$$\begin{aligned}
H &= H_0 + H_I, \\
H_0 &= H_A + H_B + H_F, \\
H_I &= -\boldsymbol{\mu}_A \cdot \mathbf{E}(\mathbf{r}_A) - \boldsymbol{\mu}_B \cdot \mathbf{E}(\mathbf{r}_B),
\end{aligned} \quad (42)$$

where H_A and H_B are respectively the Hamiltonians of atoms A and B, $\mathbf{E}(\mathbf{r})$ indicates the transverse displacement field operator (coinciding, outside the atoms, with the electric field operator, as shown by Equation (26)), and we have disregarded the term proportional to $\mathbf{P}_\perp^2(\mathbf{r})$, which does not affect interatomic interaction energies. The unperturbed ground state is $\mid g_A, g_B, \{0_{\mathbf{k}\lambda}\}\rangle$, where g_A and g_B respectively indicate the ground state of atoms A and B, and $\mid \{0_{\mathbf{k}\lambda}\}\rangle$ is the photon vacuum state. This state is not an eigenstate of the total Hamiltonian H, due to the matter–radiation interaction Hamiltonian. Thus, there is an energy shift of the bare ground state due to the interaction Hamiltonian. Evaluating this energy shift by perturbation theory, it is easily found that the second-order energy shift gives only individual single-atom shifts (the Lamb shift, after mass renormalization), not depending on the distance between the atoms. The first term yielding an interatomic-distance-dependent contribution is at fourth order in the matter–radiation coupling. It is given by

$$\begin{aligned}
\Delta E_4 = &-\sum_{I,II,III} \frac{\langle g_A, g_B, \{0_{\mathbf{k}\lambda}\} \mid H_I \mid III\rangle \langle III \mid H_I \mid II\rangle \langle II \mid H_I \mid I\rangle \langle I \mid H_I \mid g_A, g_B, \{0_{\mathbf{k}\lambda}\}\rangle}{(E_I - E_g)(E_{II} - E_g)(E_{III} - E_g)} \\
&+ \sum_{I,II} \frac{\mid \langle g_A, g_B, \{0_{\mathbf{k}\lambda}\} \mid H_I \mid II\rangle \mid^2 \mid \langle g_A, g_B, \{0_{\mathbf{k}\lambda}\} \mid H_I \mid I\rangle \mid^2}{(E_I - E_g)^2(E_{II} - E_g)},
\end{aligned} \quad (43)$$

where I, II and III indicate atoms-field intermediate states with energies E_I, E_{II} and E_{III} (with the exclusion, in the sums, of the unperturbed state), respectively; E_g is the energy of the unperturbed ground state. The second term in (43) does not contribute to the potential energy, while the first one gives twelve relevant diagrams, each representing an exchange of two virtual photons between the atoms. Using the Hamiltonian (42) and the expression (2) of the electric field, the following expression of the interaction energy between the two atoms is found from (43) [15,24]

$$\begin{aligned}
\Delta E = &-\sum_{ps}\sum_{\mathbf{k}\lambda\mathbf{k}'\lambda'} \frac{2\pi\hbar ck}{V}\frac{2\pi\hbar ck'}{V}(\hat{\mathbf{e}}_{\mathbf{k}\lambda})_i(\hat{\mathbf{e}}_{\mathbf{k}\lambda})_j(\hat{\mathbf{e}}_{\mathbf{k}'\lambda'})_\ell(\hat{\mathbf{e}}_{\mathbf{k}'\lambda'})_m \mu^{gp}_{Ai}\mu^{pg}_{A\ell}\mu^{gs}_{Bj}\mu^{sg}_{Bm} e^{i(\mathbf{k}+\mathbf{k}')\cdot\mathbf{r}} \\
&\times \frac{1}{\hbar^3 c^3}\frac{4(k_{pg} + k_{sg} + k)}{(k_{pg} + k_{sg})(k_{pg} + k)(k_{sg} + k)}\left(\frac{1}{k + k'} - \frac{1}{k - k'}\right),
\end{aligned} \quad (44)$$

where p, s indicate generic atomic states, $\mathbf{r} = \mathbf{r}_B - \mathbf{r}_A$ is the distance between the two atoms, $\hbar c k_{pg} = E_p - E_g$, and i, j, ℓ, m are Cartesian components (the Einstein convention of repeated indices has been used).

After taking the continuum limit $\sum_{\mathbf{k}} \to (V/(2\pi)^3)\int dk k^2 \int d\Omega$, next steps in the calculation are the polarization sum, done using (37), and angular integration.

We define the function

$$\begin{aligned}
G_{ij}(k,\mathbf{r}) &= \frac{1}{k^3}\left(\delta_{ij}\nabla^2 - \nabla_i\nabla_j\right)\frac{e^{ikr}}{r} \\
&= \left[(\delta_{ij} - \hat{r}_i\hat{r}_j)\frac{1}{kr} + (\delta_{ij} - 3\hat{r}_i\hat{r}_j)\left(\frac{i}{k^2r^2} - \frac{1}{k^3r^3}\right)\right]e^{ikr},
\end{aligned} \quad (45)$$

where the differential operators act on the variable r, whose real and imaginary parts are

$$\Re\{G_{ij}(k,\mathbf{r})\} = -V_{ij}(k,\mathbf{r}) = \frac{1}{k^3}\left(\delta_{ij}\nabla^2 - \nabla_i\nabla_j\right)\frac{\cos(kr)}{r} =$$
$$= (\delta_{ij} - \hat{r}_i\hat{r}_j)\frac{\cos(kr)}{kr} - (\delta_{ij} - 3\hat{r}_i\hat{r}_j)\left(\frac{\sin(kr)}{k^2r^2} + \frac{\cos(kr)}{k^3r^3}\right), \tag{46}$$

$$\Im\{G_{ij}(k,\mathbf{r})\} = \frac{1}{k^3}\left(\delta_{ij}\nabla^2 - \nabla_i\nabla_j\right)\frac{\sin(kr)}{r} = \frac{1}{4\pi}\int d\Omega\left(\delta_{ij} - \hat{k}_i\hat{k}_j\right)e^{i\mathbf{k}\cdot\mathbf{r}}$$
$$= (\delta_{ij} - \hat{r}_i\hat{r}_j)\frac{\sin(kr)}{kr} + (\delta_{ij} - 3\hat{r}_i\hat{r}_j)\left(\frac{\cos(kr)}{k^2r^2} - \frac{\sin(kr)}{k^3r^3}\right). \tag{47}$$

Using the relations above, after lengthy algebraic calculations, Equation (44) yields

$$\Delta E = -\frac{4}{\pi\hbar c}\sum_{ps}\mu_{Ai}^{gp}\mu_{A\ell}^{pg}\mu_{Bj}^{gs}\mu_{Bm}^{sg}\frac{1}{k_{pg}+k_{sg}}$$
$$\times \int_0^\infty dk k^6 \frac{k_{pg}+k_{sg}+k}{(k_{pg}+k)(k_{sg}+k)}\Re\{G_{\ell m}(k,\mathbf{r})\}\Im\{G_{ij}(k,\mathbf{r})\}. \tag{48}$$

With further algebraic manipulation, after averaging over the orientations of the atomic dipoles, Equation (48) can be cast in the following form, in terms of the atoms' dynamical polarizability (30) evaluated at the imaginary wavenumber ($k = iu$) [15,24,30,44]

$$\Delta E = -\frac{\hbar}{\pi}\int_0^\infty du\, \alpha_A(iu)\alpha_B(iu)u^6 e^{-2ur}\left(\frac{1}{u^2r^2} + \frac{2}{u^3r^3} + \frac{5}{u^4r^4} + \frac{6}{u^5r^5} + \frac{3}{u^6r^6}\right). \tag{49}$$

This expression is valid for any interatomic separation r outside the region of wavefunctions overlap, and it depends on the relevant atomic transition wavelengths from the ground state, $\lambda_{rg} = 2\pi k_{rg}^{-1}$, included in the atoms' polarizability. Two limiting cases of (49) are particularly relevant, $r \ll \lambda_{rg}$ (near zone), and $r \gg \lambda_{rg}$ (far zone).

In the near zone, Equation (49) yields

$$\delta E_{near} = -\frac{2}{3}\sum_{ps}\frac{|\mu_A^{pg}|^2|\mu_B^{sg}|^2}{E_{pg}+E_{sg}}\frac{1}{r^6}, \tag{50}$$

which coincides with the London–van der Waals dispersion interaction scaling as r^{-6} [16,45], as obtained in the minimal coupling scheme including electrostatic interactions only, i.e., neglecting contributions from the (quantum) transverse fields. For this reason, this potential is usually called the *nonretarded* London–van der Waals potential energy.

In the far zone, Equation (49) yields

$$\delta E_{far} = -\frac{23\hbar c}{4\pi}\alpha_A\alpha_B\frac{1}{r^7}, \tag{51}$$

where $\alpha_{A(B)} = \alpha_{A(B)}(0)$ is the static polarizability of the atoms [1,46]. Equation (51) shows that the dispersion energy asymptotically scales with the distance as r^{-7}, thus faster than the London potential (50). This is usually called the *retarded* or Casimir–Polder regime of the dispersion interaction energy, where retardation effects due to the transverse fields, that is, quantum electrodynamical effects, change the distance dependence of the potential energy from r^{-6} to r^{-7}, at a distance between the atoms given by the transition wavelength from the ground state to the main excited states. It should be noted that the calculation to obtain the expression (49) can be simplified by using the effective Hamiltonian introduced in Section 3 [32,34,47]. The existence of an asymptotic behaviour of the dispersion interaction falling more rapidly than the London r^{-6} dependence, was inferred for a long time from the analysis of some macroscopic properties of colloid solutions [48]. In addition, the onset of the Casimir–Polder retarded regime has been recently related to the bond length of a diatomic helium molecule [49]. Furthermore, the nonretarded interaction between two single isolated Rydberg atoms, at a distance of a few nanometers each other, has been recently measured through the measurement of the single-atom Rabi frequency [50].

We just mention that new features in the distance-dependence of the dispersion energy between ground-state atoms appear if the atoms are in a thermal bath at finite temperature. In such a case, a new length scale, related to the temperature T exists, the thermal length $\rho_{therm} = \hbar c/(2\pi k_B T)$, where k_B is the Boltzmann constant. This defines a very long-distance regime, $r \gg \rho_{therm}$, where the interaction energy is approximately proportional to T and scales with the distance as r^{-6} [51–54]. This thermal regime, usually at a much larger distance than that of the onset of the retarded regime (far zone), has the same distance-dependence of the nonretarded one.

If one (or both) atoms are in an excited state, the long-distance behaviour of the Casimir–Polder potential is different compared to that of ground-state atoms, asymptotically scaling as r^{-2}, due to the possibility of exchange of real photons between the atoms [29,55]. From a mathematical point of view, this is a consequence of the presence of a resonant pole in the energy shift. Up to very recently, a long-standing dispute in the literature has been concerned with whether a spatially oscillating modulation in the interaction energy exists or not: different results (with or without space oscillations), both mathematically correct but physically different, are obtained according to how the resonant pole is circumvented in the frequency integral. Recent results, also based on a time-dependent approach, have given a strong indication that the distance dependence of the force is monotonic, without spatial oscillations [56–60]. Dispersion interactions, between atoms involving higher-order multipoles [61–63], or chiral molecules [64–66], have been also investigated in the recent literature.

6. The Three-Body Casimir–Polder Interaction

Let us consider the case of three atoms, A, B and C, respectively located at $\mathbf{r}_A$, $\mathbf{r}_B$ and $\mathbf{r}_C$; their distances are $\alpha = |\mathbf{r}_C - \mathbf{r}_B|$, $\beta = |\mathbf{r}_C - \mathbf{r}_A|$, $\gamma = |\mathbf{r}_B - \mathbf{r}_A|$, as shown in Figure 1.

The Hamiltonian of the system, in the multipolar coupling scheme, is

$$H = H_A + H_B + H_C + H_F - \boldsymbol{\mu}_A \cdot \mathbf{E}(\mathbf{r}_A) - \boldsymbol{\mu}_B \cdot \mathbf{E}(\mathbf{r}_B) - \boldsymbol{\mu}_C \cdot \mathbf{E}(\mathbf{r}_C), \tag{52}$$

with a clear meaning of the various terms. The bare ground state is $|g_A, g_B, g_C, \{0_{\mathbf{k}\lambda}\}\rangle$ that, not being an eigenstate of the total Hamiltonian H, has an energy shift due to the atoms–field interaction. If this energy shift is evaluated up to sixth order in the interaction, considering only terms depending on the atomic coordinates, and neglecting single-atom energy shifts that do not contribute to interatomic energies, we find

$$\Delta E = \Delta E(A,B) + \Delta E(B,C) + \Delta E(A,C) + \Delta E_3(A,B,C), \tag{53}$$

where the first three terms are two-body interaction energies, depending on the coordinates of two atoms only, while $\Delta_3(A,B,C)$ is a sixth-order non-additive (three-body) contribution containing coordinates and physical parameters of all the three atoms [6,24]. For ground-state atoms, the three-body term can be expressed in terms of the dynamical polarizabilities of the three atoms or molecules, or polarizable bodies in general.

The effective Hamiltonian (29), introduced in Section 3, is particularly convenient for the calculation of the three-body interaction between three neutral ground-state atoms, allowing a third-order calculation in place of a sixth-order one, with a considerable reduction of the number of diagrams involved. In our case, the effective Hamiltonian is

$$\begin{aligned} H &= H_A + H_B + H_C + H_F + H_I, \\ H_I &= -\frac{1}{2}\sum_{\mathbf{k}\lambda}\alpha_A(k)\mathbf{E}(\mathbf{k}\lambda,\mathbf{r}_A)\cdot\mathbf{E}(\mathbf{r}_A) - \frac{1}{2}\sum_{\mathbf{k}\lambda}\alpha_B(k)\mathbf{E}(\mathbf{k}\lambda,\mathbf{r}_B)\cdot\mathbf{E}(\mathbf{r}_B) \\ &\quad -\frac{1}{2}\sum_{\mathbf{k}\lambda}\alpha_C(k)\mathbf{E}(\mathbf{k}\lambda,\mathbf{r}_C)\cdot\mathbf{E}(\mathbf{r}_C), \end{aligned} \qquad (54)$$

where $\alpha_i(k)$, with $i = A, B, C$, is the dynamical polarizability of the atoms. Because our effective Hamiltonian is quadratic in the coupling, a third-order perturbation theory allows us to obtain the non-additive term. The general expression of the third-order energy shift is

$$\begin{aligned} \Delta E_3 = &-\sum_{I,II}\frac{\langle g_A,g_B,g_C,\{0_{\mathbf{k}\lambda}\}|H_I|II\rangle\langle II|H_I|I\rangle\langle I|H_I|g_A,g_B,g_C,\{0_{\mathbf{k}\lambda}\}\rangle}{(E_I-E_g)(E_{II}-E_g)} \\ &+\sum_I\frac{\langle g_A,g_B,g_C,\{0_{\mathbf{k}\lambda}\}|H_I|g_A,g_B,g_C,\{0_{\mathbf{k}\lambda}\}\rangle\langle g_A,g_B,g_C,\{0_{\mathbf{k}\lambda}\}|H_I|I\rangle\langle I|H_I|g_A,g_B,g_C,\{0_{\mathbf{k}\lambda}\}\rangle}{(E_I-E_g)^2}, \end{aligned} \qquad (55)$$

where I and II are intermediate atoms-field states, different from the unperturbed state.

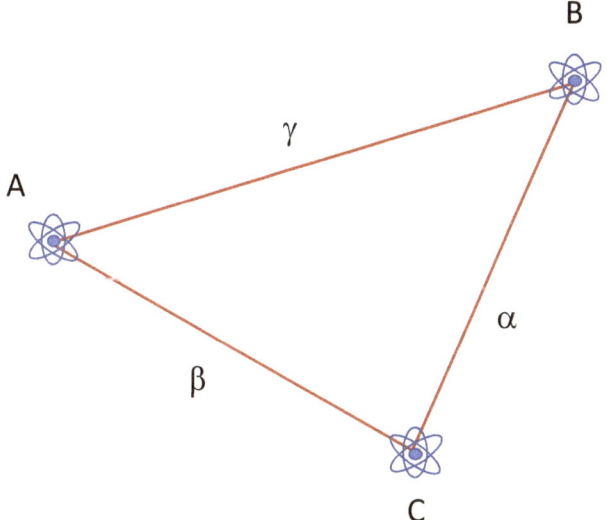

Figure 1. Geometrical configuration of the three atoms A, B and C.

Application of (55) with the interaction Hamiltonian H_I given by (54), keeping only terms containing the coordinates of all atoms, after some algebraic calculations, yields [32]

$$\Delta E_3(A,B,C) = -\frac{\hbar c}{\pi}F_{ij}^{\alpha}F_{j\ell}^{\beta}F_{\ell i}^{\gamma}\frac{1}{\alpha\beta\gamma}\int_0^{\infty}du\,\alpha_A(iu)\alpha_B(iu)\alpha_C(iu)e^{-u(\alpha+\beta+\gamma)}, \qquad (56)$$

where, in order to shorten notations, we have defined the differential operator

$$F_{mn}^r = \left(-\delta_{mn}\nabla^2 + \nabla_m\nabla_n\right)^r \qquad (57)$$

(the superscript indicates the space variable upon which the differential operators act). The expression (56), here obtained by a simpler third-order perturbative approach using the effective Hamiltonian (54), coincides with the standard sixth-order result obtained from the multipolar Hamiltonian [6].

In the far zone (retarded regime), only low-frequency field modes give a relevant contributions, and we can replace the dynamical polarizabilities with the static ones. The integral over u is then straightforward, and we get

$$\Delta E_3(A, B, C) \simeq -\frac{\hbar c}{\pi} \alpha_A \alpha_B \alpha_C F_{ij}^\alpha F_{j\ell}^\beta F_{\ell i}^\gamma \frac{1}{\alpha \beta \gamma (\alpha + \beta + \gamma)}, \quad (58)$$

where α_i $(i = A, B, C)$ are the atomic static polarizabilities.

In the specific case of three atoms in the far zone of each other, with an equilateral triangle configuration of side r $(\alpha = \beta = \gamma = r)$, Equation (58) gives [67]

$$\Delta_3(A, B, C) \simeq \frac{2^4 \cdot 79}{3^5} \frac{\hbar c}{\pi} \frac{\alpha_A \alpha_B \alpha_C}{r^{10}}. \quad (59)$$

In the same geometrical configuration, Equation (56) gives the nonretarded result (near zone) scaling as r^{-9} [5]. The r^{-10} and r^{-9} distance scaling of the three-body-component should be compared with the r^{-7} and r^{-6} scaling of the two-body component, respectively in the far and near zone. While the two-body interaction is always attractive, the three-body component can be attractive or repulsive according to the geometrical configuration of the three atoms [67]. Three-body interactions between molecules involving higher-order multipoles have been also considered in the literature [68–70].

7. Two- and Three-Body Dispersion Interactions as a Consequence of Vacuum Field Fluctuations

An important point when dealing with van der Waals and Casimir–Polder dispersion interactions is the formulation of physical models aiming to explaining their origin, stressing quantum aspects, and giving physical insights of the processes involved. This is important from two points of view: they allow for understanding what the basic origin of these interactions is, highlighting quantum and classical aspects; they can provide useful computational methods to simplify their evaluation, in particular in more complicated situations such as many-body interactions, or when boundary conditions are present, as well as when the atoms are in a noninertial motion.

In this section, we describe physical models for dispersion interactions based on the existence and properties of vacuum field fluctuations. These models clearly show how van der Waals and Casimir–Polder interactions between atoms, and polarisable bodies in general, in different conditions, can be attributed to the existence of the zero-point energy and vacuum fluctuations, specifically their change due to the presence of matter, and/or their spatial correlations. At the end of this review, we will briefly address the fundamental and conceptually subtle question if all this *proves* the real existence of vacuum fluctuations. In fact, although the results discussed in the following give strong support to the real existence of the zero-point energy, they are not a definitive confirmation of its existence because all of these effects can in principle be obtained also from considerations based on source fields, without reference to vacuum fluctuations [2,71,72].

We now introduce and apply two different models, based on the properties of vacuum field fluctuations and *dressed* vacuum field fluctuations, to obtain on a physically transparent basis the two- and three-body dispersion interactions between polarizable bodies, even in the presence of a boundary such as a reflecting mirror. The first method is based on vacuum field energy densities, and the second on vacuum field spatial correlations. These models give physical insights on the origin of the ubiquitous dispersion interactions, and also provide useful tools to calculate them in more complicated situations, which we will apply in Section 8 when a reflecting mirror close to the atoms is present.

7.1. Dressed Field Energy Densities

Let us consider two or more atoms, or in general polarizable bodies, in the unbounded space, at zero temperature, interacting with the quantum electromagnetic field in the vacuum state. The bare ground state $\mid g \rangle$, consisting of the ground state of the atoms and the vacuum state of the field, is not an eigenstate of the total Hamiltonian. The presence of matter perturbs the field [73], due to continuous emission and absorption of virtual photons [74,75], yielding the dressed ground state $\mid \tilde{g} \rangle$ of the system. In such a state, the atoms are *dressed* by the virtual photons emitted and reabsorbed by them. The presence of this *cloud* of virtual photons determines a change of the fluctuations of the electric and magnetic field around the atoms or, equivalently, of the electric and magnetic energy densities; the *renormalized* energy densities (that is, after subtraction of the spatially uniform bare field energy densities, present even in absence of the atom) are given by

$$\langle \tilde{g} \mid \mathcal{H}_{\text{el}}(\mathbf{r}) \mid \tilde{g} \rangle_{\text{ren}} = \frac{1}{8\pi} \langle \tilde{g} \mid \mathbf{E}^2(\mathbf{r}) \mid \tilde{g} \rangle - \frac{1}{8\pi} \langle g \mid \mathbf{E}^2(\mathbf{r}) \mid g \rangle, \tag{60}$$

$$\langle \tilde{g} \mid \mathcal{H}_{\text{mag}}(\mathbf{r}) \mid \tilde{g} \rangle_{\text{ren}} = \frac{1}{8\pi} \langle \tilde{g} \mid \mathbf{B}^2(\mathbf{r}) \mid \tilde{g} \rangle - \frac{1}{8\pi} \langle g \mid \mathbf{B}^2(\mathbf{r}) \mid g \rangle, \tag{61}$$

where $\mid g \rangle$ and $\mid \tilde{g} \rangle$ are respectively the bare and the dressed ground state of the system.

Let us first consider just one ground-state atom, A, located at $\mathbf{r}_A = 0$. The dressed ground state can be obtained by perturbation theory; it is convenient to use the multipolar coupling Hamiltonian obtained in Section 2, or the effective Hamiltonian (29), because in such a case we will get the transverse displacement field that, according to (26), coincides with the total (longitudinal plus transverse) electric field $\mathbf{E}(\mathbf{r})$ for $\mathbf{r} \neq 0$ (this is an essential point because it is the total electric field, and not just its transverse part, which is involved in the interaction with the other atoms). Thus, the energy density we will obtain is that associated with the total electric field and to the magnetic field. Applying first-order perturbation theory to the effective Hamiltonian (29), the dressed ground state at first order in the polarizability, that is, at second order in the electric charge, is

$$\mid \tilde{g} \rangle - \mid g_A \{0_{\mathbf{k}\lambda}\} \rangle \frac{\pi}{V} \sum_{\mathbf{k}\lambda \mathbf{k}'\lambda'} \alpha_A(k) \hat{\mathbf{e}}_{\mathbf{k}\lambda} \cdot \hat{\mathbf{e}}_{\mathbf{k}'\lambda'} \frac{kk'}{k+k'} \mid g_A, 1_{\mathbf{k}\lambda} 1_{\mathbf{k}'\lambda'} \rangle, \tag{62}$$

showing admixture with two virtual photons. We now evaluate the average value of the electric and magnetic energy density over the state (62). At first order in $\alpha_A(k)$, we get [76]

$$\langle \tilde{g} \mid \mathcal{H}_{\text{el}}(\mathbf{r}) \mid \tilde{g} \rangle_{\text{ren}} = \frac{\hbar c}{4\pi^3} \int_0^\infty dk \int_0^\infty dk' \alpha_A(k) \frac{k^3 k'^3}{k+k'} \Big[j_0(kr) j_0(k'r) \\ - \frac{1}{k'r} j_0(kr) j_1(k'r) - \frac{1}{kr} j_0(k'r) j_1(kr) + \frac{3}{kk'r^2} j_1(kr) j_1(k'r) \Big], \tag{63}$$

$$\langle \tilde{g} \mid \mathcal{H}_{\text{mag}}(\mathbf{r}) \mid \tilde{g} \rangle_{\text{ren}} = -\frac{\hbar c}{2\pi^3} \int_0^\infty dk \int_0^\infty dk' \alpha_A(k) \frac{k^3 k'^3}{k+k'} j_1(kr) j_1(k'r), \tag{64}$$

where $j_n(x)$ are spherical Bessel functions [77]. The integrals over k and k' in (63) and (64) can be decoupled by using the relation $(k + k')^{-1} = \int_0^\infty d\eta \exp(-(k+k')\eta)$. Explicit evaluation of (63) and (64) yields a positive value for the renormalized electric electric energy density, and a negative value for the magnetic one. This means that the presence of an electric dipolar source increases the electric vacuum energy density, while it decreases the magnetic vacuum energy density, with respect to their value in the absence of the source (an analogous consideration holds for the field fluctuations, of course). This is a microscopic analogue of what happens in the case of the plane conducting boundary discussed in Section 4, as Equations (40) and (41) show.

Two limiting cases of (63) and (64) are particularly relevant, according to the distance r from the atom in comparison with a relevant atomic transition wavelength λ_0 from the ground state, similarly to the dispersion energy between two ground-state atoms. In the *far zone*, $r \gg \lambda_0$, it is easy to show that only small-frequency field modes significantly contribute to the k, k' integrals in Equations (63) and (64), and we can approximate the (electric) dynamic polarizability with the static electric one, $\alpha_A(0) = \alpha_A^E$. We thus obtain

$$\langle \tilde{g} | \mathcal{H}_{\text{el}}(\mathbf{r}) | \tilde{g} \rangle_{\text{ren}} \simeq \frac{23\hbar c\alpha_A^E}{(4\pi)^2} \frac{1}{r^7}; \quad \langle \tilde{g} | \mathcal{H}_{\text{mag}}(\mathbf{r}) | \tilde{g} \rangle_{\text{ren}} \simeq -\frac{7\hbar c\alpha_A^E}{(4\pi)^2} \frac{1}{r^7}. \quad (65)$$

In the *near zone*, $r \ll \lambda_0$, we find that the renormalized electric energy density scales as $1/r^6$, and the magnetic one as $1/r^5$ [78,79]. In this limit, the electric energy density is essentially due to the electrostatic (longitudinal) field, as an explicit calculation in the minimal coupling scheme, where longitudinal and transverse contributions are separated (see Section 2), shows [78].

The results above clearly point out how the presence of a field source, a dipolar atom in our case, changes zero-point energy densities and field fluctuations in the space around it, and, more in general, that matter can influence the properties of the quantum vacuum. This is the microscopic analogue of the change of the vacuum energy densities due to a boundary discussed in Section 4, or in the two-plate case of the Casimir effect [35]; the essential difference is that in the present case, matter is described through its Hamiltonian, with its internal degrees of freedom, while in the case of macroscopic boundaries its presence is introduced by a (classical) boundary condition on the field operators.

In the far zone, the change of the zero-point energy due to the presence of atom A can be probed through a second atom, B, considered a polarizable body with static electric polarizability α_B^E and located at $\mathbf{r}_B$. Its interaction energy with field fluctuations can be written as $\Delta E = -\alpha_B^E \langle E^2(\mathbf{r}_B) \rangle / 2$, as indicated by the existence of an effective Hamiltonian of the form (31); this is also supported by the classical expression of the interaction energy of an electric field with an induced electric dipole moment. Similarly, in the far zone, the magnetic energy density can be probed through a magnetically polarizable body with static magnetic polarizability α_B^M. We stress that this is rigorously valid only in the far zone because, in the near zone, where the contribution of high-frequency photons is relevant [78], the *test* atom B responds differently to each Fourier component of the field, according to its dynamical polarizability $\alpha_B(k)$ (see Equation (29)). The distance-dependent part of the interaction energy is then

$$\Delta E^{EE}(\mathbf{r}) = -\frac{1}{2}\alpha_B^E \langle \tilde{g} | E^2(\mathbf{r}_B) | \tilde{g} \rangle_{\text{ren}} = -\frac{23\hbar c}{4\pi} \alpha_A^E \alpha_B^E \frac{1}{r^7}, \quad (66)$$

$$\Delta E^{EM}(\mathbf{r}) = -\frac{1}{2}\alpha_B^M \langle \tilde{g} | B^2(\mathbf{r}_B) | \tilde{g} \rangle_{\text{ren}} = \frac{7\hbar c}{4\pi} \alpha_A^E \alpha_B^M \frac{1}{r^7}, \quad (67)$$

where r is the distance between the two atoms, and the subscript *ren* indicates that spatially uniform bare terms have been subtracted. These expressions coincide with the dispersion interaction energy in the far-zone as obtained by a fourth-order perturbative calculation, as discussed in Section 5 (see, for example, [78,80]); from the signs of (66) and (67), it should be noted that the electric-electric dispersion force is attractive (because the presence of the electric dipolar atom decreases the electric energy density), and the electric-magnetic dispersion force is repulsive (because the presence of the electric dipolar atom increases the magnetic energy density).

This approach shows the strict relation between (renormalized) electric and magnetic vacuum energy densities and the far-zone dispersion interaction energy; moreover, it allows considerable simplification of the calculations, that in this way is split in two independent steps: evaluation of the renormalized energy densities of the field first, and then the evaluation of their interaction with the other electrically or magnetically polarizable body. This finding, besides giving a sharp insight on the origin of the retarded dispersion interaction, can be particularly useful in more complicated cases such as many-body dispersion interactions, or in the presence of boundary conditions (see Section 8).

Let us now consider the dispersion interaction between three ground-state atoms, A, B and C, respectively located at positions $\mathbf{r}_A$, $\mathbf{r}_B$ and $\mathbf{r}_C$. We concentrate on the three-body component of their interaction, discussed in Section 6. We will show that, analogously to the case of the two-body potential, the physical origin of the retarded (far zone) three-body component can be clearly attributed to vacuum fluctuations and field energy densities, as modified by the presence of the atoms [81]. The Hamiltonian of our system is given by (54); because we are limiting our considerations to the far zone only, the interaction term can be more simply expressed in terms of the Craig–Power effective Hamiltonian (31); thus, we have

$$H_I = -\frac{1}{2} \sum_{i=A,B,C} \alpha_i \mathbf{E}^2(\mathbf{r}_i)$$ (68)

(this expression can be straightforwardly obtained from (54) by replacing the dynamical polarizabilities with the corresponding static ones). The first step is to obtain the dressed ground state of the pair of atoms A and B, up to second order in the polarizabilities, i.e., up to the fourth order in the electronic charge. Writing down only terms relevant for the subsequent calculation of the three-body interaction, that is, only terms that at the end of the calculation yield contributions containing the polarizabilities of all atoms, we have

$$\begin{aligned}
|\tilde{g}_{AB}\rangle &= |g_A g_B, \{0_{\mathbf{k}\lambda}\}\rangle - \left(\frac{\pi}{V}\alpha_A \sum_{\mathbf{k}\lambda\mathbf{k}'\lambda'} (\hat{\mathbf{e}}_{\mathbf{k}\lambda}\cdot\hat{\mathbf{e}}_{\mathbf{k}'\lambda'}) \frac{\sqrt{\omega_k\omega_{k'}}}{\omega_k+\omega_{k'}} e^{-i(\mathbf{k}+\mathbf{k}')\cdot\mathbf{r}_A} |g_A g_B, \mathbf{k}\lambda\mathbf{k}'\lambda'\rangle \right.\\
&\left. + (A\to B)\right) - \left(\frac{8\pi^2}{V^2}\alpha_A\alpha_B \sum_{\mathbf{k}\lambda\mathbf{k}'\lambda'} (\hat{\mathbf{e}}_{\mathbf{k}\lambda}\cdot\hat{\mathbf{e}}_{\mathbf{k}'\lambda'})(\hat{\mathbf{e}}_{\mathbf{k}'\lambda'}\cdot\hat{\mathbf{e}}_{\mathbf{k}''\lambda''}) \frac{\omega_{k''}\sqrt{\omega_k\omega_{k'}}}{(\omega_{k''}+\omega_{k'})(\omega_k+\omega_{k'})}\right.\\
&\left. \times e^{-i(\mathbf{k}'+\mathbf{k}'')\cdot\mathbf{r}_A} e^{-i(\mathbf{k}-\mathbf{k}'')\cdot\mathbf{r}_B} |g_A g_B, \mathbf{k}\lambda\mathbf{k}'\lambda'\rangle + (\mathbf{r}_A\leftrightarrow\mathbf{r}_B)\right).
\end{aligned}$$ (69)

We now evaluate the average value of the electric field fluctuations at the position $\mathbf{r}_C$ of atom C on the dressed state (69), which is the quantity $\langle \tilde{g}_{AB} | E^2(\mathbf{r}_C) | \tilde{g}_{AB}\rangle$. After multiplying times $-\alpha_C/2$, it gives the interaction energy of atom C with the (renormalized) field fluctuations generated by atoms A and B. This quantity contains a term proportional to the atomic polarizability of the three atoms, yielding the three-body component of the dispersion interaction. After lengthy algebraic calculations, we finally obtain

$$-\frac{1}{2}\alpha_C \langle \tilde{g}_{AB} | E^2(\mathbf{r}_C) | \tilde{g}_{AB}\rangle = -\frac{\hbar c}{\pi}\alpha_A\alpha_B\alpha_C F^\alpha_{ij} F^\beta_{j\ell} F^\gamma_{\ell i} \frac{1}{\alpha\beta\gamma(\alpha+\beta+\gamma)},$$ (70)

where the distances α, β and γ have been defined in Section 6, and the superscripts indicate the variables the differential operators act on. Two analogous expressions are obtained by exchanging the role of the atoms. The total interaction energy is then obtained by averaging on these three (equal) contributions: the same expression obtained by a direct application of perturbation theory, given by (56), is thus obtained [81]. This shows that the three-body component of the retarded dispersion interaction between three atoms (far zone, Casimir–Polder regime) at sixth order in the electron charge, can be directly and more easily obtained as the interaction energy of one atom with the renormalized electric field fluctuations generated by the other two atoms, that, at the fourth order in the charge, show a sort of interference effect. In other words, at the fourth order, the fluctuations in the presence of two atoms are not simply the sum of those due to the individual atoms [82]. This approach, similarly to the two-body case previously analyzed, has two remarkable features: firstly, it gives a physically transparent insight on the origin of three-body dispersion interactions, particularly significant in the retarded regime; secondly, it allows a considerable simplification of the calculation, which is separated in two successive steps: evaluation of the renormalized field energy density due to the presence of the "source" atoms A and B, and successive interaction of them with the third atom (C).

The approach outlined in this section can be directly applied also to the retarded Casimir–Polder interaction between an atom and a perfectly reflecting wall. Renormalized electric vacuum fluctuations are given by (40) and (41). If an electrically polarizible body A, such as an atom or a molecule, is placed at a distance z from the conducting wall at $z = 0$, the second-order interaction energy in the far zone is given by

$$\Delta E_{\text{atom-wall}}(z) = -\frac{1}{2}\alpha_A \langle \{0_{\mathbf{k}\lambda}\} | E^2(z) | \{0_{\mathbf{k}\lambda}\} \rangle = -\frac{3\hbar c}{8\pi}\alpha_A \frac{1}{z^4}, \tag{71}$$

where α_A is the electric static polarizability of the atom. In a quasi-static approach (as that we are assuming for the dispersion force between atoms, too), the Casimir–Polder force on the atom is given by

$$F_{\text{atom-wall}}(z) = -\frac{d}{dz}\left(\Delta E_{\text{atom-wall}}(z)\right) = -\frac{3\hbar c}{2\pi}\alpha_A \frac{1}{z^5}, \tag{72}$$

where the minus sign implies that the force is attractive [1,83–85]. Similar considerations allow us to evaluate the far-zone magnetic interaction, starting from (41). Nonperturbative methods to evaluate the atom-wall Casimir–Polder interaction, based on Bogoliubov-type transformations and modelling the atoms as harmonic oscillators, have been also proposed in the recent literature [86,87].

7.2. Vacuum Field Correlations

A different approach showing the deep relation between dispersion interactions and zero-point field fluctuations involves the spatial correlation function of vacuum fluctuations, which, in the unbounded vacuum space is given by (35), for a single field mode, and by (36) for the field operator. Let us assume considering two isotropic ground-state atoms or molecules, A and B, respectively located at $\mathbf{r}_A$ and $\mathbf{r}_B$. Zero-point fluctuations induce instantaneous dipole moments in the two atoms according to

$$\mu_{\ell i}^{\text{ind}}(\mathbf{k}\lambda) = \alpha_\ell(k) E_i(\mathbf{k}\lambda, \mathbf{r}_\ell), \tag{73}$$

where $\mu_{\ell i}^{\text{ind}}(\mathbf{k}\lambda)$ is the $(\mathbf{k}\lambda)$ Fourier component of the i cartesian component of the induced dipole moment μ_ℓ of atom $\ell = A, B$, and $\alpha_\ell(k)$ is the isotropic dynamical polarizability of atom ℓ. Because, according to (35), vacuum fluctuations are space-correlated, the induced dipole moments in atoms A and B will be correlated too. Their dipole–dipole interaction then yields an energy shift given by [36]

$$\Delta E = \sum_{\mathbf{k}\lambda} \langle \mu_{Ai}^{\text{ind}}(\mathbf{k}\lambda) \mu_{Bj}^{\text{ind}}(\mathbf{k}\lambda) \rangle V_{ij}(k, \mathbf{r} = \mathbf{r}_B - \mathbf{r}_A), \tag{74}$$

where $V_{ij}(k, \mathbf{r})$ is the potential tensor for oscillating dipoles at frequency ck, whose expression is given by (46) [88,89]. Substitution of (73) into (74) yields

$$\Delta E = \sum_{\mathbf{k}\lambda} \alpha_A(k)\alpha_B(k) \langle \{0_{\mathbf{k}\lambda}\} | E_i(\mathbf{k}\lambda, \mathbf{r}_A) E_i(\mathbf{k}\lambda, \mathbf{r}_B) | \{0_{\mathbf{k}\lambda}\} \rangle V_{ij}(k, \mathbf{r}), \tag{75}$$

where $\mathbf{r} = \mathbf{r}_B - \mathbf{r}_A$ is the distance between the two atoms. In (75), the vacuum spatial correlation for the electric field explicitly appears. After some algebraic calculation involving polarization sum and angular integration, evaluation of (75) gives, both in the nonretarded and retarded regime, the result (49) of Section 5, as obtained by a fourth-order perturbation theory, or by a second-order perturbation theory if the effective Hamiltonian (29) is used [36,90].

In this approach, the origin of dispersion interaction is evident from a physical point of view: vacuum fluctuations, being spatially correlated, induce correlated dipole moments in the two atoms, which then interact with each other through a (classical) oscillating dipolar term. In this approach, the only quantum aspect of the electromagnetic field is the existence of the spatially correlated vacuum fluctuations. All other aspects involved are just classical ones. The interaction between the two induced dipoles, expressed by $V_{ij}(k, \mathbf{r})$, has, in fact, an essentially classical origin. The r^{-7} behaviour in the retarded Casimir–Polder regime (far zone) can be now easily understood: the spatial field correlation function behaves as r^{-4}, as Equation (36) shows, and thus the correlation function of the induced dipole moments has the same space dependence. By taking into account that the (classical) interaction energy between the induced dipoles depends as r^{-3} from r, the r^{-7} far-zone interaction energy is immediately obtained. In the near zone, the situation is a bit different and somehow more complicated because the atoms respond to the field fluctuations differently for any Fourier component, according to their frequency-dependent dynamical polarizability.

This approach, based on spatial field correlations, can be generalized to three-body van der Waals and Casimir–Polder interactions; in such case, the key element are the *dressed* vacuum fluctuations [91–93]. We consider three isotropic ground-state atoms or molecules (or polarizable bodies), A, B, and C, respectively located at r_A, r_B and r_C. In order to generalize the method outlined before to the three-body interaction, we first consider how the presence of one of the three atoms, A, changes the spatial correlation function of the electric field, yielding dressed field correlations. The multipolar-coupling interaction Hamiltonian of atom A is

$$H_I^A = -\boldsymbol{\mu}_A \cdot \mathbf{E}(\mathbf{r}_A). \tag{76}$$

The second-order dressed ground state of atom A (as atoms B and C were absent) is

$$\begin{aligned}|\tilde{g}_A\rangle &= N\,|\,g_A\{0_{\mathbf{k}\lambda}\}\rangle - \frac{1}{\hbar c}\sum_m\sum_{\mathbf{k}\lambda}\frac{\langle m,\mathbf{k}\lambda\,|\,H_I^A\,|\,g_A,\{0_{\mathbf{k}\lambda}\}\rangle}{k+k_{mg}}\,|\,m,\mathbf{k}\lambda\rangle \\ &\quad + \frac{1}{(\hbar c)^2}\sum_{mn}\sum_{\mathbf{k}\lambda\mathbf{k}'\lambda'}\frac{\langle n,\mathbf{k}\lambda\mathbf{k}'\lambda'\,|\,H_I^A\,|\,m,\mathbf{k}\lambda\rangle\langle m,\mathbf{k}\lambda\,|\,H_I^A\,|\,g_A,\{0_{\mathbf{k}\lambda}\}\rangle}{(k+k_{mg})(k+k'+k_{ng})}\,|\,n,\mathbf{k}\lambda\mathbf{k}'\lambda'\rangle, \end{aligned} \tag{77}$$

where N is a normalization factor, $\hbar c k_{mg} = E_m - E_g$, and m, n are basis states of atom A. This state has virtual admixtures with one- and two-photon states, which, in the space around atom A, modify vacuum fluctuations, specifically their spatial correlations. We can now evaluate the average value of the electric-field correlation function on the dressed ground state (77), at the positions $\mathbf{r}_B$ and $\mathbf{r}_C$ of atoms B and C, respectively. It can be expressed in the form

$$\langle \tilde{g}_A\,|\,E_i(\mathbf{k}\lambda,\mathbf{r}_B)E_j(\mathbf{k}'\lambda',\mathbf{r}_C)\,|\,\tilde{g}_A\rangle = \langle E_i(\mathbf{k}\lambda,\mathbf{r}_B)E_j(\mathbf{k}'\lambda',\mathbf{r}_C)\rangle_0 + \langle E_i(\mathbf{k}\lambda,\mathbf{r}_B)E_j(\mathbf{k}'\lambda',\mathbf{r}_C)\rangle_A^{gs}. \tag{78}$$

The first term on the right-hand-side of (78) is the same of the bare vacuum, as given by Equation (35), and it is independent from the position of atom A; thus, it does not contribute to the three-body interaction, and we disregard it. The second term is the second-order correction to the field correlation due to the presence of the ground-state atom A, and contains correlation between different modes of the field; its explicit expression is [92]

$$\begin{aligned}\langle E_i(\mathbf{k}\lambda,\mathbf{r}_B)E_j(\mathbf{k}'\lambda',\mathbf{r}_C)\rangle_A^{gs} &= \frac{4\pi^2 kk'}{V^2}\sum_m (\boldsymbol{\mu}_A^{gm}\cdot\hat{\mathbf{e}}_{\mathbf{k}'\lambda'})(\boldsymbol{\mu}_A^{mg}\cdot\hat{\mathbf{e}}_{\mathbf{k}\lambda})(\hat{\mathbf{e}}_{\mathbf{k}\lambda})_i(\hat{\mathbf{e}}_{\mathbf{k}'\lambda'})_j \\ &\quad \times \left\{\frac{e^{-i\mathbf{k}\cdot(\mathbf{r}_B-\mathbf{r}_A)}e^{i\mathbf{k}'\cdot(\mathbf{r}_C-\mathbf{r}_A)}}{(k+k_{mg})(k'+k_{mg})} + \frac{e^{i\mathbf{k}\cdot(\mathbf{r}_B-\mathbf{r}_A)}e^{i\mathbf{k}'\cdot(\mathbf{r}_C-\mathbf{r}_A)}}{k+k'}\left(\frac{1}{k+k_{mg}}+\frac{1}{k'+k_{mg}}\right) + \text{c.c.}\right\}. \end{aligned} \tag{79}$$

After some algebraic manipulation, this expression can be also cast in terms of the dynamic polarizability of atom A.

The dressed correlated vacuum field induces and correlates dipole moments in atoms B and C according to (73), and this gives an interaction energy between B and C that, for the part depending from the presence of A (see Equation (79)), is

$$\Delta E_{BC} = \sum_{\mathbf{k}\lambda\mathbf{k}'\lambda'}\alpha_B(k)\alpha_C(k')\langle E_i(\mathbf{k}\lambda,\mathbf{r}_B)E_j(\mathbf{k}'\lambda',\mathbf{r}_C)\rangle_A^{gs}V_{ij}(k\mathbf{r}_B,k'\mathbf{r}_C), \tag{80}$$

where $V_{ij}(k\mathbf{r}_B,k'\mathbf{r}_C)$ is an appropriate generalization, to the present case of dipoles oscillating at frequencies ck and ck', of the potential tensor previously introduced for the two-body potential. We assume the symmetric expression

$$V_{ij}(k\mathbf{r}_B, k'\mathbf{r}_C) = \frac{1}{2}\left(V_{ij}(k,\mathbf{r}_B) + V_{ij}(k',\mathbf{r}_C)\right). \tag{81}$$

The quantity ΔE_{BC} in (80) is the interaction between B and C in the presence of A. If we symmetrize on the role of the three atoms, after some algebraic calculations, we obtain the correct expression of the three-body potential energy in the far zone,

$$\begin{aligned}\Delta E_{ABC}^{gs} &= \frac{2}{3}\left(\Delta E_{AB} + \Delta E_{BC} + \Delta E_{AC}\right) \\ &= -\frac{\hbar c}{\pi}\alpha_A \alpha_B \alpha_C F_{ij}^\alpha F_{j\ell}^\beta F_{\ell i}^\gamma \frac{1}{\alpha\beta\gamma(\alpha+\beta+\gamma)},\end{aligned} \tag{82}$$

where the distances α, β and γ between the atoms have been defined in Section 6 (see also Figure 1). This expression coincides with the interaction energy $\Delta E_3(A,B,C)$ of Equation (58), originally obtained through a sixth-order perturbative calculation based on the multipolar Hamiltonian [6], or a third-order calculation using the effective Hamiltonian (29) [32]. This gives a new physical picture of the three-body interaction: it originates from the interaction of two atoms (B and C), whose dipole moments have been induced and correlated by the spatially correlated electric field fluctuations, as modified (*dressed*) by atom A.

The physical picture of the three-body dispersion interaction for ground-state atoms outlined above, based on dressed vacuum field correlations, can be extended to atoms in excited states too, both in static and dynamical (time-dependent) situations. Let us consider one excited atom (A), approximated as a two-level system with frequency $\omega_0 = ck_0$, and two ground-state atoms (B and C), interacting with the quantum electromagnetic field in the vacuum state. We proceed similarly to the case of three ground-state atoms, taking the atom A as a sort of source atom; firstly, we evaluate its dressed ground state, and then the dressed spatial correlation on this state of the electric field at the position of the other two atoms. We use the multipolar coupling Hamiltonian (we cannot use the effective Hamiltonian (29) for the excited atom A because it is not valid on the energy shell, and we are now dealing with an excited state). The main difference, with respect to the previous case of a ground-state atom, is that a pole at ck_0 is now present in the frequency integration, and this yields an extra (resonant) term in the correlation function. After some algebra, we obtain [94]

$$\begin{aligned}\langle E_i(\mathbf{k}\lambda,\mathbf{r}_B)E_j(\mathbf{k}'\lambda',\mathbf{r}_C)\rangle_A^{es} &= -\frac{2k_0\mu_\ell^A\mu_m^A}{\pi}F_{i\ell}^\gamma F_{jm}^\beta \frac{1}{\beta\gamma}\int_0^\infty du \frac{e^{-u(\beta+\gamma)}}{k_0^2+u^2} \\ &\quad + 2\mu_\ell^A\mu_m^A F_{i\ell}^\gamma F_{\ell j}^\beta \frac{\cos(k_0(\gamma-\beta))}{\beta\gamma},\end{aligned} \tag{83}$$

where the bare term, independent of the position of atom A, has been disregarded because it does not contribute to the three-body interaction (it only contributes to the two-body components, as previously shown in this section). In the first term of (83), we recognize the quantity $-2k_0\mu_\ell^A\mu_m^A/(\hbar c(k_0^2+u^2)) = \alpha_A^{es}(iu)$ as the dynamical polarizability for the excited state of the two-level atom A, evaluated at imaginary frequencies. It is easy to see that this term is the same of the ground-state case (see Equation (79)), except for the presence of the excited-state polarizability of atom A. The second term is new, and originates from the presence of a resonant pole at $k = k_0$ in the integration over k; it contains contributions from only the frequency ck_0. With a procedure analogue to that leading to (82) for ground-state atoms, including appropriate symmetrization over the atoms, for the three-body interaction energy of one excited- and two ground-state atoms, we finally obtain [94],

$$\begin{aligned}\delta E_{ABC}^{es} &= -\mu_n^A\mu_p^A\alpha_B(k_0)\alpha_C(k_0)F_{ij}^\alpha F_{j\ell}^\beta F_{\ell i}^\gamma \frac{1}{\alpha\beta\gamma}\left[\cos(k_0(\beta-\gamma+\alpha)) + \cos(k_0(\beta-\gamma-\alpha))\right] \\ &\quad -\frac{\hbar c}{\pi}\alpha_A^{es}\alpha_B\alpha_C F_{ij}^\alpha F_{j\ell}^\beta F_{\ell i}^\gamma \frac{1}{\alpha\beta\gamma}\int_0^\infty du\, \alpha_A^{es}(iu)\alpha_B(iu)\alpha_C(iu)e^{-u(\alpha+\beta+\gamma)}.\end{aligned} \tag{84}$$

The second term in (84) has the same structure of the ground-state three-body dispersion energy discussed previously, except for the presence of the excited-state polarizability of atom A in place of the ground-state polarizability. The first term is a new one, arising from the resonant photon exchange, characterized by a different scaling with the distance and by the presence of space oscillations with frequency ck_0, the transition frequency of the excited two-level atom [94–96]. Although this result, analogously to (82), can be also obtained by a perturbative approach, the derivation outlined in this section gives a clear and transparent physical insight of the three-body interaction as due to the modification of zero-point field correlations due to the presence of one atom, or in general a polarizable body. It should be noted that, in the present case of one excited atom, we have neglected its spontaneous decay, and thus are assuming that the times considered are shorter than its decay time. The method used can be usefully extended also to dynamical (i.e., time-dependent) situations, in order to investigate dynamical many-body dispersion (van der Waals and Casimir–Polder) interactions between atoms [97–99], as well as dynamical atom–surface interactions [100–106], for example during the dynamical dressing of one atom starting from a nonequilibrium situation.

8. Casimir–Polder Forces between Atoms Nearby Macroscopic Boundaries

In this section, we show that the methods discussed in the previous section, based on the properties of the quantum electrodynamical vacuum, can be extended to dispersion interactions between atoms or molecules in the presence of a macroscopic body, specifically an infinite plate of a perfectly reflecting material. This is relevant because it shows that dispersion interactions can be manipulated (enhanced or inhibited) through the environment—for example, a cavity [37] or a metallic waveguide [107,108], similarly also to other radiative procsses such as, for example, the radiative energy transfer [109,110].

We consider two neutral atoms, A and B, in the vacuum space at zero temperature, near a perfectly conducting infinite plate located at $z = 0$; $\mathbf{r}_A$ and $\mathbf{r}_B$ are respectively the positions of atoms A and B. We have already seen in Section 4 that the presence of the plate changes vacuum fluctuations; mathematically, this is due to the necessity of setting appropriate boundary conditions on the field operators at the plate surface. Due to the deep relation of dispersion interactions to field fluctuations, shown in Section 7, we must expect that also the van der Waals and Casimir–Polder force (and any other radiation-mediated interaction) will change due to the plate [7,43]. A fourth-order perturbative calculation has been done in Ref. [37], using the multipolar coupling Hamiltonian with the appropriate mode functions of the field in the presence of the infinite reflecting plate; in the far zone, the result is

$$\Delta E_r(r,\bar{r}) = -\frac{23\hbar c}{4\pi}\alpha_A\alpha_B\frac{1}{r^7} - \frac{23\hbar c}{4\pi}\alpha_A\alpha_B\frac{1}{\bar{r}^7} + \frac{8\hbar c}{\pi}\frac{\alpha_A\alpha_B}{r^3\bar{r}^3(r+\bar{r})^5}$$
$$\times \left[r^4\sin^2\theta + 5r^3\bar{r}\sin^2\theta + r^2\bar{r}^2(6+\sin^2\theta+\sin^2\bar\theta) + 5r\bar{r}^3\sin^2\bar\theta + \bar{r}^4\sin^2\bar\theta\right], \quad (85)$$

where $\mathbf{r} = \mathbf{r}_B - \mathbf{r}_A$ is the distance between the atoms, $\bar{\mathbf{r}} = \mathbf{r}_B - \sigma\mathbf{r}_A$ is the distance of atom B from the image of atom A (at point $\sigma\mathbf{r}_A$) with respect to the plate, having defined the reflection matrix $\sigma = \text{diag}(1,1,-1)$; θ and $\bar\theta$ are, respectively, the angles of $\mathbf{r}$ and $\bar{\mathbf{r}}$ with respect to the normal to the plate. The geometrical configuration is illustrated in Figure 2. This well-known expression shows that the retarded Casimir–Polder potential between the two atoms consists of three terms: the potential between A and B, scaling as r^{-7}, as in absence of the plate; the potential between an atom and the reflected image of the other atom, scaling as $\bar{r}^{-7}$; a term involving both distances r and $\bar{r}$.

We now show that the potential (85) can also be obtained in a simpler way through the same methods used in Section 7 for atoms in the unbounded space, based on dressed energy densities and vacuum field correleations, stressing new physical insights on the origin of this potential.

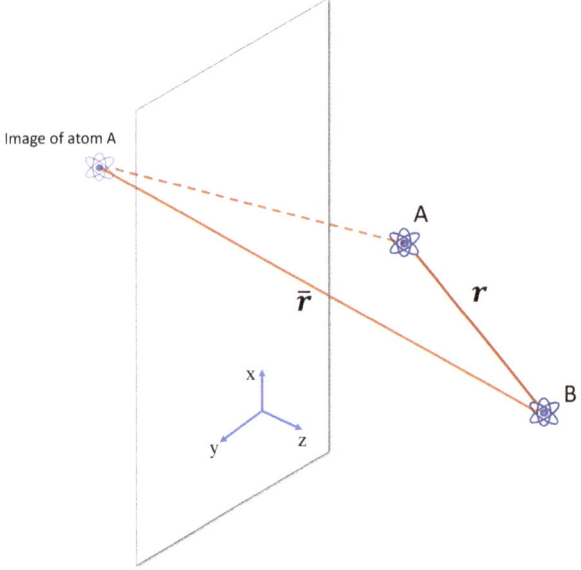

Figure 2. Two atoms, A and B, nearby a reflecting plane boundary at $z = 0$.

We use the effective Hamiltonian (29) with the electric field operator given by (38), using the appropriate mode functions $\mathbf{f}(\mathbf{k}\lambda, \mathbf{r})$ for a single infinite perfectly reflecting wall [2,12,37]. The outline of the calculation is the same as in the case of atoms in free space (see Section 7.1): we evaluate the renormalized dressed electric energy density due to one atom (A), and then its interaction energy with the other atom (B). The dressed ground state of atom A, as if atom B were absent, is given by

$$| \tilde{g}_A \rangle = | g_A, \{0_{\mathbf{k}\lambda}\} \rangle - \frac{\pi}{V} \sum_{\mathbf{k}\lambda \mathbf{k}'\lambda'} \alpha_A(k) \frac{\sqrt{kk'}}{k+k'} \mathbf{f}(\mathbf{k}\lambda, \mathbf{r}_A) \cdot \mathbf{f}(\mathbf{k}'\lambda', \mathbf{r}_A) | g_A, 1_{\mathbf{k}\lambda} 1_{\mathbf{k}'\lambda'} \rangle. \quad (86)$$

We now evaluate the average value of the effective interaction Hamiltonian relative to atom B on this state, disregarding bare terms that do not depend on the atomic distances, which is [47]

$$\Delta E_{AB} = -\frac{1}{2} \sum_{\mathbf{k}\lambda \mathbf{k}'\lambda'} \alpha_B(k) \langle \tilde{g}_A | \mathbf{E}(\mathbf{k}\lambda, \mathbf{r}_B) \cdot \mathbf{E}(\mathbf{k}'\lambda', \mathbf{r}_B) | \tilde{g}_A \rangle. \quad (87)$$

Using the expression (38) of the electric field operator with the appropriate mode functions $\mathbf{f}(\mathbf{k}\lambda, \mathbf{r})$ relative to the reflecting plate, polarization sum in (87) yields

$$\sum_\lambda f_i(\mathbf{k}\lambda, \mathbf{r}_A) f_j(\mathbf{k}\lambda, \mathbf{r}_B) = \left(\delta_{ij} - \hat{k}_i \hat{k}_j \right) e^{i\mathbf{k} \cdot (\mathbf{r}_A - \mathbf{r}_B)} - \sigma_{i\ell} \left(\delta_{\ell j} - \hat{k}_\ell \hat{k}_j \right) e^{i\mathbf{k} \cdot (\mathbf{r}_A - \sigma \mathbf{r}_B)}, \quad (88)$$

where i, j, ℓ are Cartesian components, and σ is the reflection matrix defined above. A comparison with (37) immediately shows an extra term due to the reflecting plate. Explicit evaluation of (87) in the far zone ($\alpha_{A,B}(k) \simeq \alpha_{A,B}$), using (88), after some algebra finally yields the correct result (85), originally obtained in Ref. [37] from a fourth-order calculation. The present approach, besides stressing the role of (dressed) vacuum fluctuations and their modification due to the plate, has allowed to obtain the same result through a simpler first-order calculation with the effective Hamiltonian given in Equation (29) [47].

In addition, the method based on vacuum spatial correlations of the electric field operator, outlined in Section 7.2, can be extended to the case when boundaries are present. Following the approach used in Section 7.2, we need to evaluate the following quantity

$$\Delta E_{AB}(r,\bar{r}) = \sum_{\mathbf{k}\lambda} \langle \{0_{\mathbf{k}\lambda}\} \mid E_i(\mathbf{k}\lambda, \mathbf{r}_A) E_j(\mathbf{k}\lambda, \mathbf{r}_B) \mid \{0_{\mathbf{k}\lambda}\} \rangle V_{ij}(k,r,\bar{r}), \tag{89}$$

where $\mid \{0_{\mathbf{k}\lambda}\} \rangle$ is the vacuum state of the field when the plate is present, and $V_{ij}(k,r,\bar{r})$ is the potential tensor giving the interaction energy for oscillating (induced) dipoles. In the present case, however, the expression of $V_{ij}(k,r,\bar{r})$ differs from (46), valid for dipoles in the free space, because of the effect of the image dipoles. We thus take

$$V_{ij}(k,r,\bar{r}) = V_{ij}(k,r) - \sigma_{i\ell} V_{\ell j}(k,\bar{r}), \tag{90}$$

where the second term takes into account the image dipole, reflected on the plate. Substitution of the appropriate spatial field correlation function (obtained using (88)) and (90) into (89), after algebraic calculations, finally yields the correct expression (85) for the retarded dispersion for two atoms nearby the conducting plate [47]. This finding indicates that the method based on field correlations can be directly extended to the case when boundary conditions are present, provided the appropriate field correlation function is used and the images dipole(s) are included in the classical potential tensor. It can be shown that this method is valid also for evaluating the dispersion interaction with boundary conditions and at a finite temperature [111]. All this is relevant because these methods can provide very useful computational tools to evaluate dispersion interactions in complicated geometries, also allowing the possibility to change and manipulate radiation-mediated interactions between atoms or molecules through the environment. This is still more striking for the resonance interaction, where the exchange of real photons is also present [43,112,113].

In the system considered above, we have assumed that the boundary is a perfectly reflecting one. This is an idealised situation, of course, because any real dielectric or metal material is characterised by specific magnetodielectric properies. For example, a real metal can be described with the plasma or the Drude-Lorentz model [114,115]. In the plasma model, the metal is transparent for frequencies above its plasma frequency ω_p, and this can affect renormalized vacuum fluctuations and the dispersion interaction energy between atoms nearby the boundary. It is expected that in general the corrections to vacuum fluctuations can be relevant only for short distances from the boundary [39], and the same is also expected for atom–surface and atom–atom interactions, similarly to the Casimir effect for dielectrics, as obtained through the Lifshits formula [116,117]. Indeed, the van der Waals interaction between an atom and a metallic plasma surface has been calculated, showing relevant corrections due to the finite plasma frequency in the short distance regime [118]. In addition, the dispersion interaction between a ground-state atom and a dielectric surface, and between two atoms nearby a dielectric surface, have been evaluated in terms of the dielectric constant of the surface at imaginary frequencies, using the theory of electrical images [119]. In the case of the atom–atom interaction, a strong dependence (suppression or enhancement) of the interaction on the geometry of the two atoms with respect to the surface has been found using the linear response theory [120]; enhanced dispersion interaction between two atoms in a dielectric slab between two different dielectric media has also been found [121]. The effect of real boundaries on atom–surface and atom–atom dispersion interactions can be conveniently included through appropriate body-assisted quantization of the electromagnetic field, based on the Green's functions technique [28,29,122–126]. In this approach, medium-assisted bosonic operators for the field are introduced, which take into account all magnetodielectric (dispersive and dissipative) properties of the linear macroscopic boundaries through their frequency-dependent complex electric permittivity and magnetic permeability; then, expressions of dispersion interactions for atoms or molecules nearby a generic linear environment can be obtained [10,11,28,124], as well as expressions of the intermolecular energy transfer [127].

9. Casimir–Polder and Resonance Interactions between Uniformly Accelerated Atoms

The results reviewed in the previous sections have shown the deep relation between renormalized vacuum energy densities, and zero-point field fluctuations, with dispersion interactions, in particular in the retarded Casimir–Polder regime. All these results and effects have been obtained for atoms and boundaries at rest. New phenomena appear when they are set in motion. In the case of a uniformly accelerated motion, the well-known Fulling–Davies–Unruh effect predicts that the accelerated observer perceives vacuum fluctuations as a thermal bath with the Unruh temperature $T_U = \hbar a / 2\pi c k_B$, where a is its proper acceleration and k_B is the Boltzmann constant [2,128–131]. This effect leads, for example, to a change of the Lamb shift for atoms in uniformly accelerated motion [132,133], as well as to spontaneous excitation of an accelerating atom due to its interaction with the Unruh thermal quanta [134–136]. For a boundary moving with a nonuniform acceleration (a single oscillating reflecting plate, or a metal cavity with an oscillating wall, for example), the dynamical Casimir effect occurs, consisting of the emission of pairs of real photons from the vacuum [137,138]. Theoretical analysis of the dynamical Casimir effect involves quantization of the scalar or electromagnetic field with a moving boundary [137,139,140]. Analogous effects related to Casimir energies occur in the case of a metallic cavity with a movable wall, whose mechanical degrees of freedom are described quantum-mechanically, and thus subject to quantum fluctuations of its position [41–43,141,142]. Another relevant effect involving atoms in motion, extensively investigated in the literature, is the quantum friction of an atom moving at uniform speed parallel to a surface [29,143–146]. An important question that can be asked, and that we shall address in this section, is what is the effect of a non inertial motion in the vacuum space of two or more atoms on their dispersion interaction. We should expect a change of the dispersion interaction between the atoms because of their accelerated motion; in fact, they perceive a different vacuum, equivalent to a thermal bath, and in view of the temperature dependence of the dispersion interactions, mentioned in Section 5, their interaction energy should change. However, the Unruh effect is a very tiny effect, and an acceleration of the order of 10^{20} m/s^2 is necessary to get a Unruh temperature around 1 K. Spontaneous excitation of a uniformly accelerated atom has been predicted [134], as well as changes of the Lamb shift of atomic levels and of the atom–surface Casimir–Polder force [133,147,148]. Although some experimental setups to detect the Unruh effect have been proposed [149], this effect has not been observed yet (a related effect, the Hawking radiation has been recently observed in analogue systems, specifically in an acoustical black hole [150]). The change of dispersion interactions due to a uniformly accelerated motion of the atoms, if observed, could provide a signature of the Unruh effect [151].

For the sake of simplicity, we consider two two-level atoms of frequency ω_0, separated by a distance z, and interacting with the massless relativistic scalar field. For atoms at rest and at zero temperature, the scalar Casimir–Polder interaction at zero temperature behaves as z^{-2} in the near zone ($z \ll c/\omega_0$), and as z^{-3} in the far zone ($z \gg c/\omega_0$). At a finite temperature T, after defining the thermal length $\rho^{\text{therm}} = \hbar c / (2\pi k_B T)$ and assuming $z \gg c/\omega_0$, the interaction for $z \ll \rho^{\text{therm}}$ has the same behaviour as at zero temperature, with a subleading thermal correction proportional to T^2/z, both in the near and in the far zone. At very large distances, $z \gg \rho^{\text{therm}}$, the interaction energy is proportional to T/z^2, thus scaling with the distance as in the near zone at zero temperature (this should also be compared to the case of atoms interacting with the electromagnetic field at a finite temperature, mentioned in Section 5). We note that the scaling with the distance z is different compared with the cases considered in the previous sections because we are now considering the scalar rather than the electromagnetic field.

We describe the two identical atoms, A and B, as two-level systems of frequency ω_0, using in the Dicke formalism [152]. They interact with the relativistic massless scalar field $\phi(x)$, and move in the vacuum space with the same constant proper acceleration a along the direction x, perpendicular to their distance (along z), so that their separation is constant.

The Hamiltonian of this system, in the comoving frame, which is the system in which the atoms are instantaneously at rest, is

$$H = \hbar\omega_0 S_z^A(\tau) + \hbar\omega_0 S_z^B(\tau) + \sum_{\mathbf{k}} \hbar\omega_k a_\mathbf{k}^\dagger a_\mathbf{k} \frac{dt}{d\tau} + \lambda S_x^A(\tau)\phi[x_A(\tau)] + \lambda S_x^B(\tau)\phi[x_B(\tau)], \quad (91)$$

where S_i ($i = x, y, z$) are pseudospin operators, τ is the proper time, shared by both atoms in our hypothesis, λ is a coupling constant, and $x_{A(B)}(\tau)$ are the atomic trajectories

$$t_A(\tau) = \frac{c}{a}\sinh(\frac{a\tau}{c}), \quad x_A(\tau) = \frac{c^2}{a}\cosh(\frac{a\tau}{c}); \quad y_A(\tau) = 0; \quad z_A(\tau) = z_A,$$

$$t_B(\tau) = \frac{c}{a}\sinh(\frac{a\tau}{c}), \quad x_B(\tau) = \frac{c^2}{a}\cosh(\frac{a\tau}{c}); \quad y_B(\tau) = 0; \quad z_A(\tau) = z_A + z, \quad (92)$$

with $\tau_A = \tau_B = \tau$ the common proper time of the atoms, and z the (constant) distance between them [153,154]. The massless scalar field operator is

$$\phi(\mathbf{r},t) = \sum_\mathbf{k} \sqrt{\frac{\hbar}{2V\omega_k}}\left[a_\mathbf{k} e^{i(\mathbf{k}\cdot\mathbf{r}-\omega_k t)} + a_\mathbf{k}^\dagger e^{-i(\mathbf{k}\cdot\mathbf{r}-\omega_k t)}\right], \quad (93)$$

with the dispersion relation $\omega_k = c|\mathbf{k}|$. The Casimir–Polder energy is the distance-dependent energy shift due to the atoms-field interaction, and it is a fourth-order effect in the coupling constant λ. The calculation of the relevant part of the energy shift can be done using a method, originally introduced by Dupont-Roc et al. to separate vacuum fluctuations and radiation reaction contributions to second-order radiative shifts [155,156], even in accelerated frames [157,158], and recently generalized to fourth-order processes [159,160]. We find that a new distance scale, related to the acceleration, $z_a = c^2/a$, appears. For $z \ll z_a$, the dispersion interaction for atoms with the same uniform acceleration a is the same as that for atoms at rest in a thermal bath with the Unruh temperature $T_U = \hbar a/2\pi c k_B$. In this regime, our extended system of two atoms exhibits the Unruh equivalence between acceleration and temperature, similarly to a point-like detector. However, for larger distances, $z \gg z_a$, we have

$$\Delta E^{accel} \simeq -\frac{\lambda^4}{512\pi^4 \hbar c \omega_0^2 a}\frac{1}{z^4}. \quad (94)$$

In this regime, the interaction energy decays with the distance faster than in the near and far zone, showing lack of the equivalence between acceleration and temperature [159]. In addition, the dispersion interaction between accelerating atoms has qualitative features, specifically its distance dependence, different from those of inertial atoms; measure of the dispersion force between atoms subjected to a uniform acceleration could thus give an indirect signature of the Unruh effect.

The non-thermal behaviour of acceleration at large distances is obtained also for the resonant interaction energy between two identical accelerating atoms, one excited and the other in the ground state, prepared in an entangled symmetric or antisymmetric Bell-type state

$$|\phi_\pm\rangle = \frac{1}{\sqrt{2}}(|e_A g_B, \{0_{\mathbf{k}\lambda}\}\rangle \pm |g_A e_B, \{0_{\mathbf{k}\lambda}\}\rangle), \quad (95)$$

where $e_{A(B)}$ and $g_{A(B)}$ respectively represent the excited and the ground state of the atoms. In this case, a *resonance* interaction energy between the two correlated atom exists, due to the exchange of one real or virtual photon between them.

We consider two identical two-level atoms, A and B, with frequency $\omega_0 = ck_0$, and interacting with the electromagnetic field through the multipolar coupling Hamiltonian. μ is the dipole moment operator. In the unbounded space, at zero temperature, the resonance interaction energy for atoms at rest, obtained by evaluating the second-order energy shift due to the interaction Hamiltonian (42), is given by

$$\Delta E_{\pm}^{\text{res}} = \pm \mu_{Ai}^{eg} \mu_{Bj}^{ge} \left[\left(\delta_{ij} - 3\hat{r}_i \hat{r}_j \right) \left(\cos(k_0 r) + k_0 r \sin(k_0 r) \right) - \left(\delta_{ij} - \hat{r}_i \hat{r}_j \right) k_0^2 r^2 \cos(k_0 r) \right] \frac{1}{r^3}, \quad (96)$$

where the $+$ or $-$ sign respectively refers to the symmetric or antisymmetric state in (95), r is the distance between the atoms, and $\mu_{A(B)i}^{eg}$ is the matrix element of $(\mu_{A(B)})_i$ between the excited and the ground state, assumed real. In the near zone ($k_0 r \ll 1$), the interaction energy (96) scales as r^{-3}, while in the far zone ($k_0 r \gg 1$) it scales as r^{-1} with space oscillations [15,112].

We now assume that the two atoms are moving with a uniform proper acceleration along x, and that their distance is along z, so that it is constant in time; their trajectory is given by Equation (92). The Hamiltonian, similarly to Equation (91) of the scalar field case, within dipole approximation and in the comoving frame, is

$$H = H_A + H_B + \sum_{\mathbf{k}\lambda} \hbar \omega_k a_{\mathbf{k}\lambda}^\dagger a_{\mathbf{k}\lambda} \frac{dt}{d\tau} - \boldsymbol{\mu}_A(\tau) \cdot \mathbf{E}(x_A(\tau)) - \boldsymbol{\mu}_B(\tau) \cdot \mathbf{E}(x_B(\tau)), \quad (97)$$

where $H_{A(B)}$ is the Hamiltonian of atom A (B), and $\boldsymbol{\mu}_{A(B)}$ is their dipole moment operator.

After lengthy algebraic calculations, for $z \gg z_a = c^2/a$, the following expression for the resonant interaction energy between the two accelerating atoms, in the comoving frame, is obtained

$$\Delta E_{\pm}^{\text{accel}} \simeq \pm \mu_{A\ell}^{eg} \mu_{Bm}^{ge} \frac{1}{z^3} \left\{ \left(\delta_{\ell m} - q_\ell q_m - 2n_\ell n_m \right) \left[\frac{2\omega_0 z}{c} \sin \left(\frac{2\omega_0 c}{a} \log \left(\frac{az}{c^2} \right) \right) \right. \right.$$
$$\left. \left. - \frac{\omega_0^2 z^2}{c^2} \left(\frac{2c^2}{za} \right) \cos \left(\frac{2\omega_0 c}{a} \log \left(\frac{az}{c^2} \right) \right) \right] + q_\ell q_m \left(\frac{8c^2}{az} \right) \cos \left(\frac{2\omega_0 c}{a} \log \left(\frac{az}{c^2} \right) \right) \right\}, \quad (98)$$

where ℓ, m are Cartesian components, $\mathbf{n} = (0,0,1)$ is a unit vector along z (the direction of the distance between the atoms), $\mathbf{q} = (1,0,0)$ is a unit vector along x (the direction of the acceleration), and $\pm$ refers to the symmetric or antisymmetric superposition in (95) [161]. Comparison of (98) with (96) shows that the acceleration of the atoms can significantly change the distance-dependence of the resonance interaction, with respect to inertial atoms. For atoms at rest, Equation (96) gives a far-zone dependence as z^{-1}, while for accelerating atoms Equation (98) asymptotically gives a more rapid decrease of the interaction with the distance: z^{-2} if the two dipoles are along z or y, and z^{-4} for dipole oriented along the direction of the acceleration, x. This change in the dependence of the interaction energy from the interatomic distance is a signature of the noninertial motion of the atoms, and ultimately related to the Unruh effect (even if, in this case, the change of the interaction energy is not equivalent to that due to a thermal field). Moreover, Equation (98) shows striking features of the accelerated-atoms case with respect to the orientation of the atomic dipole moments. For example, if the two dipoles are orthogonal to each other, with one along z and the other in the plane (x, y), the interaction energy vanishes for inertial atoms, as immediately follows from (96); on the other hand, Equation (98) shows that it is different from zero when $a \neq 0$. In such a case, the resonance interaction is a unique signature of the accelerated motion of the atoms [161]. Similar results are also found in the coaccelerated frame [162], and for atoms nearby a reflecting boundary [163]. Finally, we wish to mention that very recently the Casimir–Polder and resonance interactions between atoms has been also investigated in the case of a curved-spacetime background [164–166].

In conclusion, the results outlined in this section give a strong indication that the radiation-mediated interactions, specifically dispersion and resonance interactions, between uniformly accelerated atoms could provide a promising setup to detect the effect of a noninertial motion on radiative processes, possibly allowing an indirect detection of the Unruh effect through the measurement of dispersion or resonance interaction between accelerating atoms [151,159,161,163].

10. Conclusions

In this review, we have discussed recent developments on radiation-mediated interactions between atoms or molecules, in the framework of nonrelativistic quantum electrodynamics. Major emphasis has been on stressing the role of zero-point field fluctuations and (dressed) vacuum field energy densities and spatial correlations. This has allowed us to give a transparent physical interpretation of dispersion (van der Waals and Casimir–Polder) and resonance interactions, as well as useful computational tools for their evaluation, even in more complicated situations, for example in the presence of boundaries or for uniformly accelerating atoms. We have shown that dispersion interactions can be seen as a direct consequence of the existence of vacuum fluctuations, with features directly related to their physical properties. We have also discussed how dispersion and resonance interactions could provide an experimental setup to get an indirect evidence of the Unruh effect, also testing the Unruh acceleration-temperature equivalence.

A fundamental final question arises: are van der Waals and Casimir–Polder dispersion interactions a definitive proof of the real existence of vacuum fluctuations and zero-point energy? In the author's opinion, the answer is yes and no. As clearly shown by all physical systems and situations discussed in this review, the results here reviewed are certainly fully consistent with the existence of vacuum fluctuations. Assuming the existence of vacuum fluctuations of the electromagnetic field, as predicted by nonrelativistic quantum electrodynamics, we can derive these observable phenomena, using clear and transparent physical models. However, the same results can be also obtained in a different way, that is, from source fields, without invoking vacuum fluctuations [71]. A similar consideration applies to the Lamb shift, for example [2]. This approach is based on the solution of the Heisenberg equations of motion for the field operators, which contains a free (vacuum) term and a source term, the latter depending on the presence of matter, while the former is related to the vacuum field. Using a specific ordering, specifically the normal ordering, between field and atomic operators (while the atomic operators commute with the full field operators, in general they do not commute with the single parts of field operators related to vacuum and source terms [167,168]); it is possible to show that only the radiation reaction term contributes to dispersion interactions when this ordering of operators is chosen [71,169]. Thus, it seems that a deep dichotomy between vacuum fluctuations and source fields exists in nature. As pointed out and stressed by P.W. Milonni, the two physical models and interpretations of several radiative processes, in terms of vacuum fluctuations or source fields (radiation reaction), should be considered as the *two sides of a coin* [72]. In conclusion, we just wish to mention that, probably, a definitive and unambiguous confirmation of the existence of the zero-point energy (density) can be obtained only from its gravitational effects, it being a component of the energy-momentum tensor, and thus a source term for the gravitational field [170–172]. The essential point is that Casimir and Casimir–Polder forces are always related to *differences* of vacuum energies for different configurations of the system; necessarily, this involves the presence of matter and thus of both vacuum and source fields. On the contrary, gravitational effects are related to the absolute value of the vacuum energy density [173]. In any case, as we have pointed out above, dispersion interactions between atoms, also in the presence of macroscopic bodies or for accelerated systems, are observable physical effects fully consistent with the real existence of the zero-point energy of the quantum electromagnetic theory, even if they cannot be considered as a definitive proof of the existence of the vacuum energy.

Funding: This research received no external funding.

Acknowledgments: The content of this review, and many ideas therein, owe much to the long-standing past collaboration of the author with Franco Persico, Edwin A. Power and Thiru Thirunamachandran, to whose memory this paper is dedicated. The author also wishes to thank Lucia Rizzuto for a careful and critical reading of the manuscript. The author gratefully acknowledges financial support from the Julian Schwinger Foundation.

Conflicts of Interest: The author declares no conflict of interest.

References

1. Casimir, H.B.G.; Polder, D. The Influence of Retardation on the London–van der Waals Forces. *Phys. Rev.* **1948**, *73*, 360. [CrossRef]
2. Milonni, P.W. *The Quantum Vacuum. An Introduction to Quantum Electrodynamics*; Academic Press: San Diego, CA, USA, 1994.
3. Power, E.A. Casimir–Polder potential from first principles. *Eur. J. Phys.* **2001**, *22*, 453. [CrossRef]
4. Barnett, S.M.; Aspect, A.; Milonni, P.W. On the quantum nature of the Casimir–Polder interaction. *J. Phys. B* **2000**, *33*, L143. [CrossRef]
5. Axilrod, B.M.; Teller, E. Interaction of the van der Waals type between three atoms. *J. Chem. Phys.* **1943**, *11*, 299. [CrossRef]
6. Aub, M.R.; Zienau, S. Studies on the retarded interaction between neutral atoms I. Three-body London–van der Waals interaction of neutral atoms. *Proc. R. Soc. A* **1960**, *257*, 464. [CrossRef]
7. Milton, K.A.; Abalo, E.; Parashar, P.; Shajesh, K.V. Three-body Casimir–Polder interactions. *Nuovo Cimento C* **2013**, *36*, 183. [CrossRef]
8. Brevik, M.; Marachevsky, V.N.; Milton, K.A. Identity of the van der Waals Force and the Casimir Effect and the Irrelevance of These Phenomena to Sonoluminescence. *Phys. Rev. Lett.* **1999**, *82*, 3948. [CrossRef]
9. Barcellona, P.; Passante, R. A microscopic approach to Casimir and Casimir–Polder forces between metallic bodies. *Ann. Phys.* **2015**, *355*, 282. [CrossRef]
10. Buhmann, S.Y.; Welsch, D.-K. Dispersion forces in macroscopic quantum electrodynamics. *Prog. Quantum Electron.* **2007**, *31*, 51. [CrossRef]
11. Scheel, S.; Buhmann, S.Y. Macroscopic quantum electrodynamics—Concepts and applications. *Acta Phys. Slov.* **2008**, *58*, 675. [CrossRef]
12. Compagno, G.; Passante, R.; Persico, F. *Atom-Field Interactions and Dressed Atoms*; Cambridge University Press: Cambridge, UK, 1995.
13. Power, E.A. *Introductory Quantum Electrodynamics*; Longmans: London, UK, 1964.
14. Power, E.A.; Thirunamachandran, T. On the nature of the Hamiltonian for the interaction of radiation with atoms and molecules: $(e/mc)\mathbf{p} \cdot \mathbf{A}, -\boldsymbol{\mu} \cdot \mathbf{E}$, and all that. *Am. J. Phys.* **1978**, *46*, 370. [CrossRef]
15. Craig, D.P.; Thirunamachandran, T. *Molecular Quantum Electrodynamics*; Dover Publ.: Mineola, NY, USA, 1998.
16. London, F. Zur Theorie und Systematik der Molekularkräfte. *Z. Phys.* **1930**, *63*, 245. [CrossRef]
17. Power, E.A.; Zienau, S. On the radiative contribution to the van der Waals Force. *Nuovo Cim.* **1957**, *6*, 7. [CrossRef]
18. Power, E.A.; Zienau, S. Coulomb gauge in non-relativistic quantum electrodynamics and the shape of spectral lines. *Philos. Trans. R. Soc. A* **1959**, *251*, 427. [CrossRef]
19. Woolley, R.G. Molecular quantum electrodynamics. *Proc. R. Soc. Lond. A* **1971**, *321*, 557. [CrossRef]
20. Cohen-Tannoudji, C.; Dupont-Roc, J.; Grynberg, G. *Photons and Atoms: Introduction to Quantum Electrodynamics*; Wiley: New York, NY, USA, 1989.
21. Atkins, P.W.; Woolley, R.G. The interaction of molecular multipoles with the electromagnetuc field in the canonical formulation of non-covariant quantum electrodynamics. *Proc. R. Soc. Lond. A* **1970**, *319*, 549. [CrossRef]
22. Babiker, M.; Loudon, R. Derivation of the Power–Zienau–Woolley Hamiltonian in quantum electrodynamics by gauge transformation. *Proc. R. Soc. Lond. A* **1983**, *385*, 439. [CrossRef]
23. Andrews, D.L.; Jones, G.A.; Salam, A.; Woolley, R.G. Perspective: Quantum Hamiltonians for optical interactions. *J. Chem. Phys.* **2018**, *148*, 040901. [CrossRef]
24. Salam, A. *Molecular Quantum Electrodynamics: Long-Range Intermolecular Interactions*; Wiley: Hoboken, NJ, USA, 2010.
25. Bykov, V.P. *Radiation of Atoms in a Resonant Environment*; World Scientific: Singapore, 1993.
26. Biswas, A.K.; Compagno, G.; Palma, G.M.; Passante, R.; Persico, F. Virtual photons and causality in the dynamics of a pair of two-level atoms. *Phys. Rev. A* **1990**, *42*, 4291. [CrossRef]
27. Compagno, G.; Palma, G.M.; Passante, R.; Persico, F. Relativistic causality and quantum-mechanical states in the Fermi problem. *Chem. Phys.* **1995**, *198*, 19. [CrossRef]
28. Buhmann, S.Y. *Dispersion Forces I: Macroscopic Quantum Electrodynamics and Ground-State Casimir, Casimir–Polder and van der Waals forces*; Springer: Berlin, Germany, 2012.

29. Buhmann, S.Y. *Dispersion Forces II: Many-Body Effects, Excited Atoms, Finite Temperature and Quantum Friction*; Springer: Berlin, Germany, 2012.
30. Salam, A. Molecular quantum electrodynamics in the Heisenberg picture: A field theoretic viewpoint. *Int. Rev. Phys. Chem.* **2008**, *27*, 405. [CrossRef]
31. Salam, A. *Non-Relativistic QED Theory of the Van Der Waals Dispersion Interaction*; Springer: Cham, Switzerland, 2016.
32. Passante, R.; Power, E.A.; Thirunamachandran, T. Radiation-molecule coupling using dynamic polarizabilities: Application to many-body forces. *Phys. Lett. A* **1998**, *249*, 77. [CrossRef]
33. Passante, R.; Power, E.A. The Lamb shift in non-relativistic quantum electrodynamics. *Phys. Lett. A* **1987**, *122*, 14. [CrossRef]
34. Craig, D.P.; Power, E.A. The asymptotic Casimir–Polder potential from second-order perturbation theory and its generalization for anisotropic polarizabilities. *Int. J. Quantum Chem.* **1969**, *3*, 903. [CrossRef]
35. Casimir, H.B.G. On the attraction between two perfectly conducting plates. *Proc. Kon. Ned. Akad. Wet.* **1948**, *51*, 793.
36. Power, E.A.; Thirunamachandran, T. Casimir–Polder potential as an interaction between induced dipoles. *Phys. Rev. A* **1993**, *48*, 4761. [CrossRef]
37. Power, E.A.; Thirunamachandran, T. Quantum electrodynamics in a cavity. *Phys. Rev. A* **1982**, *25*, 2473. [CrossRef]
38. Ford, L.H.; Svaiter, N.F. Vacuum energy density near fluctuating boundaries. *Phys. Rev. D* **1998**, *58*, 065007. [CrossRef]
39. Bartolo, N.; Passante, R. Electromagnetic-field fluctuations near a dielectric-vacuum boundary and surface divergences in the ideal conductor limit. *Phys. Rev. A* **2012**, *86*, 012122. [CrossRef]
40. Bartolo, N.; Butera, S.; Lattuca, M.; Passante, R.; Rizzuto, L.; Spagnolo, S. Vacuum Casimir energy densities and field divergences at boundaries. *J. Phys. Condens. Matter* **2015**, *27*, 214015. [CrossRef]
41. Butera, S.; Passante, R. Field Fluctuations in a One-Dimensional Cavity with a Mobile Wall. *Phys. Rev. Lett.* **2013**, *111*, 060403. [CrossRef] [PubMed]
42. Armata, F.; Passante, R. Vacuum energy densities of a field in a cavity with a mobile boundary. *Phys. Rev. D* **2015**, *91*, 025012. [CrossRef]
43. Armata, F.; Butera, S.; Fiscelli, G.; Incardone, R.; Notararigo, V.; Palacino, R.; Passante, R.; Rizzuto, L.; Spagnolo, S. Effect of boundaries on vacuum field fluctuations and radiation mediated interactions between atoms. *J. Phys. Conf. Ser.* **2017**, *880*, 012064. [CrossRef]
44. Shahmoon, E. Van der Waals and Casimir–Polder dispersion forces. In *Forces of the Quantum Vacuum. An Introduction to Casimir Physics*; Simpson, W.M.R., Leonhardt, U., Eds.; World Scientific Publ. Co.: Singapore, 2015; p. 61.
45. Margenau, H. Van der Waals forces. *Rev. Mod. Phys.* **1939**, *11*, 1. [CrossRef]
46. Babb, J.F. Casimir effects in atomic, molecular, and optical physics. In *Advances in Atomic, Molecular, and Optical Physics*; Arimondo, E., Berman, P.R., Lin, C.C., Eds.; Elsevier Inc.: London, UK, 2010; Volume 59, p. 1.
47. Spagnolo, S.; Passante, R.; Rizzuto, L. Field fluctuations near a conducting plate and Casimir–Polder forces in the presence of boundary conditions. *Phys. Rev. A* **2006**, *73*, 062117. [CrossRef]
48. Verwey, E.J.W.; Overbeek, J.T. *Theory of the Stability of Lyophobic Colloids*; Dover Publ.: Mineola, NY, USA, 1999.
49. Przybytek, M.; Jeziorski, B.; Cencek, W.; Komasa, J.; Mehl, J.B.; Szalewicz, K. Onset of Casimir–Polder Retardation in a Long-Range Molecular Quantum State. *Phys. Rev. Lett.* **2012**, *108*, 183201. [CrossRef] [PubMed]
50. Béguin, L.; Vernier, A.; Chicireanu, R.; Lahaye, T.; Browaeys, A. Direct measurement of the van der Waals Interaction between Two Rydberg Atoms. *Phys. Rev. Lett.* **2013**, *110*, 263201. [CrossRef]
51. McLachlan, A.D. Retarded dispersion forces in dielectrics at finite temperatures. *Proc. R. Soc. Lond. A* **1963**, *274*, 80. [CrossRef]
52. Boyer, T.H. Temperature dependence of Van der Waals forces in classical electrodynamics with classical electromagnetic zero-point radiation. *Phys. Rev. A* **1975**, *11*, 1650. [CrossRef]
53. Goedecke, G.H.; Wood, R.C. Casimir–Polder interaction at finite temperature. *Phys. Rev. A* **1999**, *11*, 2577. [CrossRef]
54. Barton, G. Long-range Casimir–Polder-Feinberg-Sucher intermolecular potential at nonzero temperature. *Phys. Rev. A* **2001**, *64*, 032102. [CrossRef]

55. Power, E.A.; Thirunamachandran, T. Quantum electrodynamics with nonrelativistic sources. V. Electromagnetic field correlations and intermolecular interactions between molecules in either ground or excited states. *Phys. Rev. A* **1993**, *47*, 2593. [CrossRef]
56. Rizzuto, L.; Passante, R.; Persico, F. Dynamical Casimir–Polder energy between an excited- and a ground-state atom. *Phys. Rev. A* **2004**, *70*, 012107. [CrossRef]
57. Berman, P.R. Interaction energy of nonidentical atoms. *Phys. Rev. A* **2015**, *91*, 042127. [CrossRef]
58. Donaire, M.; Guérout, R.; Lambrecht, A. Quasiresonant van der Waals Interaction between nonidentical atoms. *Phys. Rev. Lett.* **2015**, *115*, 033201. [CrossRef]
59. Milonni, P.W.; Rafsanjani, S.M.H. Distance dependence of two-atom dipole interactions with one atom in an excited state. *Phys. Rev. A* **2015**, *92*, 062711. [CrossRef]
60. Barcellona, P.; Passante, R.; Rizzuto, L.; Buhmann, S.Y. Van der Waals interactions between excited atoms in generic environments. *Phys. Rev. A* **2016**, *94*, 012705. [CrossRef]
61. Power, E.A.; Thirunamachandran, T. Dispersion interactions between atoms involving electric quadrupole polarizabilities. *Phys. Rev. A* **1996**, *53*, 1567. [CrossRef]
62. Salam, A.; Thirunamachandran, T. A new generalization of the Casimir–Polder potential to higher electric multipole polarizabilities. *J. Chem. Phys.* **1996**, *104*, 5094. [CrossRef]
63. Salam, A. A general formula obtained from induced moments for the retarded van derWaals dispersion energy shift between two molecule with arbitrary electric multipole polarizabilities: I. Ground state interactions. *J. Phys. B* **2006**, *39*, S651. [CrossRef]
64. Jenkins, J.K.; Salam, A.; Thirunamachandran, T. Retarded dispersion interaction energies between chiral molecules. *Phys. Rev. A* **1994**, *50*, 4767. [CrossRef] [PubMed]
65. Salam, A. On the effect of a radiation field in modifying the intermolecular interaction between two chiral molecules. *J. Chem. Phys.* **2006**, *124*, 014302. [CrossRef] [PubMed]
66. Barcellona, P.; Passante, R.; Rizzuto, L.; Buhmann, S.Y. Dynamical Casimir–Polder interaction between a chiral molecule and a surface. *Phys. Rev. A* **2016**, *93*, 032508. [CrossRef]
67. Power, E.A.; Thirunamachandran, T. The non-additive dispersion energies for *N* molecules: A quantum electrodynamical theory. *Proc. R. Soc. Lond. A* **1985**, *401*, 167. [CrossRef]
68. Salam, A. Higher-order electric multipole contributions to retarded non-additive three-body dispersion interaction energies between atoms: Equilateral triangle and collinear configurations. *J. Chem. Phys.* **2013**, *139*, 244105. [CrossRef] [PubMed]
69. Salam, A. Dispersion potential between three-bodies with arbitrary electric multipole polarizabilities: Molecular QED theory. *J. Chem. Phys.* **2014**, *140*, 044111. [CrossRef]
70. Buhmann, S.Y.; Salam, A. Three-Body Dispersion Potentials Involving Electric Octupole Coupling. *Symmetry* **2018**, *10*, 343. [CrossRef]
71. Milonni, P.W. Casimir forces without the vacuum radiation field. *Phys. Rev. A* **1982**, *25*, 1315. [CrossRef]
72. Milonni, P.W. Different ways of looking at the electromagnetic vacuum. *Phys. Scr.* **1988**, *T21*, 102. [CrossRef]
73. Power, E.A. Zero-point energy and the Lamb shift. *Am. J. Phys.* **1966**, *34*, 516. [CrossRef]
74. Compagno, G.; Passante, R.; Persico, F. The role of the cloud of virtual photons in the shift of the ground state energy of a hydrogen atom. *Phys. Lett. A* **1983**, *98*, 253. [CrossRef]
75. Passante, R.; Compagno, G.; Persico, F. Cloud of virtual photons in the ground state of the hydrogen atom. *Phys. Rev. A* **1985**, *31*, 2827. [CrossRef]
76. Passante, R.; Rizzuto, L.; Spagnolo, S. Vacuum local and global electromagnetic self-energies for a point-like and an extended field source. *Eur. Phys. J. C* **2013**, *73*, 2419. [CrossRef]
77. *NIST Handbook of Mathematical Functions*; Olver, F.W.J., Lozier, D.W., Boisvert, R.F., Clark, C.W., Eds.; NIST and Cambridge University Press: Cambridge, UK, 2010.
78. Passante, R.; Power, E.A. Electromagnetic-energy-density distribution around a ground-state hydrogen atom and connection with van der Waals forces. *Phys. Rev. A* **1987**, *35*, 188. [CrossRef]
79. Compagno, G.; Palma, G.M.; Passante, R.; Persico, F. Atoms dressed and partially dressed by the zero-point fluctuations of the electromagnetic field. *J. Phys. B* **1995**, *28*, 1105. [CrossRef]
80. Feinberg, G.; Sucher, J. General theory of the van der Waals Interaction: A model-independent Approach. *Phys. Rev. A* **1970**, *2*, 2395. [CrossRef]
81. Passante, R.; Persico, F. Virtual photons and three-body forces. *J. Phys. B* **1999**, *32*, 19. [CrossRef]

82. Compagno, G.; Persico, F.; Passante, R. Interference in the virtual photon clouds of two hydrogen atoms. *Phys. Lett. A* **1985**, *112*, 215. [CrossRef]
83. Hinds, E.A.; Sandoghdar, V. Cavity QED level shifts of simple atoms. *Phys. Rev. A* **1991**, *43*, 398. [CrossRef] [PubMed]
84. Messina, R.; Passante, R.; Rizzuto, L.; Spagnolo, S.; Vasile, R. Casimir–Polder forces, boundary conditions and fluctuations. *J. Phys. A* **2008**, *41*, 164031. [CrossRef]
85. Barton, G. Frequency shifts near an interface: Inadequacy of two-level atomic models. *J. Phys. B* **1974**, *29*, 1871. [CrossRef]
86. Passante, R.; Rizzuto, R.; Spagnolo, S.; Petrosky, T.Y.; Tanaka, S. Harmonic oscillator model for the atom–surface Casimir–Polder interaction energy. *Phys. Rev. A* **2012**, *85*, 062109. [CrossRef]
87. Ciccarello, F.; Karpov, E.; Passante, R. Exactly solvable model of two three-dimensional harmonic oscillators interacting with the quantum electromagnetic field: The far-zone Casimir–Polder potential. *Phys. Rev. A* **2005**, *72*, 052106. [CrossRef]
88. Born, M.; Wolf, E. *Principles of Optics*; Pergamon Press: Oxford, UK, 1980.
89. McLone, R.R.; Power, E.A. On the Interaction between two identical neutral dipole systems, one in an excited state and the other in the ground state. *Mathematika* **1964**, *11*, 91. [CrossRef]
90. Passante, R.; Persico, F.; Rizzuto, L. Spatial correlations of vacuum fluctuations and the Casimir–Polder potential. *Phys. Lett. A* **2003**, *316*, 29. [CrossRef]
91. Cirone, M.; Passante, R. Vacuum field correlations and the three-body Casimir–Polder potential. *J. Phys. B* **1996**, *29*, 1871. [CrossRef]
92. Cirone, M.; Passante, R. Dressed zero-point field correlations and the non-additive three-body van der Waals potential. *J. Phys. B* **1997**, *30*, 5579. [CrossRef]
93. Cirone, M.A.; Mostowski, J.; Passante, R.; Rzążewsi, K. The concept of vacuum in nonrelativistic QED. *Recent. Res. Devel. Physics* **2001**, *2*, 131.
94. Passante, R.; Persico, F.; Rizzuto, L. Vacuum field correlations and three-body Casimir–Polder potential with one excited atom. *J. Mod. Opt.* **2005**, *52*, 1957. [CrossRef]
95. Power, E.A.; Thirunamachandran, T. Dispersion forces between molecules with one or both molecules excited. *Phys. Rev. A* **1995**, *51*, 3660. [CrossRef]
96. Power, E.A.; Thirunamachandran, T. Two- and three-body dispersion forces with one excited molecule. *Chem. Phys.* **1995**, *198*, 5. [CrossRef]
97. Passante, R.; Persico, F.; Rizzuto, L. Causality, non-locality and three-body Casimir–Polder energy between three ground-state atoms. *J. Phys. B* **2006**, *39*, S685. [CrossRef]
98. Passante, R.; Persico, F.; Rizzuto, L. Nonlocal field correlations and dynamical Casimir–Polder forces between one excited- and two ground-state atoms. *J. Phys. B* **2007**, *40*, 1863. [CrossRef]
99. Rizzuto, L.; Passante, R.; Persico, F. Nonlocal Properties of Dynamical Three-Body Casimir–Polder Forces. *Phys. Rev. Lett.* **2007**, *98*, 240404. [CrossRef]
100. Vasile, R.; Passante, R. Dynamical Casimir–Polder force between an atom and a conducting wall. *Phys. Rev. A* **2008**, *78*, 032108. [CrossRef]
101. Shresta, S.; Hu, B.L.; Phillips, N.G. Moving atom-field interaction: Correction to the Casimir–Polder effect from coherent backaction. *Phys. Rev. A* **2003**, *68*, 062101. [CrossRef]
102. Hu, B.L.; Roura, A.; Shresta, S. Vacuum fluctuations and moving atoms/detectors: From the Casimir–Polder to the Unruh–Davies–DeWitt–Fulling effect. *J. Opt. B Quantum Semiclass. Opt.* **2004**, *6*, S698. [CrossRef]
103. Messina, M.; Vasile, R.; Passante, R. Dynamical Casimir–Polder force on a partially dressed atom near a conducting wall. *Phys. Rev. A* **2010**, *82*, 062501. [CrossRef]
104. Messina, R.; Passante, R.; Rizzuto, L.; Spagnolo, S.; Vasile, R. Dynamical Casimir–Polder potentials in non-adiabatic conditions. *Phys. Scr.* **2014**, *T160*, 014032. [CrossRef]
105. Haakh, H.R.; Henkel, C.; Spagnolo, S.; Rizzuto, L.; Passante, R. Dynamical Casimir–Polder interaction between an atom and surface plasmons. *Phys. Rev. A* **2014**, *89*, 022509. [CrossRef]
106. Armata, F.; Vasile, R.; Barcellona, P.; Buhmann, S.Y.; Rizzuto, L.; Passante, R. Dynamical Casimir–Polder force between an excited atom and a conducting wall. *Phys. Rev. A* **2016**, *94*, 042511. [CrossRef]
107. Haakh, H.R.; Scheel, S. Modified and controllable dispersion interaction in a one-dimensional waveguide geometry. *Phys. Rev. A* **2015**, *91*, 052707. [CrossRef]

108. Dung, H.T. Interatomic dispersion potential in a cylindrical system: Atoms being off axis. *J. Phys. B* **2016**, *49*, 165502. [CrossRef]
109. Weeraddana, D.; Premaratne, M.; Gunapala, S.D.; Andrews, D.L. Controlling resonance energy transfer in nanostructure emitters by positioning near a mirror. *J. Chem. Phys.* **2017**, *147*, 074117. [CrossRef] [PubMed]
110. Fiscelli, G.; Rizzuto, L.; Passante, R. Resonance energy transfer between two atoms in a conducting cylindrical waveguide. *Phys. Rev. A* **2018**, *98*, 013849. [CrossRef]
111. Passante, R.; Spagnolo, S. Casimir–Polder interatomic potential between two atoms at finite temperature and in the presence of boundary conditions. *Phys. Rev. A* **2007**, *76*, 042112. [CrossRef]
112. Incardone, R.; Fukuta, T.; Tanaka, S.; Petrosky, T.; Rizzuto, L.; Passante, R. Enhanced resonant force between two entangled identical atoms in a photonic crystal. *Phys. Rev. A* **2014**, *89*, 062117. [CrossRef]
113. Notararigo, V.; Passante, R.; Rizzuto, L. Resonance interaction energy between two entangled atoms in a photonic bandgap environment. *Sci. Rep.* **2018**, *8*, 5193. [CrossRef]
114. Bordag, M.; Klimchitskaya, G.L.; Mohideen, U.; Mostepanenko, V.M. *Advances in Casimir Physics*; Oxford Science Publications: Oxford, UK, 2009.
115. Kittel, C. *Introduction to Solid State Physics*; John Wiley & Sons: Hoboken, NJ, USA, 2004.
116. Lifshits, E.M. The theory of molecular attractive fiorces between solids. *Sov. Phys. JETP* **1956**, *2*, 73.
117. Lifshits, E.M.; Pitaevskii. *Landau and Lifshits Course of Theoretical Physics, Vol. 9: Statistical Physics, Part 2*; Pergamon Press: Oxford, UK, 1980.
118. Babiker, M.; Barton, G. Quantum frequency shifts near a plasma surface. *J. Phys. A* **1976**, *9*, 129. [CrossRef]
119. McLachlan, A.D. Van der Waals forces between an atom and a surface. *Mol. Phys.* **1964**, *7*, 381. [CrossRef]
120. Cho, M.; Silbey, R.J. Suppression and enhancement of van der Waals interactions. *J. Chem. Phys.* **1996**, *104*, 8730. [CrossRef]
121. Marcovitch, M.; Diamant, H. Enhanced dispersion interaction in confined geometry. *Phys. Rev. Lett.* **2005**, *95*, 223203. [CrossRef] [PubMed]
122. Matloob, R.; Loudon, R.; Barnett, S.M.; Jeffers, J. Electromagnetic field quantization in absorbing dielectrics. *Phys. Rev. A* **1995**, *52*, 4823. [CrossRef]
123. Gruner, T.; Welsch, D.-G. Green-function approach to the radiation-field quantization for homogeneous and inhomogeneous Kramers-Kronig dielectrics. *Phys. Rev. A* **1996**, *53*, 1818. [CrossRef]
124. Dung, H.T.; Knöll, L.; Welsch, D.-G. Three-dimensional quantization of the electromagnetic field in dispersive and absorbing inhomogeneous dielectrics. *Phys. Rev. A* **1998**, *57*, 3931. [CrossRef]
125. Buhmann, S.Y.; Butcher, D.T.; Scheel, S. Macroscopic quantum electrodynamics in nonlocal and nonreciprocal media. *New J. Phys.* **2012**, *65*, 032813. [CrossRef]
126. Scheel, S. The Casimir stress in real materials. In *Forces of the Quantum Vacuum. An Introduction to Casimir Physics*; Simpson, W.M.R., Leonhardt, U., Eds.; World Scientific Publ. Co.: Singapore, 2015; p. 107.
127. Dung, H.T.; Knöll, L.; Welsch, D.-G. Intermolecular energy transfer in the presence of dispersing and absorbing media. *Phys. Rev. A* **2002**, *14*, 083034. [CrossRef]
128. Unruh, W.G. Notes on black-hole evaporation. *Phys. Rev. D* **1976**, *14*, 870. [CrossRef]
129. Fulling, S.A. Nonuniqueness of canonical field quantization in Riemannian space-time. *Phys. Rev. D* **1973**, *7*, 2850. [CrossRef]
130. Davies, P.C.W. Scalar production in Schwarzschild and Rindler metrics. *J. Phys. A* **1973**, *8*, 609. [CrossRef]
131. Crispino, L.C.B.; Higuchi, A.; Matsas, G.E.A. The Unruh effect and its applications. *Rev. Mod. Phys.* **2008**, *80*, 787. [CrossRef]
132. Audretsch, G.; Müller, R. Radiative energy shifts of an accelerated two-level system. *Phys. Rev. A* **1995**, *52*, 629. [CrossRef] [PubMed]
133. Passante, R. Radiative level shifts of an accelerated hydrogen atom and the Unruh effect in quantum electrodynamics. *Phys. Rev. A* **1998**, *57*, 1590. [CrossRef]
134. Audretsch, G.; Müller, R. Spontaneous excitation of an accelerated atom: The contributions of vacuum fluctuations and radiation reaction. *Phys. Rev. A* **1994**, *50*, 1755. [CrossRef] [PubMed]
135. Zhu, A.; Yu, H. Fulling-Davies-Unruh effect and spontaneous excitation of an accelerated atom interacting with a quantum scalar field. *Phys. Lett. B* **2006**, *645*, 459. [CrossRef]
136. Calogeracos, A. Spontaneous excitation of an accelerated atom: (i) acceleration of infinite duration (the Unruh effect), (ii) acceleration of finite duration. *Res. Phys.* **2016**, *6*, 377. [CrossRef]

137. Moore, G.T. Quantum theory of the electromagnetic field in a variable-length one-dimensional cavity. *J. Math. Phys.* **1970**, *11*, 2679. [CrossRef]
138. Dodonov, V.V. Current status of the dynamical Casimir effect. *Phys. Scr.* **2010**, *82*, 038105. [CrossRef]
139. Dodonov, V.V.; Klimov, A.B. Generation and detection of photons in a cavity with a resonantly oscillating boundary. *Phys. Rev. A* **1996**, *53*, 2664. [CrossRef]
140. Mundarain, D.F.; Maia Neto, P.A.M. Quantum radiation in a plane cavity with moving mirrors. *Phys. Rev. A* **1998**, *57*, 1379. [CrossRef]
141. Law, C.K. Effective Hamiltonian for the radiation in a cavity with a moving mirror and a time-varying dielectric medium. *Phys. Rev. A* **1994**, *49*, 433. [CrossRef]
142. Law, C.K. Interaction between a moving mirror and radiation pressure: A Hamiltonian formulation. *Phys. Rev. A* **1995**, *51*, 2537. [CrossRef] [PubMed]
143. Dalvit, D.A.R.; Maia Neto, P.A.; Mazzitelli, D. Fluctuations, dissipation and the dynamical Casimir effect. In *Casimir Physics*; Dalvit, D., Milonni, P., Roberts, D., Rosa, F., Eds.; Springer: Berlin, Germany, 2011; p. 419.
144. Barton, G. On van der Waals friction. I. Between two atoms. *New J. Phys.* **2010**, *10*, 113044. [CrossRef]
145. Barton, G. On van der Waals friction. II: Between atom and half-space. *New J. Phys.* **2010**, *10*, 113045; Corrigendum in *New J. Phys.* **2012**, *14*, 079502, doi:10.1088/1367-2630/14/7/07950. [CrossRef]
146. Intravaia, F.; Mkrtchian, V.E.; Buhmann, S.Y.; Scheel, S.; Dalvit, D.A.R.; Henkel, C. Friction forces on atoms after acceleration. *J. Phys. Condens. Matter* **2015**, *27*, 214020. [CrossRef] [PubMed]
147. Rizzuto, L.; Spagnolo, S. Lamb shift of a uniformly accelerated hydrogen atom in the presence of a conducting plate. *Phys. Rev. A* **2009**, *79*, 062110. [CrossRef]
148. Rizzuto, L. Casimir–Polder interaction between an accelerated two-level system and an infinite plate. *Phys. Rev. A* **2007**, *76*, 062114. [CrossRef]
149. Schützhold, R.; Schaller, G.; Habs, D. Signatures of the Unruh Effect from Electrons Accelerated by Ultrastrong Laser Fields. *Phys. Rev. Lett.* **2006**, *97*, 121302. [CrossRef]
150. Steinhauer, J. Observation of quantum Hawking radiation and its entanglement in an analogue black hole. *Nature Phys.* **2016**, *12*, 959. [CrossRef]
151. Noto, A.; Passante, R. Van der Waals interaction energy between two atoms moving with uniform acceleration. *Phys. Rev. D* **2013**, *88*, 025041. [CrossRef]
152. Dicke, R.H. Coherence in spontaneous radiation processes. *Phys. Rev.* **1954**, *93*, 99. [CrossRef]
153. Rindler, W. *Relativity. Special, General, and Cosmological*; Oxford Univ. Press: Oxford, UK, 2006.
154. Birrell, N.D.; Davies, P.C.W. *Quantum Fields in Curved Space*; Cambridge Univ. Press: Cambridge, UK, 1982.
155. Dalibard, J.; Dupont-Roc, J.; Cohen-Tannoudji, C. Vacuum fluctuations and radiation reaction: Identification of their respective contributions. *J. Phys. (Fr.)* **1982**, *43*, 1617. [CrossRef]
156. Dalibard, J.; Dupont-Roc, J.; Cohen-Tannoudji, C. Dynamics of a small system coupled to a reservoir: reservoir fluctuations and self-reaction. *J. Phys. (Fr.)* **1984**, *45*, 637. [CrossRef]
157. Menezes, G.; Svaiter, N.F. Radiative processes of uniformly accelerated entangled atoms. *Phys. Rev. A* **2016**, *93*, 052117. [CrossRef]
158. Zhou, W.; Yu, H. Spontaneous excitation of a uniformly accelerated atom coupled to vacuum Dirac field fluctuations. *Phys. Rev. A* **2012**, *86*, 033841. [CrossRef]
159. Marino, J.; Noto, A.; Passante, R. Thermal and Nonthermal Signatures of the Unruh Effect in Casimir–Polder Forces. *Phys. Rev. Lett.* **2014**, *113*, 020403. [CrossRef]
160. Noto, A.; Marino, J.; Passante, R. A fourth–order statistical method for the calculation of dispersion Casimir–Polder interactions. **2018**, in preparation.
161. Rizzuto, L.; Lattuca, M.; Marino, J.; Noto, A.; Spagnolo, S.; Zhou, W.; Passante, R. Nonthermal effects of acceleration in the resonance interaction between two uniformly accelerated atoms. *Phys. Rev. A* **2016**, *94*, 012121. [CrossRef]
162. Zhou, W.; Passante, R.; Rizzuto, L. Resonance interaction energy between two accelerated identical atoms in a coaccelerated frame and the Unruh effect. *Phys. Rev. D* **2016**, *94*, 105025. [CrossRef]
163. Zhou, W.; Passante, R.; Rizzuto, L. Resonance Dipole–Dipole Interaction between Two Accelerated Atoms in the Presence of Reflecting Plane Boundary. *Symmetry* **2018**, *10*, 185. [CrossRef]
164. Menezes, G.; Kiefer, C.; Marino, J. Thermal and nonthermal scaling of the Casimir–Polder interaction in a black hole spacetime. *Phys. Rev. D* **2016**, *95*, 085014. [CrossRef]

165. Zhou, W.; Yu, Y. Resonance interatomic energy in a Schwarzschild spacetime. *Phys. Rev. D* **2017**, *96*, 045018. [CrossRef]
166. Zhou, W.; Yu, Y. Boundarylike behaviors of the resonance interatomic energy in a cosmic string spacetime. *Phys. Rev. D* **2018**, *97*, 045007. [CrossRef]
167. Senitzky, I.R. Radiation-reaction and vacuum-field effects in Heisenberg-picture quantum electrodynamics. *Phys. Rev. Lett.* **1973**, *31*, 955. [CrossRef]
168. Milonni, P.W.; Ackerhalt, J.R.; Smith, W.A. Interpretation of radiative corrections in spontaneous emission. *Phys. Rev. Lett.* **1973**, *31*, 958. [CrossRef]
169. Milonni, P.W. Semiclassical and quantum-electrodynamical approaches in nonrelativistic radiation theory. *Phys. Rep.* **1976**, *25*, 1. [CrossRef]
170. Adler, R.J.; Casey, B.; Jacob, O.C. Vacuum catastrophe: An elementary exposition of the cosmological constant problem. *Am. J. Phys.* **1995**, *63*, 620. [CrossRef]
171. Cree, S.S.; Davis, T.M.; Ralph, T.C.; Wang, Q.; Zhu, Z.; Unruh, W.G. Can the fluctuations of the quantum vacuum solve the cosmological constant problem? *Phys. Rev. D* **2018**, *98*, 063506. [CrossRef]
172. Solá, J. Cosmological constant and vacuum energy: Old and new ideas. *J. Phys. Conf. Ser.* **2013**, *453*, 012015. [CrossRef]
173. Carroll, S.M. *Spacetime and Geometry: An Introduction to General Relativity*; Pearson Education Limited: Harlow, UK, 2014.

© 2018 by the author. Licensee MDPI, Basel, Switzerland. This article is an open access article distributed under the terms and conditions of the Creative Commons Attribution (CC BY) license (http://creativecommons.org/licenses/by/4.0/).

Review

Symmetries, Conserved Properties, Tensor Representations, and Irreducible Forms in Molecular Quantum Electrodynamics

David L. Andrews

University of East Anglia, Norwich Research Park, Norwich NR4 7TJ, UK; david.andrews@physics.org

Received: 13 June 2018; Accepted: 11 July 2018; Published: 23 July 2018

Abstract: In the wide realm of applications of quantum electrodynamics, a non-covariant formulation of theory is particularly well suited to describing the interactions of light with molecular matter. The robust framework upon which this formulation is built, fully accounting for the intrinsically quantum nature of both light and the molecular states, enables powerful symmetry principles to be applied. With their origins in the fundamental transformation properties of the electromagnetic field, the application of these principles can readily resolve issues concerning the validity of mechanisms, as well as facilitate the identification of conditions for widely ranging forms of linear and nonlinear optics. Considerations of temporal, structural, and tensorial symmetry offer significant additional advantages in correctly registering chiral forms of interaction. More generally, the implementation of symmetry principles can considerably simplify analysis by reducing the number of independent quantities necessary to relate to experimental results to a minimum. In this account, a variety of such principles are drawn out with reference to applications, including recent advances. Connections are established with parity, duality, angular momentum, continuity equations, conservation laws, chirality, and spectroscopic selection rules. Particular attention is paid to the optical interactions of molecules as they are commonly studied, in fluids and randomly organised media.

Keywords: symmetry; parity; quantum electrodynamics; optics; nanophotonics; chirality; helicity; optical activity; optical angular momentum; dual transform; electromagnetic duality; irreducible tensor; multiphoton process; quantum information

1. Introduction

It is a truism that principles of symmetry lie at the heart of modern physics. Indeed, it is perhaps to be expected, when scientific reductionism demands fundamental theory to be valid at every level from the smallest subatomic particle upwards. A well-known illustration is afforded by the symmetry principles associated with spherical geometry, which largely determine the character of electronic transitions in atoms—and thereby the form of each atomic spectrum. By contrast, it might be supposed that in the realm of the larger agglomerations we designate as molecules, with a vast multitude of shapes and structures, the operation of symmetry principles would be less prominent. Yet, a moment's reflection tells us this is not so. Consider, for example, the lowly water molecule: it is only because its three atoms, as a result of their intrinsic electronic structure, form a bent rather than a linear arrangement (Schoenflies point group C_{2v} as opposed to $D_{\infty h}$) that H_2O possesses an electric dipole moment—and every life as we know it could not exist otherwise.

The interactions of light with matter exhibit a range of especially puissant symmetry principles, many owing their origin to the intrinsic features of electromagnetism as one of the four fundamental forces of nature. Just as the atoms in a molecule are primarily held together by electrical forces, molecules engage with light by primarily electrical—and to some extent, magnetic—forms of coupling.

At this level, where the quantum nature of the world is very evident, the one theory that correctly accounts for the optical properties of both molecules and light, in their mutual interactions, is molecular quantum electrodynamics (QED) [1–8]. It draws on principles that operate at the deepest fundamental level; QED is widely known as the most successful theory in physics, unsurpassed in the precision of its agreement with experiment. However, it is not simply quantitative accuracy for which the theory has value; it has a robust character of immense power for determining absolutely the validity, viability, and necessary conditions for optical interactions of a wide-ranging nature, based on principles of structural and mathematical symmetry.

In the concise overview that follows, a range of such principles is drawn out with reference to some of the most recent applications. Connections are established with continuity equations, conservation laws, and spectroscopic selection rules, and particular attention is paid to the optical interactions of molecules in fluids (gas, liquid, or solution phases) or randomly organised media—which together account for most common molecular systems. Although, with relatively little reformulation, almost the same framework has been shown to apply to dielectric solids, quantum dots, and quantum wells, those are excluded from the analysis that follows, simply for the sake of brevity. The article substantially builds upon a recent review of the role of symmetry in the quantum theory of nanoscale optical and material chirality [9]; the expanded scope of the present work more widely addresses optical phenomena in molecular systems, with a particular focus on optical transitions. A differently formulated group theoretical approach is necessary to address non-molecular, effectively continuous materials [10].

The following analysis begins in Section 2 with a brief review of charge-parity-time ($\mathcal{CPT}$) symmetry with conservation laws and electro-magnetic dual symmetry discussed in Section 3. (By happy coincidence, the initials $\mathcal{CPT}$ are shared by three pioneers in this field: Craig, Power, and Thiru (as Thirunamachandran insisted he be called, for ease to Westerners). The three worked together extensively, though seldom publishing as a threesome; they were very well known to both the present author and the editor of this special issue. Sadly, all three have departed this life since the millennium). The subsequent Section 4 introduces the full foundation for a detailed analysis of various forms of photon–molecule interaction—those explicitly involving real quantum transitions in the material medium, with directly associated selection rules. In Section 5, the further development of the theory for multiphoton processes then introduces the construction of a convenient representation for radiation and molecular tensors, whose structure and permutational symmetry receives detailed attention in Section 6. After a focus on the general form of observables in Section 7, Section 8 introduces Cartesian tensors of irreducible form, facilitating identification of the dependencies of multiphoton processes on experimental configurations—such as beam geometries and polarization—and on molecular structure. Here, the group theoretical connection with angular momentum coupling comes to the fore. On this basis, Section 9 develops a symmetry categorisation of transition classes, establishing a connection to information content. Then, in Section 10, it is shown how, as a result of effecting isotropic or axial averaging procedures, dramatic simplification ensues when the theory is further developed for application to measurements on fluids, or indeed any substantially or partially disordered molecular system. To illustrate the application and significance of several principles outlined within this paper, Section 11 provides a concise illustration of how they apply to the elucidation of some interesting dichroic effects in the simple case of single-photon absorption. The paper ends in Section 12 with a brief discussion of recent applications.

2. Charge-Parity-Time Symmetry in Molecular Electrodynamics

Symmetry principles are powerfully operative in determining the allowed or forbidden character of optical processes in molecular systems. To fully appreciate the origin of the detailed rules that emerge from such considerations, in each form of interaction, it is necessary to formulate theory that treats both matter and light with full quantum rigour. The wide variety of symmetry types into which molecules fall affords an even greater variety regarding their individual electronic and vibrational quantum states.

The realm of optics and electrodynamics generally addresses mechanisms that fundamentally involve the positions and motions of electrical charges. Accordingly, it is the symmetry laws associated with charge, spatial position, and time that are of primary significance, that is, the operations of charge, space, and time inversion denoted by $\mathcal{C}$, $\mathcal{P}$, and $\mathcal{T}$, respectively [11–13]. Each is formally represented by the Abelian group Z_2, whose ±1 eigenvalues signify even or odd parity. All optical phenomena preserve symmetry under the product operation $\mathcal{CPT}$—a proof of this universality and analysis of its implications has been authoritatively presented in a recent review by Lehnert [12], and a broad spectroscopic perspective on the topic has been given by Lazzeretti [14]. Nonetheless, considerations of charge conjugation symmetry are seldom relevant for conventional electrodynamic phenomena, as the mathematical operation $\mathcal{C}$ is never physically realized; clouds of negative charge always surround positively charged nuclei. Accordingly, in the consideration of optical effects, it is usually sufficient to restrict consideration to the $\mathcal{PT}$ product, which, through the constraints of Lorentz invariance, ensures Hamiltonian operators of Hermitian form. Moreover, $\mathcal{PT}$-symmetric quantum theory has been shown to be exactly equivalent to standard (Hermitian) quantum mechanics in terms of all observables [15].

It is worth emphasizing that applying the symmetry operation of time reversal to any mathematical representation both changes the sign of any explicitly occurring time variable, and it effects Hermitian conjugation—which also subsumes complex variable conjugation. In terms of relativity theory, this is consistent with the four-vector symmetry for Lorentz transformations on the Minkowski space (ict, x, y, z); in the sphere of quantum mechanics, it also ensures, for example, that the Hamiltonian operator $i\hbar\partial/\partial t$ is itself time-even [16,17]. An extensive summary of the properties, physical significance, and interpretation of P and T within the framework of molecular QED is given in the literature [9]. Other issues of non-Hermitian photonics and $\mathcal{PT}$ symmetry, which specifically relate to non-molecular media, and are therefore beyond the scope of the present article, are notably discussed in two other recent references [18,19].

3. Dual Symmetry and Conservation Laws in Quantum Electromagnetism

To proceed, it is appropriate to recall that for the constituent fields of electromagnetic radiation, the electric field e is formally of odd parity under $\mathcal{P}$ as well as under $\mathcal{T}$; the converse applies to the magnetic induction field b. This symmetry is indeed required by the structure of the Faraday and Ampère Laws. Nonetheless, these and the other two Maxwell's equations also support another well-known, fundamental symmetry, registering a dual complementarity between the electric and magnetic fields of optical radiation in free space. It is a symmetry that is compromised in the presence of electric charge, owing to the asymmetry in existence of electric but not magnetic monopoles; for the electric field, a charge-driven source term accordingly appears in Gauss's Law, but there is no counterpart in the expression for divergence of the magnetic field. Nonetheless, there is sufficient interest and power in the underlying free-space symmetry that there is recurrent attention in electromagnetic duality. Indeed, much of the recent interest—largely centred on structured and singular light, with associated momentum and angular momentum issues—does concern essentially free-space propagation.

In a range of acceptable formulations for the Poynting vector, for example, Berry advocates an 'electric–magnetic democracy' [20]. This is a feature that is evident not only the classical formulation, but also in the operator formulation due to Power and Thirunamachandran;

$$p(r,t) = \frac{1}{2}\varepsilon_0[e(r,t) \times b(r,t) - b(r,t) \times e(r,t)] \quad (1)$$

which is Equation (3.1) in the literature [21], here recast in SI units: r and t are space and time coordinates. For more general application, it is the transverse electric displacement field $d^\perp$ that should feature in (1), rather than the electric field e, but in source- and current-free regions, there is no physical distinction (the symbol d is also commonly used in entirely different connections). Here, too, with a view to the microscopic formulation that is appropriate for application to systems on the molecular scale, the standard lower-case symbols are used; the context will generally make it clear if quantum operators are signified. Notably, in the above Equation (1), symmetrisation is necessary to ensure Hermiticity, because of the non-commutativity of the electric and magnetic field operators at a common point in space [4,22]:

$$[e_i(r), b_j(r')] = \frac{i\hbar}{\varepsilon_0} \varepsilon_{ijk} \frac{\partial}{\partial r'_k} \delta(r - r') \quad (2)$$

here and henceforth, there is implied summation over repeated tensor indices; ε_{ijk} is the Levi–Civita antisymmetric tensor. Equation (2) immediately exhibits quantum uncertainty in optics, manifested at the photon level; it signifies that the electric and magnetic fields cannot be simultaneously determined at any single position.

Another aspect of the free-space relationship between e and b is manifested in the form of the duality transformation under which Maxwell's source-free equations prove invariant:

$$(e, cb) \to (e\cos\theta + bc\sin\theta, \ bc\cos\theta - e\sin\theta) \quad (3)$$

where the brackets simply denote the orthogonally paired fields. Here, θ is an arbitrary pseudoscalar, signifying that it changes sign under spatial parity inversion. The odd parity of the electric field, and the even parity of the magnetic field, both under $\mathcal{P}$, are thus preserved in the transformation; temporal parity is compromised, except in the case of $\theta = \pi/2$ (or, trivially, multiples of $\pi/2$). In anticipation of later details, it is worth noting that casting equations in units such as the commonly used $c = \hbar = 1$ can obscure any connection between the transformation properties under $\mathcal{P}$ and $\mathcal{T}$. For example, in the above equation, every element necessarily has the same units, but c clearly does not change under any such transformation; the e and b fields exhibit different spatial and temporal parities because they have different physical dimensions.

Equation (3), known as a Heaviside–Larmor transformation, has the specific form of a 2D rotation, with symmetry SO(2). In some accounts, it is misleadingly described as a Lorentz boost, because an expression of Lorentz transformation equations in terms of hyperbolic (cosh and sinh) functions of *rapidity* has a similar cast [23], and those functions convert to their trigonometric counterparts when their argument is imaginary. However, the signs in (3) are not consistent with this interpretation; moreover, the Lorentz transform necessarily engages time with one physical dimension. A useful account of the Lorentz transforms of electric and magnetic fields is given by Ivezić [24].

The textbook compartmentalisation of optical angular momentum j into spin and orbital parts, s and l, respectively, proceeds along the following lines [25]:

$$j = \int r \times p \, d^3 r \equiv l + s \quad (4)$$

$$l = \varepsilon_0 \hat{r}_i \int e_j (r \times \nabla)_i a_j d^3 r \quad (5)$$

$$s = \varepsilon_0 \hat{r}_i \int (e \times a)_i d^3 r \quad (6)$$

where ε_0 is the vacuum permittivity and a is the vector potential field. Quite apart from the gauge-dependence of a, it is well known that this separation is beset with problems; the spin operator s does not satisfy the necessary commutation relations amongst its Cartesian components, to be acceptable as a true quantum mechanical operator [26]. As pointed out by Barnett et al., the same conclusion therefore necessarily applies to the counterpart orbital angular momentum l, as the sum of the two does constitute a mathematically correct formulation of the orbital momentum from the vector product $r \times p$. [27]. Their work nonetheless exhibits the dual transformation as essentially consistent, within the paraxial approximation, to the rotations generated by treating l and s as infinitesimal angle generators.

In an incisive analysis by Cameron et al. [28], it has been shown how, through application of Noether's theorem [29] to the appropriate symmetries of Maxwell's equations, it is possible to secure a refined form of angular momentum conservation laws, revealing a subtle interplay of spin and orbital angular momentum features. Further manipulation of the Maxwell equations in terms of vector and scalar potentials, exploiting dual symmetry, has also been shown to reveal a cascade of mathematically equivalent formulations, [30] though with a caveat that application in the presence of charge will introduce complications. For any non-paraxial beam, it is possible to secure exact, self-consistent operator descriptions of the spin and orbital angular momenta in both quantum and classical energy-flow formulations [31]. The ultimately incomplete separability of spin and orbital angular momenta for structured light is essentially connected to the spatial variation of momentum flux, which undermines canonical separation [32]. Accordingly, there is a host of optical phenomena that manifest optical spin-orbit coupling, as shown and summarised in a commendable review [33]. Not surprisingly, the extent and nature of such coupling is compounded when knotted fields are entertained [34].

Bliokh et al. have highlighted problems with exploiting electromagnetic duality in standard electromagnetic field theory, as its association with an incontrovertibly non-dual Lagrangian [35] leads to conflicts in the associated conservation laws. Of course, for any dynamic system, the Lagrangian is not unique; observables relate to equations of motion that are at least invariant to the addition to the Lagrangian of any total time-derivative. However, by recasting the formulation in terms of a dual field tensor, Bliokh's work has shown how it is possible to resolve the issues, and also to afford a more robust method for separately identifying spin and orbital parts of the angular momentum. The analysis engages another field vector with a significant symmetry role, now usually known as the Riemann–Silberstein vector f [36–38]:

$$f(r,t) = e(r,t) + icb(r,t) \tag{7}$$

together with its Hermitian conjugate, this field also serves as a suitable basis for representing electromagnetic fields. Fernandez-Corbaton and Molina-Terriza favour the Riemann–Silberstein (RS) formulation in their account of duality symmetry in transformation optics [39], for the transformation Equation (3) can then be cast as follows:

$$f(r) \rightarrow \exp(+i\theta)f(r) \tag{8}$$

In a detailed analysis of the parity and general symmetry properties of dual symmetry, duality transformations, and helicity density associated with electromagnetic waves in widely-ranging dispersive media, it has recently been noted that the generator of the dual transformation has eigenmodes that are fields of well-defined ($\pm$) helicity [40]. In earlier work, Bialynicki-Birula proposed that the RS field vector fulfils the function of a photon wavefunction. [41] However, there are obstacles to such an interpretation. Consider any specific radiation mode $(\mathbf{k}, \eta)$, for wave-vector k and polarization η; there is no way to represent the wavefunction for the two-photon state, $|2(\mathbf{k}, \eta)\rangle$, as any kind of combination or product of one-photon $|1(\mathbf{k}, \eta)\rangle$ state functions (just as it is not possible to represent the wavefunction for a 2s electron in hydrogen in simple terms of 1s wavefunctions). The notion of a photon wavefunction can serve as a workable pragmatism when single photons are involved, and the distinction from a state vector poses less of a problem, but for states with two or more identical photons, there is no conventional sense in which any one photon can be considered to have its own wavefunction [42].

A range of conservation principles also relates to the symmetry properties of electromagnetic radiation. However, the engagement of light with matter undermines the applicability in most cases. For example, although energy is conserved between matter and radiation as an overall quantity in any optical interaction, for any measurement that is made beyond a near-field region of quantum uncertainty, the same cannot necessarily be asserted for all other quantities conserved in freely propagating radiation. A key illustration, to be examined below, is afforded by an optical chirality measure known after its originator as the *Lipkin zilch*. Associated with conservation of polarization [43–46], this is just one of a group of properties that is conserved in free electromagnetic fields, representative of a group whose invariance under the space–time Poincaré group is associated with an eight-dimensional Lie algebra of non-geometric symmetry transformations.

Work by Bliokh and Nori has uncovered close connections between the optical chirality density and such measures as polarization helicity and energy density [47], and the results have been shown to have a direct dependence on a difference in the photon number operators for left- and right- handed modes [48]. For example, the helicity operator for the free field, defined by the volume integral of $\boldsymbol{a}.\boldsymbol{b}$ emerges as follows:

$$\kappa \equiv \int \boldsymbol{a}.\boldsymbol{b}\, d^3\boldsymbol{r} = \frac{\hbar}{c\varepsilon_0} \sum_k \left[n^{(L)}(\boldsymbol{k}) - n^{(R)}(\boldsymbol{k}) \right] \quad (9)$$

where the brackets on the right contain a difference of the number operators for left- and right-handed circularly polarized photons. Locally, the appropriate operators for measures of radiation helicity are a chirality flux $\boldsymbol{\varphi}(\mathbf{r},t)$ and chirality density $\chi(\mathbf{r},t)$, which together satisfy a continuity (helicity conservation) equation [48];

$$\frac{\partial \chi}{\partial t} + \nabla.\boldsymbol{\varphi} = 0 \quad (10)$$

with the defining equations

$$\chi = \frac{\varepsilon_0}{2} \left[\boldsymbol{e}.(\nabla \times \boldsymbol{e}) + c^2 \boldsymbol{b}.(\nabla \times \boldsymbol{b}) \right] \quad (11)$$

$$\boldsymbol{\varphi} = \frac{c^2 \varepsilon_0}{2} \left[\boldsymbol{e} \times (\nabla \times \boldsymbol{b}) - \boldsymbol{b} \times (\nabla \times \boldsymbol{e}) \right] \quad (12)$$

in terms of fundamental symmetries, the matrix elements of χ are pseudoscalar fields, odd with respect to the operator for space inversion (or parity), $\mathcal{P}$, but even under time reversal, $\mathcal{T}$; $\boldsymbol{\varphi}$ is a polar vector field, even under $\mathcal{P}$ and odd under $\mathcal{T}$. Together, the operators defined by Equations (11) and (12) represent components of a four-vector $(c\chi, \boldsymbol{\varphi})$ in Minkowski space [49], signifying the conserved Lipkin 'zilch' [43].

The issues of electromagnetic helicity become considerably more intricate for radiation passively propagating within complex media; the subject is thoroughly explored in a recent paper by Alpeggiani et al. [50]. However, the pursuit of conservation laws in connection with *active* processes, where real electronic transitions occur and energy is exchanged between radiation and matter, is a fundamentally different proposition [51]. In this respect, helicity-related aspects of optical radiation behave quite differently from energy, linear momentum, and angular momentum, to which global conservation laws apply. For example, when any molecule absorbs a circularly polarized photon, it does not thereby gain in helicity, nor does circularly polarized emission deplete any measurable chiral character in the emitter. Fundamentally, there is no quantum operator for helicity/chirality in a material system—nor can there be. The spectroscopic study of circular dichroism (CD), that is, circularly differential single-photon absorption, manifests the implausibility of any such measure, for quantitative measurements exhibit a dependence on optical wavelength that is far from absolute; generally, the CD rate differential is a sensitive and intricate function of the electronic wavefunctions, excited state energies, and transition dipoles for each material.

4. Symmetry Principles for Photon–Molecule Interactions

Having outlined the symmetry principles that apply for free radiation, we can now undertake a review of the principles that apply to photon–molecule interactions, with a particular view to electronic transitions. Accordingly, this section begins with a concise summary of key equations that will underpin any conventional deployment of QED, in deriving expressions for the observables in optical transitions. The generic framework described in previous work [9,52], which provides a basis for describing both processes and electrodynamic properties based on the Power–Zienau–Woolley (PZW) Hamiltonian [1,53–55], is here consolidated for specific application to electronic transitions—facilitating simplification by excluding features that would only feature in the theory of optically parametric processes, or in the representation of static quantities such as permanent dipoles or susceptibilities.

The approach to be taken allows provision for the full symmetry of the free electromagnetic field to engage with the full symmetry of the molecular system. In this respect, it substantially differs in its approach from complementary forms of analysis considering direct reductions of electromagnetic field symmetry through propagation within gyrotropic media, for example [56].

Although the principles to be enunciated are in principle applicable to 'photonic molecules'—a term that was introduced to highlight a similarity between the optical modes in simple cavity photonics and those of diatomic molecules [57]—the former constructs generally lack the rich diversity of symmetry elements afforded by real molecular systems, and in this respect, a more general use of the term is perhaps misleading. Equally, the fabrication of structures to impose excitation symmetry in surface plasmonics has no real connection with the intrinsic selection rules for electronic transitions [58]. However, applications of the symmetry framework to be developed in the following analysis do invite extension to nanofabricated dielectric structures, where transition processes such as circular dichroism occur, even as specifically quantum aspects of the radiation become less prominent. An example of recent work in this area is a fine combination of theory and experimental work on dichroism in chiral sculptured thin films [59].

The key observable for optical transitions is a signal proportional to the rate Γ—which may directly represent a rate of transition, or equally a rate of change in the energy, linear or angular momentum content of the radiation fields that are responsible. Unless saturation occurs, the rate is usually cast in terms of Fermi's Golden Rule:

$$\Gamma = \frac{2\pi\rho}{\hbar}|M_{FI}|^2 \tag{13}$$

if either saturation or oscillatory behaviour occurs, the detailed dynamics is still essentially determined by the matrix element for the process The density of states ρ exhibited in this equation is in principle a convolution of functions representing the number of states per unit energy interval for each of the light-matter system components; in practice, one component usually dominates, and for the many applications (including almost all multiphoton processes) that involve narrow linewidth lasers, it is usually the molecular excited state whose linewidth effectively determines the value. The core of Equation (13) is M_{FI}, the *matrix element* of an operator M that couples an initial state $|I\rangle$ to a final state $|F\rangle$ in a composite *system* (i.e., molecule plus radiation). In the present connection, with a focus on processes in which energy is exchanged between the radiation and matter, the final state is presumed to be measurably different from, though necessarily isoenergetic with, the initial state of energy E_I. The operator M may itself be cast in the following resolvent operator form [22]:

$$M = \sum_{p=0}^{\infty} H_{\text{int}}(G_0 H_{\text{int}})^p \tag{14}$$

where the propagator is given by

$$G_0 \simeq (E_I - H_0)^{-1} \tag{15}$$

$$H_0 = H_{\text{mol}} + H_{\text{rad}} \tag{16}$$

Here, H_0 is the basis Hamiltonian, comprising the unperturbed molecular and radiation operators. Implementing the completeness relation delivers the system matrix element $(M_{FI})_{\text{sys}}$ in the form of a familiar expansion in the light-matter interaction operator H_{int}, representing a time-dependent perturbation:

$$(M_{FI})_{\text{sys}} = \langle F|H_{\text{int}}|I\rangle + \sum_R \frac{\langle F|H_{\text{int}}|R\rangle\langle R|H_{\text{int}}|I\rangle}{(E_I - E_R)} + \sum_{R,S} \frac{\langle F|H_{\text{int}}|S\rangle\langle S|H_{\text{int}}|R\rangle\langle R|H_{\text{int}}|I\rangle}{(E_I - E_R)(E_I - E_S)} + \ldots \tag{17}$$

the intermediate states $|R\rangle$, $|S\rangle$ associated with energies E_R, E_S, and so on, are also cast in the system basis. Each Dirac bracket featured in the numerators of terms in Equation (17), and thus entails both matter and radiation components—and to identify symmetry aspects, both must be brought into explicit consideration.

It is the structure of the operator M, given by Equation (14), that proves a key to unlocking the symmetry aspects of light-matter interactions examined in the following section. As the system basis comprises products of molecular and radiation states, the symmetry of the propagator G_0 is separable in each component. Clearly, all energies are scalar quantities, and H_{mol} is invariant under the same full set of symmetry operations as the molecule, whose symmetry class is always identified with the ground state (or higher, in the case of chiral species [60]—where the ground state wavefunction lacks a two-fold permutational symmetry that is present in the molecular Hamiltonian).

As noted above, it is most expedient to deploy the PZW form of interaction Hamiltonian, cast as multipolar coupling in terms of a transverse electric field operator $e^\perp$ and a counterpart magnetic induction operator b. This affords major calculational advantages and insights; expressing the couplings between the optical fields and charges directly in terms of experimentally meaningful electric and magnetic fields also highlights their involvement with corresponding multipole moments and optical response tensors in Cartesian form, thus elucidating their connection to molecular symmetry. Strictly, when casting theory in terms of a PZW Hamiltonian formulation, all orders of the electric multipole coupling should be cast in terms of a transverse electric displacement $d^\perp$, rather than the electric field $e^\perp$ [4,61]. However, in isotropic media such as gases, and all conventional liquids and solutions, the two quantities are related by a scalar, so precisely the same symmetry arguments apply. The equations here are written in terms of the electric field for simplicity, and consistency with previous work. The leading terms of H_{int} are thus expressible as follows:

$$H_{\text{int}} = -\mu_i e_i^\perp - Q_{ij} \nabla_j e_i^\perp - \ldots - m_i b_i - \ldots \tag{18}$$

where μ is the electric dipole operator, Q is the (second rank tensor) electric quadrupole operator, and m is the magnetic dipole operator. The first and third of these are vectors; the quadrupole operator is a second rank tensor; and the indices i, j represent coordinates in any consistent frame of spatial reference with orthonormal axes—usually Cartesian, but not necessarily so (see Section 11). Every index that is repeated signifies an implied summation over the 3D basis set.

For concise reference in the text, the three consecutive terms in the above Equation (18) will be referred to as E1, E2, and M1, respectively, and as a rule, the first of these represents a coupling that is significantly stronger than the other two—where selection rules permit them all to occur (see later). It is important to recognize that the E2 and M1 forms of coupling may in principle constitute equally significant contributions to the light-matter coupling, together representing a leading correction to the E1 term. The proof of this connection is straightforward; both terms emerge from the same level of approximation in transforming between minimal coupling and multipolar Hamiltonian forms [62].

In addition to the terms explicitly exhibited in Equation (18), there are further terms of higher order—which generally indicates that they will be responsible for much weaker effects. These include a diamagnetisation contribution that has recently attracted fresh interest. As this term is quadratic in the optical magnetic field, it is of even parity with respect to both space and time, and may therefore in principle be considered pervasive (in this respect it is like the conventional polarizability, which is non-zero for every material). However, the same property also renders this form of coupling less potentially useful as a tool of symmetry analysis. Thus, although it is now recognized that in some connections, diamagnetisation coupling may prove quantitatively more significant than warrants its usual disregard [63–65], it is not a concern in a primary focus on symmetry features.

For most optical applications—the majority, which do not specifically concern the confined geometries of a fabricated microcavity, or an exotic beam structure as such may be imparted by optical elements—electromagnetic fields are most expediently commonly cast in terms of plane waves; these represent propagation modes whose wave-fronts are perpendicular to a single director in all cases. Moreover, in order to accommodate multimode radiation fields, the field operators are best expressed as mode expansions in the form of Fourier representations. Promoting the two fields to operator status leads to the following standard expansions [4]:

$$e^{\perp}(r) = \sum_{k,\eta} \left\{ i \left(\frac{\hbar c k}{2\varepsilon_0 V} \right)^{\frac{1}{2}} \varepsilon^{(\eta)}(k) a^{(\eta)}(k) \exp(i k \cdot r) + h.c. \right\} \quad (19)$$

$$b(r) = \sum_{k,\eta} \left\{ i \left(\frac{\hbar k}{2\varepsilon_0 c V} \right)^{\frac{1}{2}} \left(\hat{k} \times \varepsilon^{(\eta)}(k) \right) a^{(\eta)}(k) \exp(i k \cdot r) + h.c. \right\} \quad (20)$$

where h.c. denotes Hermitian conjugate. The above equations express the fields at position r, within an arbitrary quantization volume V, in terms of sums over wave-vector k and polarization state η. The latter sum may in principle be taken on a basis comprising any two states that are represented by opposing points on the Poincaré sphere; [66] commonly, those chosen are either left and right circular polarizations, or *horizontal* and *vertical* plane polarizations. The circularly polarized basis can in fact be expressed in terms of the following unit vectors:

$$\varepsilon^{(L)}(k) = \frac{1}{\sqrt{2}}(\hat{i} + i\hat{j}); \; \varepsilon^{(R)}(k) = \frac{1}{\sqrt{2}}(\hat{i} - i\hat{j}) \quad (21)$$

where $\hat{i}$ and $\hat{j}$ are Cartesian unit vectors such that $(\hat{i}, \hat{j}, \hat{k})$ comprise a right-handed orthogonal group. The quantum optical features of Equations (19) and (20) reside in the photon annihilation operators $a^{(\eta)}(k)$ for each mode (k, η), and in their counterpart creation operators $a^{\dagger(\eta)}(k)$ implicit in the Hermitian conjugate part of each expression. In passing, it is interesting to observe that the RS field vectors, constructed from (19) and (20) using the defining Equation (7), have the particular property that f annihilates a left-handed photon and creates a right-handed photon, whereas its Hermitian conjugate $f^{\dagger}$ does the opposite [37]. An important corollary follows; noting the linearity of the electromagnetic fields in H_{int} (a feature that also carries through to the RS expression of coupling, see below), it becomes evident from the above sequence of expressions that the nth term in the matrix element M_{FI}, Equation (17), delivers the leading contribution for any process involving n photons.

We can now introduce symmetry principles—but first, a caveat. A difference in the symmetry behaviour of electric and magnetic transition moments is sometimes expressed in terms of their being orthogonal to each other—presumably an inference derived from that feature of the relationship between the vector characters of the electric and magnetic fields, exhibited by the cross-product in Equation (20). Others write of the difference as signifying the two kinds of moments are out of phase, as indeed the counterpart fields are out of phase in circularly polarised radiation. In certain applications to atoms, such essentially classical arguments may appear superficially credible, but in general, such inferences are very misleading—not least, because *transition* moments are very different from *induced* moments. Moreover, quadrupole and higher moments cease to be amenable to such straightforward unidirectional interpretations. In molecules, more significantly, both static and transition moments are quantities whose vector or tensor components relate specifically to directions with a fixed and specific relation to the internal molecular geometry.

To establish the ensuing analysis on a firm footing, we first recall that the electric field of the radiation is formally odd with respect to parity $\mathcal{P}$, and even with respect to $\mathcal{T}$; the magnetic field has the opposite character in both respects. Individual modes of the radiation field need not conform to either parity, but in the sum over all modes, this is the definitive character [17]. Clearly, since H_{int} is an energy operator, and therefore even in both space and time, the electric dipole operator μ is necessarily also odd with respect to parity $\mathcal{P}$, and even with respect to $\mathcal{T}$, its magnetic counterpart m is even in $\mathcal{P}$ and odd in $\mathcal{T}$. Accounting for the gradient operator featured in the second term of (18), the electric quadrupole operator Q has to be even in both forms of parity. To illustrate the significance of a difference in spatial parity, it emerges that the difference between electric and magnetic transition moments is the key to most common forms of chirality-sensitive response. As the former are polar vectors (odd in $\mathcal{P}$), and the latter are axial vectors (even in $\mathcal{P}$), it takes a molecule with no center of symmetry—that is, one that is *not* invariant under $\mathcal{P}$, such as a chiral molecule—to support an electronic transition involving both electric and magnetic transition moments. It is indeed an interference of these two kinds of coupling that proves to supply the main mechanism for chiroptical differentiation—see the literature for an example [67].

Before proceeding further, it is worth returning to the Riemann–Silberstein formulation introduced in Section 3, to observing a superficial appeal in connecting creation and annihilation operations with photons of a specific handedness. To this end, for processes of potentially chiroptical significance, the interaction Hamiltonian is, in some accounts, written as follows:

$$H_{int} \simeq -\frac{1}{2}\left[d^\dagger \cdot f + d \cdot f^\dagger\right] \qquad (22)$$

where

$$d = \mu + ic^{-1}m \qquad (23)$$

this cast of the interaction operator is readily shown to precisely replicate the E1 and M1 terms in Equation (18). Although electric quadrupole interactions are thereby excluded from consideration, it transpires for phenomena such as circular dichroism and optical rotation that the absent E2 term in fact plays no role in randomly oriented media; in conjunction with E1 coupling, it generates only terms that vanish on orientational averaging (see Section 10).

The combination of electric and magnetic dipole operators in (23) is real (the former involves only charge positions, and the latter only the corresponding angular momenta operators) and it is of even temporal parity, but it is not an eigenfunction of $\mathcal{P}$; spatial parity is not a good quantum number. The same, of course, is true for f. So although, for chiral molecules, transition dipoles based on Equation (23) may comprise non-vanishing contributions from both its electric and magnetic components, the difference in selection rules that applies for most other materials means that d itself cannot be regarded as a secure gauge of chiral propensity. Moreover, for many chiroptical processes, E2 contributions do not indeed disappear on orientational averaging; Raman optical activity is a familiar example [68,69]. Any advantage of deploying the RS formulation for light-matter coupling is therefore circumscribed; the representation is not generally applicable.

5. The Coupling of Radiation and Molecular Tensors

When we consider any multiphoton process involving $n \geq 2$ photons, the detailed structure of the relevant term in Equation (17) generates tensorial forms of interaction, coupling the material response to elements of the optical fields. Because the denominators of each term in (18) are scalars, symmetry aspects of the result are entirely associated with the products of Dirac brackets in numerator expressions. The rule for each Dirac bracket is that the product of the irreducible representations (*irreps*) of the states of the molecule at each end of the bracket must be spanned by one or more components of the appropriate multipole operator. With regard to the initial and final states for the overall process, the same rule applies with respect to the operator M, introduced in Equation (14). From earlier observations on the symmetry of the associated propagator G_0, it follows that the irrep $\chi(M)$ for M is a product of the individual irreps for each of the multipoles involved in the whole process. Attending to the leading multipole terms given in Equation (18), we can write the following:

$$\chi(M) = \prod_{i=1}^{e}\prod_{j=1}^{m}\prod_{k=1}^{q} \chi_i(e)\chi_j(m)\chi_k(q) \qquad (24)$$

where labels e, m, and q represent the number of E1, M1, and E2 interactions, respectively, whose sum $n = (e + m + q)$ is the total number of photon interactions. For most absorption or scattering processes—and also emission to the ground state—the irreducible representation of the transition specifies the extent of symmetry difference between the relevant molecular excited state and the stable, ground state.

Commonly, excited state wavefunctions lack invariance under the full set of operations corresponding to symmetry elements of the ground state function. For example, in centrosymmetric molecules, whose equilibrium nuclear coordinates from a suitable point of origin represent a set that is even under parity $\mathcal{P}$, some excited states will also be even; others will display odd parity. Often, under C_n rotations to which a ground state is invariant, excited states acquire an integer power of the phase factor $\exp(2\pi i/n)$. Consider, for example, each term of the matrix element for a two-interaction process (noting that more than one term will usually arise, because all sequences of interaction are accommodated in the theory). Each term may entail one Dirac bracket of E1 form and the other of M1 form; all combinations of multipoles are possible in principle, though not all will necessarily be symmetry-allowed. Nonetheless, a first step is to consider what constraints are imposed on each individual interaction, as a result of the group theoretical rules imposed by molecular symmetry [70].

The matrix element M_{FI} for any specific n-photon interaction now emerges in the form of a linear combination of terms, each of which entails vector and tensor interactions between molecule-based and radiation-based properties. The molecular system is cast in terms of products of transition moments, and the corresponding radiation constructs comprise products of components of the field vectors. The general form can be expressed as follows:

$$M_{FI} \sim \sum_{e=0}^{n} \sum_{q=0}^{n} \sum_{m=n-e-q}^{n} S_{e;m;n-e-m}^{(e+m+2q)} \odot^{(e+m+2q)} T_{e;m;n-e-m}^{(e+m+2q)} \qquad (25)$$

which is Equation (25) in the literature [9], without the phase factor that becomes redundant for transition processes—where it disappears in the Fermi rate equation. Here, the result comprises the inner product, signified by $\odot$, of a radiation tensor S and a molecular response tensor T. Specifically, $S^{(r)} \equiv S_{i_1 i_2 \ldots i_r}$ comprises an outer product of components of the electric field and the magnetic field (and in addition, where quadrupoles are involved, the field wave-vector); the corresponding molecular tensor $T^{(r)} = T_{i_1 i_2 \ldots i_r}$ entails products of n Dirac brackets, and its spatial symmetry properties are determined by Equation (24). Each tensor has a rank r given by $r = (e + m + 2q)$ so that the inner product contrasts this number of indices; the molecular tensor $T^{(r)}$ specifically incorporates $(e + m + q)$ products of transition multipole moments.

Because their product M_{FI} has the physical dimensions of energy, the $S^{(r)}$ and $T^{(r)}$ tensors must have identical signatures of parity for each separate parity operation, $\mathcal{P}$ and $\mathcal{T}$. The respective eigenvalues are $(-1)^e$ and $(-1)^m$, as determined by the space-odd, time-even character of the electric field, and the space-even, time-odd character of the magnetic field. Any electric quadrupole, having even parity under both $\mathcal{P}$ and $\mathcal{T}$, plays no part in this determination. If, for example, the $S^{(r)}$ and $T^{(r)}$ tensors are odd with respect to both parity operations, their product will remain the same if both radiation and matter are inverted in space, physically representing opposite parity enantiomers, and also opposite helicity radiation.

In this connection, it is worth briefly noting certain aspects of the physics relating to molecular orientation, with an important bearing on chirality principles. The angular disposition of molecules with respect to any propagating stimulus can play a role in the exhibition of chiral differentiation; the commonly long lifetime for quantum tunneling between oppositely handed enantiomeric forms (which are usually high orders of magnitude greater than optical interaction times) may also be a significant factor. Consider, as a counterexample, a molecule of hydrogen peroxide, H_2O_2; in its ground electronic state, it has only C_2 rotational symmetry and is therefore chiral in principle, but it is not normally regarded as such—because at common ambient temperatures, where the substance is a liquid, thermal energy is sufficient to provide equilibration between the two oppositely handed forms. Relatively low potential energy barriers must be surmounted for interconversion to occur [71]; in this case, evidence is readily afforded by the significant energy splitting between even and odd parity combinations of the two enantiomeric state functions [72].

Conversely, consider a molecule such as boric acid, $B(OH)_3$, which possesses, in addition to a pure rotational (C_3) axis, a plane of symmetry (it belongs to the C_{3h} point group); it is not intrinsically chiral, but if the molecule is held at a fixed angle with respect to any transversely propagating signal beam of light, it has the capacity to differentiate between circular polarizations. This type of effect—essentially 2D chirality—is more commonly encountered (and more easily registered) in the surface features of suitably fabricated metamaterials—gammadion structures are a well-studied example—where even in the absence of an external stimulus, there is a clear disparity across the planar interface between physically dissimilar regions. In this way, effects more commonly associated with optical activity may be exhibited by an intrinsically achiral material or metamaterial [73]. Nonetheless, consideration of the complete light-matter system reveals that chiroptical differentiation will only be manifest in optical fields with a helical character—either through circular polarizations, in chirally configured beams, or within optical nanofibres [74]. When circularly polarized light impinges upon a suitably nanostructured surface, propagation by reflection or transmission may exhibit differences according to direction of travel, as opposite directions are not equivalent under the operations of spatial parity $\mathcal{P}$ [75].

6. Structure and Permutation Symmetry in Material and Radiation Tensors

It is easy to recognise, in the general tensor form of light-matter coupling for nonlinear optical interactions, a potential for the theory to deliver expressions of great complexity, rapidly increasing with the number of photons involved. It will emerge that three-photon absorption, for example, in its most general formulation, leads to a rate equation with 225 independent terms; for four-photon absorption, the figure is 8281 (the explanation of these figures will emerge in Section 10). Such complicated results are of little practical value, and only narrow academic interest. However, a raft of symmetry considerations dramatically redeems the situation. The features discussed below will often reduce the number of independent parameters to a mere handful. The implementation of symmetry principles thus not only lends important physical insights, it also leads to equations that are realistic for experimental application and data interpretation.

There are three distinct structure and geometry-related symmetry properties that can produce major simplifications; in each case, considerations of symmetry lead to a reduction in the number of independent variables. One aspect concerns the inherent photonic character of the nonlinear process itself, reflected in a permutational symmetry between equivalent photon interactions. Another is the possibility of polarization-configured symmetry, which is directly under experimental control. Finally, there are symmetry features determined by the intrinsic symmetry of the molecular component, dependent upon the geometry of its nuclear framework and the spatial symmetry of the transition taking place. We are now in a position to address the first two of these, and in the following section, each feature is illustrated by a specific, typical case: the hyper-Raman effect. Issues associated with molecular structural symmetry are deferred to Sections 8 and 9, pending the further development of the tensor formulation that next ensues.

First, we consider the photonic symmetry that may be intrinsic in the nature of any optical process. The hyper-Raman effect [76] is an inelastic scattering effect in which an intense input beam of optical frequency ω produces scattering, Stokes-shifted (slightly lowered in frequency) from the second harmonic 2ω by a vibrational frequency ω_{vib} for one of the normal modes of the molecule. Thus, it is a three-photon process, detectable in the optical output of a frequency $\omega' = 2\omega - \omega_{\text{vib}}$. Recognising that the leading form of coupling is associated with E1 transitions alone, Equation (25) casts the matrix element as $S^{(3)}_{3;0;0} \odot^3 T^{(3)}_{3;0;0}$. The detailed structure of the molecular tensor $T^{(3)}_{3;0;0}$—a form of transition hyperpolarizability—is usually determined through the construction of time-ordered diagrams [77], which represent every topologically distinct sequence of the individual photon interactions—three in this case; see Figure 1. The same information is in fact conveyed by a single state-sequence diagram, Figure 2 [78,79]. Each path in a state-sequence diagram is in a *topological* sense a dual transform of one of the time-ordered diagrams, interchanging vertices with line segments. The complications that arise in this case, when other multipoles are entertained, will be considered subsequently. The explicit expression for the E1^3 molecular tensor, written as a sum of three corresponding terms, accounting for overall energy conservation in each case, is as follows:

$$\beta_{\lambda\mu\nu}^{nm} = \sum_{r,s} \left[\frac{\mu_\lambda^{ns} \mu_\mu^{sr} \mu_\nu^{rm}}{(E_{sm} - 2\hbar\omega)(E_{rm} - \hbar\omega)} + \frac{\mu_\mu^{ns} \mu_\lambda^{sr} \mu_\nu^{rm}}{(E_{sn} + \hbar\omega)(E_{rm} - \hbar\omega)} + \frac{\mu_\mu^{ns} \mu_\nu^{sr} \mu_\lambda^{rm}}{(E_{sn} + \hbar\omega)(E_{rn} + 2\hbar\omega)} \right] \quad (26)$$

where $\hbar$ is the reduced Planck's constant $h/2\pi$, subscript Greek indices denote Cartesian indices referring to a molecule-fixed reference frame, vector components of the form μ_λ^{ab} and so on refer to electric dipole transition moments for transition $a \leftarrow b$, and E_{ab} denotes an energy difference $E_a - E_b$. Three terms arise because this is the order of index permutations given by the symmetric group product $S_3 \times S_2$.

It will be evident on inspection that Equation (26) does not exhibit the permutational symmetry between the indices μ and ν connected with the two physically indistinguishable input photons (vertices coloured red in Figure 1). However, the radiation tensor with which it forms an inner product, does so as follows:

$$S_{\lambda\mu\nu} = \bar{\varepsilon}'_\lambda \varepsilon_\mu \varepsilon_\nu \quad (27)$$

where an overbar (on the polarization vector for the emitted photon) denotes complex conjugation. This same permutational symmetry can therefore be accommodated in a symmetrized tensor, expediently identified by bracketing the relevant index pair:

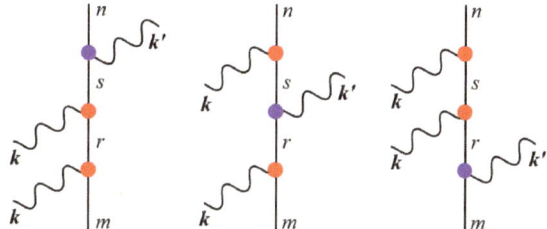

Figure 1. Three topologically distinct time-ordered diagrams (time progressing upwards) for hyper-Raman scattering from an input mode k into an output mode k': the molecule undergoes a transition $n \leftarrow m$ via two virtual intermediate states r and s.

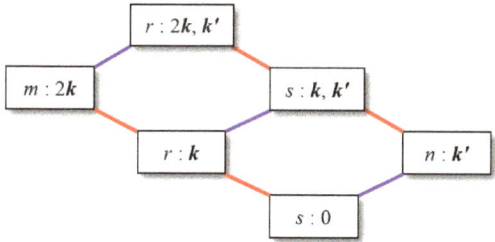

Figure 2. State-sequence diagram (time progressing to the right) for hyper-Raman scattering, subsuming all three pathways exhibited in Figure 1. Here, the interactions denoted by line sequences are colour-coded to highlight the input and output modes.

$$\beta_{\lambda(\mu\nu)}^{nm} \equiv \frac{1}{2}\left(\beta_{\lambda\mu\nu}^{nm} + \beta_{\lambda\nu\mu}^{nm} \right) \quad (28)$$

it is readily shown that this serves to deliver the same completely correct result when it is implemented as $T_{3;0;0}^{(3)}$ in Equation (25).

In general, for any multiphoton process in which two or more of the photons derive from the same monochromatic beam, a corresponding permutational symmetry will be latent in the radiation tensor. This symmetry is ensured if the corresponding photon interactions engage the same level if multipole interaction—usually E1; it is potentially compromised by admitting mixtures of different multipoles. For example, in any form of frequency-doubling conversion—including second harmonic generation (SHG) and elastic second harmonic (hyper-Rayleigh) scattering, as well as the hyper-Raman effect—the substitution of one E1 interaction by M1 or E2 gives the leading correction terms of the form E1^{2}M1 and E1^{2}E2. The associated 'J' and 'K' tensors molecular tensors retain index permutational symmetry if the M1 or E2 interaction is involved in the output emission, but not if it is linked with one of the two input photon annihilation events [80,81].

The second form of index symmetry is now easily identified. Again, consider the hyper-Raman case, exhibited in (27). There need not be any correlation between the polarizations of the two input and single output photons—but in an experiment, it may prove useful to make a measurement (using polarizing optics) in which the polarizations are identical. One example for the commonly studied case of right-angle scattering is if all polarizations are linear and perpendicular to the scattering plane; another is if a forward-scattering geometry is used and the input and output photons are circularly polarized with opposite helicity, as follows from the form of Equation (21). In either case, the S tensor acquires full permutational symmetry amongst all three of its indices—and by similar arguments to those presented above, the same index symmetry is effectively conferred upon the molecular tensor.

To address The third symmetry feature, associated with molecular symmetry and the nature of transitions, will become more accessible on the introduction of an irreducible form of tensor analysis in Section 8. As will emerge, there are further considerations that can serve to very substantially ameliorate the complexity of analysis in the case of more complex forms of optical interaction; to secure their application, there are additional symmetry principles that first need to be developed.

7. Observables

At this stage, it becomes important to return to the generalized matrix elements M_{FI} to distinguish expectation values (signifying identical initial and final system states) from the off-diagonal matrix elements that feature as modulus squares in process observables. The distinction, recently re-emphasized by Stokes [82], becomes especially important when physically identifiable effects arise from the interference between terms involving different kinds of multipolar coupling—chiral and mechanical effects in particular, as shown in other recent work [83–85] To secure an expression for the rate of an observable transition process, we now work from Equation (13) to arrive at the following:

$$\Gamma \sim \sum_{e=0}^{n}\sum_{q=0}^{n}\sum_{m=n-e-q}^{n}\sum_{e'=0}^{n}\sum_{q'=0}^{n}\sum_{m'=n-e'-q'}^{n} \Sigma^{(e+m+2q+e'+m'+2q')} \odot^{(e+m+2q+e'+m'+2q')} \Pi^{(e+m+2q+e'+m'+2q')} \quad (29)$$

where

$$\Pi^{(e+m+2q+e'+m'+2q')} \equiv \left(T^{(e+m+2q)}_{e;m;n-e-m} \otimes^t \overline{T}^{(e'+m'+2q')}_{e';m';n-e'-m'} \right) \quad (30)$$

$$\Sigma^{(e+m+2q+e'+m'+2q')} = \left(S^{(e+m+2q)}_{e;m;n-e-m} \otimes^t \overline{S}^{(e'+m'+2q')}_{e';m';n-e'-m'} \right) \quad (31)$$

Here, the superscript t denotes 'total'—signifying that in the outer product, $\otimes^t$ effects no index contractions and therefore generates a tensor product whose rank is the sum of ranks for its tensor multiplicands. In Equation (30), the shorter representation of the outer product $T^{(r)}\overline{T}^{(r')}$ may be regarded as a material tensor $\Pi^{(r+r')}$; the $S^{(r)}\overline{S}^{(r')}$ construct in (31) may equally be regarded as a radiation tensor $\Sigma^{(r+r')}$. It is evident that for the terms with $r = r'$, each $\Sigma^{(r+r')}$ and counterpart $\Pi^{(r+r')}$ tensor product in (29) will have even parity with respect to both $\mathcal{P}$ and $\mathcal{T}$. However, in the quantum interference terms, $r \neq r'$, some products may have odd parity.

The alarming complexity of the above equations primarily reflects the generality of form in which they are cast; major simplifications arise in almost every specific application. Consider, for example, a single-photon transition $\alpha \leftarrow 0$. In the leading E1^2 contribution to the rate equation, $\Gamma = \left| M_{FI}^{(E1)} \right|^2$, is expressible in terms of the product $S^{(1)}_{1;0;0} \otimes \overline{S}^{(1)}_{1;0;0} \odot^2 T^{(1)}_{1;0;0} \otimes \overline{T}^{(1)}_{1;0;0}$. Here, the material and radiation tensor constructs, as defined above, take the form of a transition electric dipole product $\mu_\lambda^{\alpha 0}\overline{\mu}_\mu^{\alpha 0}$, and a polarization component product $e_\lambda \overline{e}_\mu$ (where these subscript indices imply components in principle referred to the molecule-fixed Cartesian frame—with implications to be addressed in the following sections). This rate contribution, which even for chiral molecules retains its sign irrespective of the enantiomeric form or the circular handedness of the input radiation, is almost invariably the term that generates the largest contribution to the absorption rate. However, attending to the terms beyond E1 in the coupling delivers a corrected rate equation of the form

$$\Gamma = \left| M_{FI}^{(E1)} \right|^2 + M_{FI}^{(E1)}\overline{M}_{FI}^{(M1)} + \overline{M}_{FI}^{(E1)} M_{FI}^{(M1)} + \ldots \quad (32)$$

in which the E1M1 correction terms—which may be non-zero for transitions in chiral media—deliver odd-parity $\Sigma^{(r+r')} \equiv S_{1;0;0}^{(1)} \otimes \overline{S}_{1;0;0}^{(1)} \odot^2 T_{1;0;0}^{(1)} \otimes \overline{T}_{0;1;0}^{(1)}$, and its conjugate, both of which clearly change sign either on substituting enantiomers (necessarily changing the sign of $T_{1;0;0}^{(1)} \otimes \overline{T}_{0;1;0}^{(1)}$)—or, alternatively, by inverting the circularity of the input (producing the same effect on $S_{1;0;0}^{(1)} \otimes \overline{S}_{0;1;0}^{(1)}$)). In either case, the absolute value of the sum (32) changes, resulting in circular dichroic absorption. [84] In the less familiar case, of the hyper-Raman effect, Equation (29) delivers the product $S_{1;0;0}^{(3)} \otimes \overline{S}_{1;0;0}^{(3)} \odot^6 T_{1;0;0}^{(3)} \otimes \overline{T}_{1;0;0}^{(3)}$.

8. Irreducible Cartesian Tensor Framework for Multiphoton Interactions

Molecules, necessarily of less than spherical symmetry, may possess no other symmetry elements than those that can together constitute a subset of the orthogonal group O(3) [86]. Mapping the irreducible representations of this group onto any lower symmetry is *surjective*, and the physical consequence is to permit transitions to occur between states of more than one symmetry class. The irreducible representations of any molecular point group are therefore related through chain decomposition to irreps of O(3) associated with odd or even parity representations of quantum angular momentum states S, P, D, and so on [87]. Functional expressions of the latter are, of course, the vector spherical harmonics, which thereby constitute a natural choice for describing atomic transitions [88,89]. However, with the lower symmetry of molecules, at least one direction within the structure is commonly distinct—usually an axis with the highest level rotational symmetry. The nature of most commonly arising symmetry elements then generally favors a representation of molecular vector or tensor properties in terms of a Cartesian basis. In fact, the advantage often carries over to the representation of the radiation field too.

In principle, the derivation and expression of selection rules for molecular transition moments and other properties can therefore benefit from expression in either a spherical tensor or a Cartesian basis. The deployment of spherical tensors [90] most clearly exhibits angular momentum aspects; it can, for example, elicit important physical insights with regard to angular momentum transfer and multipolar forms of interaction in single-photon processes [91,92]. However, developing an equivalent *irreducible* Cartesian basis can also establish connections with aspects of beam geometry and molecular shape; it is much more directly suited to analyzing multiphoton processes with regard to optical selection rules, because molecular symmetry properties are also usually registered in Cartesian form [70, 93–97]. This is especially important because, in the nonlinear optical spectroscopy of molecules, different lines or bands in the spectrum will commonly be associated with transitions of disparate symmetry, and it is possible, by judicious experimentation based on multiple polarization studies, to elicit their individual character. It is also notable that optical beam configurations are most readily specified in an (x, y, z) form. For example, in a conventional geometry optical table set up for scattering or fluorescence measurements, a Cartesian basis is the standard for describing the orientations of beam propagation and polarization vectors. It has furthermore been shown how the applicability of Cartesian bases extends to beams of complex Gaussian-weighted structure [98].

The two distinct formalisms, spherical and Cartesian—which are, of course, rigorously equivalent—both have an intricacy that rapidly escalates with the tensor rank. However, while conversion between the two forms is not simple [90,93,94,96,99], it is rarely necessary, and the power of analysis that they afford proves its value in processes of more complex photonic interactions. Because the detailed forms of results for arbitrary tensors of up to fourth rank have been calculated, the necessary formulae can be directly deployed [95,97,100,101]. Most of the original QED work on multiphoton electronic processes in molecules was conducted using such an irreducible Cartesian framework [102–106], as well as subsequent studies based on a semiclassical formulation [107]; it is this Cartesian cast of theory that most readily establishes connections between the angular momentum properties of electromagnetic radiation and the multipoles or nonlinear response tensors of molecules, with which the light engages in any particular form of interaction.

The starting point for constructing an irreducible Cartesian calculus is to identify a *natural tensor basis*. Natural tensors are tensors that are fully symmetric under interchange of any (Cartesian) index pair, and are also fully traceless with respect to each such pair. Any such tensor of rank n can be expressed in terms of exactly $(2n + 1)$ linearly independent, non-zero components. In the more general case, a *reducible* Cartesian tensor of a given rank n comprises a sum of irreducible constituent tensors of the same rank n, individually distinguished by *weights* $j = 0 \ldots n$. Each irreducible tensor of weight j and rank n thus represents a natural tensor of rank j embedded in a space of rank n. [70] The advantages of connection to an angular momentum basis are retained in the Cartesian basis, as the coupling between irreducible tensors follows the usual rules of angular momentum coupling. For tensors or rank greater than two, individual weights may have a degenerate representation; in general, the multiplicity of weights j for a tensor of rank n is given by the following [108,109];

$$Q_n^{(j)} = \sum_{k=0}^{\lfloor (n-j)/3 \rfloor} (-1)^k \frac{n(n-1)(2n-3k-j-2)!}{k!(n-3k-j)!(n-k)!} \tag{33}$$

in which the upper limit on the summation is cast in terms of the floor function (signifying the highest integer no greater than the argument). The above result necessarily satisfies the following formula for the total number of independent components:

$$\sum_{j=0}^{n}(2j+1)\, Q_n^{(j)} = 3^n \tag{34}$$

the factor of $(2j + 1)$ accounts for the fact that, for even parity tensors, any $j = 0$ term transforms under the symmetry operations of the molecular point group as a scalar, any $j = 1$ term as a vector (three independent components), $j = 2$ as a *deviator* (a traceless symmetric second rank tensor with five independent components), and so on. For odd parity tensors, $j = 0$ represents a *pseudoscalar* (odd under $\mathcal{P}$) and so on. However, when any degree of index symmetry is present in the tensor, the number of independent components is obviously decreased, and accordingly the multiplicities in representation of each weight are also subject to reductions.

Table 1 lists the structure of decomposition into weights relevant for the most extensively studied kinds of optical interaction; the most prevalent forms of index symmetry are accommodated in tensors up to rank $n = 6$ (which, though less familiar, arise, for example, for six-wave mixing). [110–114] In this table, the tilde on $\widetilde{Q}_n^{(j)}$ indicates that index symmetry is taken into account. Other cases of index permutational symmetry are possible, and most can be identified from the distinct partitions of n; although additional possibilities such as $T_{((\lambda\mu)(\nu\pi))}$ are possible in principle, no experimental implementations of such cases are evident in the optics literature as yet. The parameters t_n and $\widehat{t}_n$ listed in the right-hand pair of columns will be introduced in Section 10.

Table 1. Maximum number of independent components for the tensors $T^{(n)}$ that most commonly arise in n-photon molecular interactions, brackets embracing indices with permutational symmetry. Illustrative examples: Abs—single photon absorption; nPA—n-photon absorption (single-beam); RRE—resonance Raman effect; HR—hyper-Raman effect; HS—second harmonic scattering; SFG—sum-frequency generation; SFS—sum-frequency scattering; 4WM—four-wave mixing; OKE—optical Kerr effect; THS—third harmonic scattering; SWM—six-wave mixing.

$T^{(n)}$	Effect	N	$\widetilde{Q}_n^{(0)}$	$\widetilde{Q}_n^{(1)}$	$\widetilde{Q}_n^{(2)}$	$\widetilde{Q}_n^{(3)}$	$\widetilde{Q}_n^{(4)}$	$\widetilde{Q}_n^{(5)}$	$\widetilde{Q}_n^{(6)}$	t_n	$\widehat{t}_n$
$n=1$											
T_λ	Abs	3	0	1						1	1
$n=2$											
$T_{\lambda\mu}$	RRE	9	1	1	1					3	3
$T_{(\lambda\mu)}$	2PA	6	1	0	1					2	2
$n=3$											
$T_{\lambda\mu\nu}$	SFG/SFS	27	1	3	2	1				15	11
$T_{\lambda(\mu\nu)}$	HR/SHS	18	0	2	1	1				6	5
$T_{(\lambda\mu\nu)}$	3PA	10	0	1	0	1				2	2
$n=4$											
$T_{\lambda\mu(\nu\pi)}$	4WM	54	2	3	4	2	1			34	23
$T_{(\lambda\mu)(\nu\pi)}$	OKE	36	2	1	3	1	1			16	12
$T_{\lambda(\mu\nu\pi)}$	THS	30	1	1	2	1	1			8	7
$T_{(\lambda\mu\nu\pi)}$	4PA	15	1	0	1	0	1			3	3
$n=5$											
$T_{(\lambda\mu\nu\pi\rho)}$	5PA	21	0	1	0	1	0	1		3	3
$n=6$											
$T_{(\lambda\mu\nu\pi)(\rho\sigma)}$	SWM	90	2	1	4	2	3	1	1	36	25
$T_{(\lambda\mu\nu\pi\rho\sigma)}$	6PA	28	1	0	1	0	1	0	1	4	4

Returning once again to the hyper-Raman effect to provide an example, it is immediately evident from the above that considerable simplification ensues in the response tensor on taking account of the pair index symmetry in $\beta_{\lambda(\mu\nu)}^{nm}$, observed in Section 6. The number of independent tensor elements is reduced from 27 to 18; just as significantly, weight 0 contributions fall away entirely; weights 1, 2, and 3 are sustained. So the conclusion is that transitions are only allowed when the product of irreducible representations for the initial and final state—which in the hyper-Raman case, equates to the symmetry of the molecular vibration excited in its course—must span one or more of the irreps for weights 1, 2, and/or 3 in the relevant molecular point group. The odd parity of the E1^3 coupling also applies.

It is relatively straightforward to derive the transformation properties for successive weights of either even or odd parity, and an extensive tabulation of the results is available in the literature. [17,70] Earlier work identified specific components rather than weights, [115] but these prove unnecessary for effective conclusions to be drawn on the simple basis of considering symmetry. Consider, for example, the case of the octahedral molecule sulfur hexafluoride, SF$_6$; the Schoenflies point group is O$_h$ and the odd-parity representations of weights 1, 2, and 3 are T$_{1u}$, (E$_u$+T$_{2u}$), (A$_{2u}$+T$_{1u}$+T$_{2u}$), respectively. This signifies that only vibrations of A$_{2u}$, E$_u$, T$_{1u}$, or T$_{2u}$ symmetry can produce a hyper-Raman signal. For vibrations of all other symmetries, the process is forbidden. It is to be emphasized that the symmetry properties of the *transition* are key here—not the permanent properties of the molecule itself. Again, taking the instance of SF$_6$; because it is octahedral, it has no *permanent* hyperpolarizability—and as such, it cannot exhibit the *elastic* frequency doubling process of second harmonic generation (SHG). Nonetheless, the molecule can produce a hyper-Raman spectrum.

In other connections, decomposition into irreducible terms still has considerable value and power when it is applied to static tensor properties—in which case the rule for a non-vanishing response is simply that one or more of the irreps for non-vanishing weights must transform under the totally symmetric representation of the relevant molecular point group. In a classic paper, Zyss showed in clear and elaborate detail how such principles provide a basis for the molecular engineering of nonlinear optical materials [116]. (In that and subsequent work, the term with weight j is referred to as a 2^j-pole, e.g., a deviator is identified as quadrupolar. In its own specific context, where it is implicit that every photon interaction in fact has E1 form, there is no likelihood of confusion, but the potential ambiguity is noted.)

At this juncture, however, it needs to be pointed out that erroneous deductions can be (and some studies have been) made if complete tensor index symmetry is assumed. Such an approach, which became widespread owing to its appealing simplicity, is largely credited to Kleinman [117], whose expressly limited intention was indeed to make the interpretation of early experiments in nonlinear optics more tractable. The slender argument, not to be pursued in detail here, is based on a case that in expressions such as Equation (26), photon energy terms such as $\hbar\omega$ and $2\hbar\omega$ are small compared with the electronic energy differences that arise in the sum over states. Appeals to such arguments led to a supposition that the hyper-Raman and analogous tensors could effectively be treated as fully index-symmetric. As Table 1 shows, in its entry for $T_{(\lambda\mu\nu)}$, one hidden implication was that weight 2 contributions could not arise. In the SF_6 case examined above, this would wrongly suggest that E_u vibrations are also forbidden. The essential flaws and general inapplicability of Kleinman symmetry were in fact quickly pointed out by Wagnière [118]. Recent work on third harmonic scattering has again shown that emphatic differences arise, according to whether or not full index symmetry is assumed [119]. As a corollary to all such cases, however, it is of interest that in a specific case where all the photons involved in the interaction have identical polarization, then, for the same reasons discussed in Section 6, the results will indeed be consistent with Kleinman symmetry.

9. Transition Classes and Information Content

The various combinations of weight that are possible for each order n have been used to designate *classes* of transition symmetry, which are individually discernible with suitably configured polarization measurements in principle. For $n \geq 2$, the permissible classes are essentially the partitions of n, subject to the exclusion rules: $2 \Rightarrow 4$, and the combination $p1 \Rightarrow p+1$ for any integer p. For example, the pairing 01, equivalent to 10, implies the additional presence of weight 2. Then, allowing weight 2 serve to introduce the pair 21, which in turn implies weight 3 (rank allowing), and so forth. While not excluded by these rules, in rank 4, there are no known occurrences of 41 or 30. Classes up to $n = 4$, with known implementations among the commonly listed molecular point groups, are shown in Table 2.

Table 2. Combinations of weight that arise in processes involving up to four photons, in all common molecular point groups (those with up to six-fold rotational symmetry, and also the linear groups).

$T_{\lambda\mu}$	210	21	20	2	1	0								
$T_{(\lambda\mu)}$			20	2		0								
$T_{\lambda\mu\nu}$	3210	321	320	32	31	30	20	3	2	1	0			
$T_{\lambda(\mu\nu)}$		321		32				3	2	1				
$T_{(\lambda\mu\nu)}$					31			3		1				
$T_{\lambda(\mu\nu\pi)}$	43210	4321	4320	432	431	430	420	43	42	40	4	3	1	0
$T_{(\lambda\mu\nu\pi)}$							420		42		4			0

Every one of the classes exhibited in Table 2 is represented in different point group/irrep combinations. More strikingly, any transition, in a molecule of any known symmetry, must conform to one of them; extensive listings are given elsewhere [70]. There are no known materials in which *every* class arises, however. In the octahedral group O_h, for example, the following classes arise for any even-parity, fourth rank tensor lacking full index symmetry: (432)—T_{2g}; (431)—T_{1g}; (42)—E_g; (40)—A_{1g}; and (3)—A_{2g}. As shown in the Table, the number of classes is generally diminished by any admission of tensor index symmetry. Specific processes for which classification schemes based on these principles have been introduced are hyper-Raman scattering, [84] multiphoton absorption [85–88], and third harmonic scattering [96].

It is interesting to observe the growth in order of the number of classes, which can be considered physical (i.e., excited state) implementations of symmetry properties conferred through multiphoton excitation. Given the associated experimental difficulty, it is evident that there is only a marginal advantage to be gained in progressing from $n = 3$ to 4. The number of distinctly identifiable classes in either instance might nonetheless appear to present a potential prospect for high-dimensional data, with a capacity to exceed the second order of a simple binary basis per photon. However, a single n-photon experiment cannot provide an unequivocal basis for class assignment. To achieve that end, in general, requires a *complete polarization study*—a term and concept enunciated by McClain [120,121]. As will become evident in the next section, the number of such experiments required always exceeds the number of distinct classes.

Before moving on to consider fluid media, it is noteworthy that casting electrodynamic theory in terms of irreducible Cartesian tensors proves its value in a variety of other connections. One clever example is afforded by Bancewicz's work on two-centre (collisional) corrections to molecular hyperpolarizabilities, [122] and there are several applications connected with multipole coupling in intermolecular energy transfer [123–125]. The same formalism also facilitates the derivation of analytically tractable formulations for the properties of optically ordered anisotropic nanoparticles [126].

10. Isotropic and Axial Invariants and Ensemble Averages

The majority of optical phenomena in molecules are registered in liquid or solution, where individual molecular constituents are orientationally unconstrained over the timespan for most experimental measurements. The molecules' effective symmetry can then accurately be identified with the properties of their intrinsic nuclear framework in the ground state equilibrium. To secure the appropriate forms of results for experiments on such systems, it therefore becomes necessary to account for an optical response whose time-average, for any individual molecule, will equate to the ensemble average, based on the ergodic theorem. Moreover, the distribution of orientations within the ensemble is usually isotropic (unless orienting fields are present; a case to be considered shortly). The analysis that follows, pursuing the irreducible tensor formulation, represents an alternative perspective to the one given in detail in Section 9 of Ruggenthaler [9].

To begin, a general result can be noted. In general, the product of two irreducible tensors $A^{(n_1)}_{j_1}$ and $B^{(n_2)}_{j_2}$ may entail a fully outer product, in which case it generates a result of the highest possible rank, or at the other extreme, a fully inner product (if the two have the same rank), thus generating a tensor of rank 0—that is, a scalar. In the most general case comprising p inner products (tensor index contractions, $p \leq \min(n_1, n_2)$), the result may be expressed as follows:

$$A^{(n_1)}_{j_1} \overset{\otimes^{n_1+n_2-2p}}{\odot^p} B^{(n_2)}_{j_2} = \sum_{r=0}^{r_{max}} C^{(n_1+n_2-2p)}_{|j_1-j_2|+r} \tag{35}$$

where $r_{max} = \min[2j_1, 2j_2, (n_1 + n_2 - 2p - |j_1 - j_2|)]$. The principles involved in this coupling are illustrated in Figure 3. Relation (35) proves to be extremely important for the simplifications that it can effect as we consider isotropic fluids. To this end, consider the constructs for the product tensors Π as given by Equation (30). To most simply illustrate the implementation of an orientational average, let us restrict consideration to dipole (allowing for both E1 and M1) coupling—that is, the representation of E2 couplings, $q = 0$. The product tensor thus has rank $e + m + e' + m'$, which equates to $2n$. Again, one example from hyper-Raman scattering is the sixth rank term $\beta^{nm}_{\lambda(\mu\nu)} \overline{\beta}^{nm}_{\pi(\pi\rho)}$.

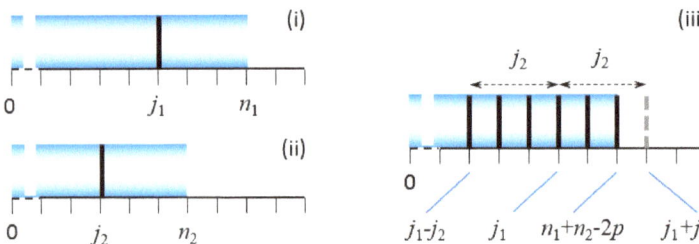

Figure 3. Illustration of coupling weights j_1 and j_2, in a partially inner, partially outer products of two tensors, of respective ranks n_1 and n_2, as given by Equation (35). Assuming $j_1 > j_2$, the span of weights in the product may range from $j_1 - j_2$ to $j_1 + j_2$, capped by an upper limit $n_1 + n_2 - 2p$ that is the rank of the product tensor.

Now, as follows from a theorem by Jeffreys [127], orientation-averaged parameters (in the present application both molecular and radiation forms) must have the transformation properties of scalars under the full rotation group SO(3). As such, they can only be cast as scalar multiples of isotropic tensors g with the same, even rank ($r + r' = 2n$), which comprise products of Kronecker deltas. Averaging can proceed on this basis using Equation (32) in the literature [9]—which also provides for more complicated cases—but by utilising irreducible forms, we now take a different tack. The inner product of the Π and g tensors generates results of the following form, utilizing Equation (35) from the present section and the defining Equation (30) for the explicit form of Π (while the Σ tensors are treated in the same way):

$$\Pi^{(2n)} \odot^{2n} g^{(2n)} = \sum_{j_1=0}^{n}\sum_{j_2=0}^{n} T^{(n)}_{j_1}\overline{T}^{(n)}_{j_2} \odot^{2n} g^{(2n)} = \sum_{j=0}^{n} T^{(n)}_{j}\overline{T}^{(n)}_{j} \odot^{2n} g^{(2n)} \qquad (36)$$

the simplification in the second step, which enforces $j_1 = j_2$, is a consequence of the range for the coupled weights being subject to an upper limit of 0—as the isotropic tensors are weight 0 alone, and the whole expression (which results in a scalar, i.e., a tensor of rank zero) must itself result in weight 0.

Averaging can now proceed on the basis of the above Equation (36), as shown by Andrews and Blake [128]. It then follows that the emerging rate equation will be cast in terms of molecular invariants generated by this equation. These comprise a set of t_n linearly independent set of parameters whose number follows from the multiplicity $\widetilde{Q}^j_n$ of each weight, as listed in Table 1; each weight only couples with itself, and hence we have the following:

$$t = \sum_{j=0}^{n} \left(\widetilde{Q}^j_n\right)^2 \qquad (37)$$

this is the result for the general case (complex T); if the molecular tensor can be treated as real (which generally applies for E1 coupling in regions far from optical resonance), then it follows that the number of invariants reduces to the following:

$$\widehat{t} = \frac{1}{2}\sum_{j=0}^{n}\left[\left(\widetilde{Q}^j_n\right)^2 + \widetilde{Q}^j_n\right] \qquad (38)$$

therefore, for example, the orientationally averaged rate equation for a three-photon process entailing the index non-symmetric tensor $T_{\lambda\mu\nu}$ is cast in terms of $t_n = 15$ molecular invariants. Because, in general, the radiation tensor is subject to the same development, the rate (29) in principle accommodates $t_3^2 = 15^2 = 225$ terms (for four-photon processes lacking permutational symmetry, the corresponding number is $t_4^2 = 8281$). However, if the molecular tensor is real, $\widehat{t}_3 = 11$ and the number of terms in the rate is almost halved.

In practice, because the set of molecular invariants in any case constitutes a complete, linearly independent set, it is equally possible to express results in terms of any other set obtained by injective linear transform—and these are the invariants that emerge from the direct averaging process [127,129]. McClain's approach to securing the maximum information from multiphoton studies of fluids is fundamentally based on the premise of conducting a number $\widehat{t}_n$ of experiments with polarization conditions ensuring linearly independent radiation terms. In general, it is not possible to configure the radiation tensor constructs Σ, using electromagnetic fields of conventional, plane wave form to only comprise any chosen, arbitrary combination of weights. By exploiting the orbital angular momentum of structured beams, Molina-Terriza et al. have shown that it is in fact possible to prepare photons in multidimensional vector states of angular momentum [130]—but orbital angular momentum is known not to engage with the leading E1 form of coupling for electronic transitions [131]. In consequence, as observed earlier, to secure the fullest information from separate experiments with different polarization conditions, the required number of studies always exceeds the number of distinct symmetry classes. Whichever method of tensor representation is deployed (reducible or irreducible), it is noteworthy that it is unnecessary to derive expressions for individual tensor components; they are not required, nor are they measurable in fluid media.

The same kinds of approaches as those considered above can be applied to molecular systems with partial orientational order—one familiar example being cholesteric liquid crystals under the influence of a static electric field acting as a *director* vector. In the most general case, the distribution of molecular orientations is no longer isotropic, but residual degrees of freedom provide for an axially weighted distribution. In the dipole approximation, the associated orientational averaging procedure then requires contracting the radiation and material constructs Π and Σ with isotropic tensors of rank $(2n + 1)$, as shown in Andrews [9] and detailed in reference Wagnière [132].

Although the focus through much of this account is on processes involving up to four photons, it is interesting to note that some theory has been developed for absorption processes of higher order—notably by Wagnière [133] and Friese [134], the latter recently providing remarkable results for five-, six-, and seven-photon excitations. Those calculations make explicit use of index symmetry from the outset, so the results are not claimed to have general applicability for multiphoton interactions. It does not appear that group theoretical methods have as yet been applied to such cases—but the degree of experimental challenge in resolving the associated spectra suggests that such a symmetry analysis would not serve any immediately practicable purpose.

11. Intricate Aspects of Dichroism

To illustrate the principles, and to highlight the powerful significance of orientational averaging, it proves salutary to consider some potentially circular dichroic aspects of single photon absorption—where, despite the simplicity, some recent developments invite such a perspective. At its simplest, there is only a single interaction to consider, as given by Equation (18); the interference terms between forms of coupling with different parity signatures can only be elicited in chiral materials, and using radiation with a degree of helicity—which, as we observed earlier, generally indicates the use of circular polarizations. The original case of circular dichroism (CD), introduced at the end of Section 7, has been addressed in numerous works—see, for example, the treatment given in Craig and Thirunamachandran [4].

One of the features of conventional CD is that in leading order, it entails E1–M1, but not E1–E2, interference terms. The reason for the exclusion of the latter, which also satisfies the spatial symmetry criterion for eliciting chiral response, is that in a fluid, the associated orientational average entails isotopic tensors of rank 3—which are scalar multiples of a Levi–Civita antisymmetric tensor. The inner product of this tensor with the E1–E2 molecular tensor product vanishes, because of the index symmetry in the quadrupole transition moment. However, the fact that this situation changes when orientational order is present, has invited study of the possible engagement of E2 terms in chirality associated with vortex beams (whose photons convey the orbital angular momentum introduced in Section 3). A recent analysis has shown that this is indeed the case, leading to a distinctive circular vortex dichroism (CVD) provided orientational order is present [135]. It nonetheless emerges, by applying the orientational average in a cylindrical coordinate basis (in which the local orthonormality of the basis unit vectors still applies), that the effect once more disappears in isotropic media.

Now, returning to the case of conventional radiation, let us consider that a static magnetic field is introduced. As discussed by Andrews [9], the presence of a static magnetic field is often described as 'symmetry breaking'. When it engages with any optical interaction linearly (or indeed in any odd power), its time-odd character imparts a propensity to undermine Helmholtz reciprocity (i.e., forward-backward equivalence), as, for example, in the familiar Faraday effect. However, due inclusion of the field as a full component of the light-matter system confirms that its involvement is entirely consistent with $\mathcal{PT}$ symmetry. In principle, a static magnetic field might engage with electron spin, in molecules or radicals with one (or more) unpaired electron. In such cases, strong magnetic fields can lift the degeneracy of spin states to a photophysically significant degree, and the result is to permit circularly polarized photons of opposite handedness to excite each component of any resulting spin doublet. However, there are other more interesting, and more general mechanisms that may come into play where magnetic fields are involved, where the significance of both molecular symmetry and rotational averaging become especially evident. In the following section, we assume 'closed-shell' states of time-even parity, for simplicity excluding states with unpaired electron spin. By far the majority of stable molecules and larger assemblies are known to be of this kind.

To suitably develop the theory, we now extend the single-interaction Equation (18) by writing the following:

$$M_{FI} = \sum_{\Omega} M_{FI}^{(\Omega)} \quad (\Omega \equiv \text{E1, E2, M1, E1M01, E2M01, M1M01}, \ldots) \tag{39}$$

where the first three terms can be identified in explicit form with those given in Equation (18), and the second three are double-interaction terms engaging each multipolar form of photon interaction, along with a static magnetic field dipole interaction denoted M01. Figure 4 shows the salient forms of time-ordered diagram, for the influence of the magnetic field on single-photon absorption, with the static (i.e., non-propagating) field depicted by a horizontal line; Figure 5 is the corresponding state-sequence diagram, accommodating all time-orderings to this level of interaction.

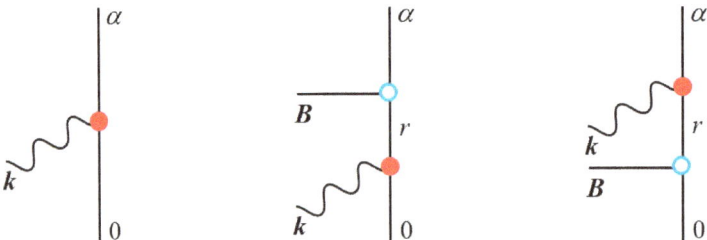

Figure 4. Key time-ordered diagrams for engaging a static magnetic field B in the absorption of a single photon of wave-vector k. The diagram on the left represents the leading term, where the red circles denote E1, E2, or M1 coupling. Additional coupling with the static field (empty blue circle) engages two distinct time-orderings.

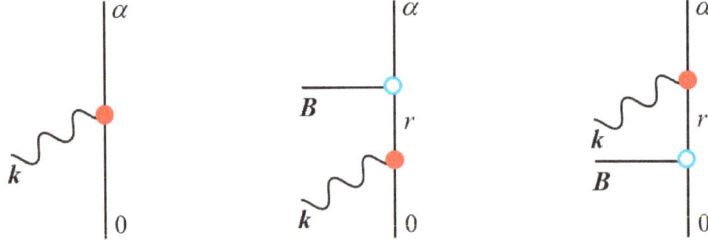

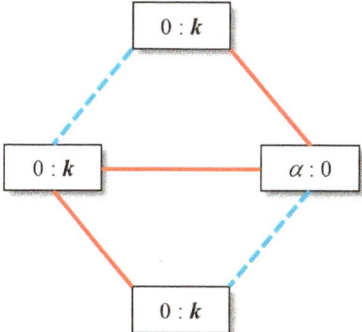

Figure 5. State-sequence diagram for magnetic field engagement in single-photon absorption, connectors coloured to match the time-ordered representations of Figure 4.

Before proceeding further, it is worth noting that the present method of dealing with the engagement in an optical process of any static field, whether magnetic or electric, is a pragmatic shortcut to correct results. Formally, although static fields are absent in the PZW Hamitonian, their effect on a system of interest can be introduced either by applying time-independent perturbation theory to establish static field-modified basis states for a standard time-dependent perturbation theoretic treatment of the optical process [136], or equally by including, as a source, a static dipole whose influence on the system is mediated by E1–E1 or M1–M1 virtual photon coupling; the former has been demonstrated in several connections—see the literature for an example [137].

Now, applying Equation (13) to secure an observable, the rate of absorption, Equation (39) yields a series of terms; details have been reported elsewhere [138]. The leading term is quadratic in E1 (assuming the transition is conventionally allowed); then follow cross-terms such as E1–E1M01, and so on. Suppose we look for terms that will exhibit involvement with the magnetic field, but which are allowed only by non-centrosymmetric molecules. As E1 is of odd parity under $\mathcal{P}$, but E2, M1, and M01 are even, the leading terms of interest are E1–E2M01 and E1–M1M01. The significance of these was first considered by Wagnière and Meier [139], whose depiction of the former cross-term deploys another diagrammatic form shown in Figure 6.

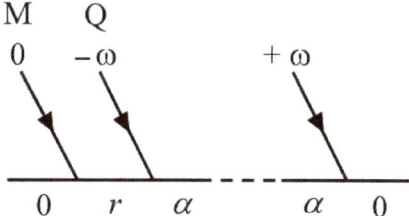

Figure 6. Cross-term for magnetic field engagement in single-photon absorption (Wagnière and Meier depiction). Such diagrams signify partial rate contributions, as compared with the individual matrix element depictions in Figures 4 and 5.

Consider first the E1–E2M01 term. This generates fourth rank Π and Σ tensors constructed according to Equations (30) and (31), each to be contracted with an isotropic tensor of the same rank that is, a product of two Kronecker deltas. In particular, the field tensor Σ comprises products of components of $\varepsilon^{(\eta)}(\mathbf{k})$, $\bar{\varepsilon}^{(\eta)}(\mathbf{k})$, $\mathbf{k}$, and $\mathbf{B}$ (one component of the polarization vector, one of its complex conjugate, one of the wave-vector, and one of the magnetic field). Therefore, the result of contraction with two deltas, which produces two scalar products, may appear to be non-zero and acquire its maximum value if the static field is aligned with the direction of beam input; because $\varepsilon^{(\eta)}(k) \cdot \bar{\varepsilon}^{(\eta)}(k) = 1$ for any polarization, the result is ostensibly non-zero. An interesting aspect for chiral molecules is that the two opposite enantiomeric forms would appear to produce opposite E1–E2M01 contributions of opposite signs, even when linearly polarized light is deployed. However, the molecular part Π of this result involves two terms, each one entailing a transition magnetic dipole moment—one with the transition dipole $m^{\alpha r}$ and the other m^{r0}, corresponding to the interaction vertices denoted by empty blue circles in the middle and right-hand diagrams in Figure 4. With real wavefunctions, the values of these moments are imaginary, because the angular momentum operator implicit in a magnetic moment operator is itself imaginary; hence, the associated rate contribution in fact vanishes (the rate equation entails twice the real part of this imaginary cross-term contribution).

A similar logic, but with a different outcome, applies to the E1–M1M01 term. Here, the Π and Σ tensors that arise are third rank, and accordingly, each demands contraction with the isotropic tensor of rank 3, that is, the Levi–Civita tensor. For Σ, comprising a product of components of $\varepsilon^{(\eta)}(k)$, $\hat{k} \times \bar{\varepsilon}^{(\eta)}(k)$, and B, this generates a vector triple product that can again be non-zero if B is aligned with k. In this case, the molecular part Π of the result again entails two terms, from the middle and right-hand diagrams in Figure 4, but now each one comprises *two* magnetic moments, so that the molecular part of the rate contribution is real. The result persists for both linearly and circularly polarized light; the vector triple product entails the cross-product of $\varepsilon^{(\eta)}(k)$ with $\hat{k} \times \bar{\varepsilon}^{(\eta)}(k)$, which equals $\hat{k}$ for any polarization state—which, therefore, also includes the case of unpolarized light. This distinct difference in physical significance, compared with E1–E2M01, appears not to have been noted before.

It can, therefore, be directly concluded that the presence of a static magnetic field, with any non-zero component along the propagation axis of incident light, enables chiral molecules to exhibit a differential response according to the handedness of the enantiomer. The involvement of this phenomenon, which has been categorised as one aspect of a magnetochiral anisotropy [140], has been suggested as being responsible for magneto-chiral enantioselective photochemistry [141]. It is now conjectured that it may also play a role in recent reports of enantioselective adsorption onto a magnetic substrate [142], interpreted using CD measurements.

12. Discussion

This account has aimed to exemplify symmetry principles that can profitably be applied to secure information of various kinds, in the context of molecular photonics. Over and above its well-known relevance to material properties, symmetry considerations most obviously provide a basis for determining whether any specific form of optical effect is allowed or forbidden. This kind of criterion has comprehensive application to optical phenomena of all kinds; furthermore, it extends to individual optical transitions. Using a quantum electrodynamical basis for the physics provides a framework of equations in which the interplay of radiation and material symmetries becomes especially lucid, as the molecules and radiation are treated as twin components of a fully quantized system. QED methods also facilitate the identification of information content relating to transition symmetry classes, and they provide a framework for devising or interpreting the relevant optical experiments. In this connection, an emphasis on observables—generally process rates and signal intensities—has additional impact; it affords advantages over semiclassical equations whose potentiality in representing observables can be obscure. The distinction is especially important in dealing with theory for systems in which the molecules are randomly oriented, as the implementation of orientational averaging can itself have major implications for the viability of the mechanism. The results of averaging provide a means for establishing rigorous conditions under which processes may be detected. The spheres of optical and material chirality provide numerous examples of how all these principles lend insights into the prospects and possibilities for achieving chiroptical differentiation.

Acknowledgments: The author is grateful for comments from Kayn Forbes.

Conflicts of Interest: The author declares no conflict of interest.

References

1. Woolley, R. The electrodynamics of atoms and molecules. *Adv. Chem. Phys.* **1975**, *33*, 153–233.
2. Cohen-Tannoudji, C.; Dupont-Roc, J.; Grynberg, G. *Photons and Atoms: Introduction to Quantum Electrodynamics*; Wiley: New York, NY, USA, 1989; pp. xviii, 486.
3. Andrews, D.L.; Craig, D.P.; Thirunamachandran, T. Molecular quantum electrodynamics in chemical physics. *Int. Rev. Phys. Chem.* **1989**, *8*, 339–383. [CrossRef]
4. Craig, D.P.; Thirunamachandran, T. *Molecular Quantum Electrodynamics: An Introduction to Radiation-Molecule Interactions*; Dover Publications: Mineola, NY, USA, 1998.
5. Woolley, R.G. Gauge invariance in non-relativistic electrodynamics. *Proc. R. Soc. A* **2000**, *456*, 1803–1819. [CrossRef]
6. Salam, A. Molecular quantum electrodynamics in the Heisenberg picture: A field theoretic viewpoint. *Int. Rev. Phys. Chem.* **2008**, *27*, 405–448. [CrossRef]
7. Białynicki-Birula, I.; Białynicka-Birula, Z. *Quantum Electrodynamics*; Elsevier: New York, NY, USA, 2013.
8. Ruggenthaler, M.; Tancogne-Dejean, N.; Flick, J.; Appel, H.; Rubio, A. From a quantum-electrodynamical light–matter description to novel spectroscopies. *Nat. Rev. Chem.* **2018**, *2*, 0118. [CrossRef]
9. Andrews, D.L. Quantum formulation for nanoscale optical and material chirality: Symmetry issues, space and time parity, and observables. *J. Opt.* **2018**, *20*, 033003. [CrossRef]
10. Rodríguez-Lara, B.M.; El-Ganainy, R.; Guerrero, J. Symmetry in optics and photonics: A group theory approach. *Sci. Bull.* **2018**, *63*, 244–251. [CrossRef]
11. Greenberg, O. Why is CPT Fundamental? *Found. Phys.* **2006**, *36*, 1535–1553. [CrossRef]
12. Lehnert, R. CPT symmetry and its violation. *Symmetry* **2016**, *8*, 114. [CrossRef]
13. Kaplan, A.D.; Tsankov, T.D. CPT invariance in classical electrodynamics. *Eur. J. Phys.* **2017**, *38*, 065205. [CrossRef]
14. Lazzeretti, P. The abstract GPT and GCPT groups of discrete C, P and T symmetries. *J. Mol. Spectrosc.* **2017**, *337*, 178–184. [CrossRef]
15. Brody, D.C. Consistency of PT-symmetric quantum mechanics. *J. Phys. A Math. Theor.* **2016**, *49*, 10LT03. [CrossRef]
16. Stedman, G.E. *Diagram Techniques in Group Theory*; Cambridge University Press: Cambridge, UK, 1990.
17. Andrews, D.L.; Allcock, P. *Optical Harmonics in Molecular Systems*; Wiley-VCH: Weinheim, Germany, 2002.
18. Gbur, G.; Makris, K. Introduction to non-Hermitian photonics in complex media: PT-symmetry and beyond. *Photonics Res.* **2018**, *6*, PTS1–PTS3. [CrossRef]
19. El-Ganainy, R.; Makris, K.G.; Khajavikhan, M.; Musslimani, Z.H.; Rotter, S.; Christodoulides, D.N. Non-Hermitian physics and PT symmetry. *Nat. Phys.* **2018**, *14*, 11. [CrossRef]
20. Berry, M.V. Optical currents. *J. Opt. A Pure Appl. Opt.* **2009**, *11*, 094001. [CrossRef]
21. Power, E.A.; Thirunamachandran, T. Quantum electrodynamics with nonrelativistic sources. II. Maxwell fields in the vicinity of a molecule. *Phys. Rev. A* **1983**, *28*, 2663–2670. [CrossRef]
22. Cohen-Tannoudji, C.; Dupont-Roc, J.; Grynberg, G. *Atom-Photon Interactions: Basic Processes and Applications*; Wiley: New York, NY, USA, 1992; pp. xxii, 656.
23. Taylor, E.F.; Wheeler, J.A. *Spacetime Physics*, 2nd ed.; W.H. Freeman: New York, NY, USA, 1992.
24. Ivezić, T. Lorentz transformations of the electric and magnetic fields according to Minkowski. *Phys. Scr.* **2010**, *82*, 055007. [CrossRef]
25. Mandel, L.; Wolf, E. *Optical Coherence and Quantum Optics*; Cambridge University Press: Cambridge, UK; New York, NY, USA, 1995.
26. van Enk, S.J.; Nienhuis, G. Commutation Rules and Eigenvalues of Spin and Orbital Angular-Momentum of Radiation-Fields. *J. Mod. Opt.* **1994**, *41*, 963–977. [CrossRef]
27. Barnett, S.M.; Allen, L.; Cameron, R.P.; Gilson, C.R.; Padgett, M.J.; Speirits, F.C.; Yao, A.M. On the natures of the spin and orbital parts of optical angular momentum. *J. Opt.* **2016**, *18*, 064004. [CrossRef]
28. Cameron, R.P.; Speirits, F.C.; Gilson, C.R.; Allen, L.; Barnett, S.M. The azimuthal component of Poynting's vector and the angular momentum of light. *J. Opt.* **2015**, *17*, 125610. [CrossRef]
29. Noether, E. Invariante variationsprobleme. *Nachr. d. König. Gesellsch. d. Wiss. zu Göttingen, Math-phys. Klasse* (1918) 235–257. *Physics/0503066* **1918**, *57*. [CrossRef]
30. Cameron, R.P. On the 'second potential' in electrodynamics. *J. Opt.* **2014**, *16*, 015708. [CrossRef]

31. Bliokh, K.Y.; Alonso, M.A.; Ostrovskaya, E.A.; Aiello, A. Angular momenta and spin-orbit interaction of nonparaxial light in free space. *Phys. Rev. A* **2010**, *82*, 063825. [CrossRef]
32. Bialynicki-Birula, I.; Bialynicka-Birula, Z. Canonical separation of angular momentum of light into its orbital and spin parts. *J. Opt.* **2011**, *13*, 064014. [CrossRef]
33. Bliokh, K.Y.; Rodríguez-Fortuño, F.; Nori, F.; Zayats, A.V. Spin-orbit interactions of light. *Nat. Photonics* **2015**, *9*, 796–808. [CrossRef]
34. Arrayás, M.; Trueba, J. Spin-Orbital Momentum Decomposition and Helicity Exchange in a Set of Non-Null Knotted Electromagnetic Fields. *Symmetry* **2018**, *10*, 88. [CrossRef]
35. Bliokh, K.Y.; Bekshaev, A.Y.; Nori, F. Dual electromagnetism: Helicity, spin, momentum and angular momentum. *New J. Phys.* **2013**, *15*, 033026. [CrossRef]
36. Silberstein, L. Nachtrag zur Abhandlung über, Elektromagnetische Grundgleichungen in bivektorieller Behandlung". *Ann. Phys. (Berl.)* **1907**, *329*, 783–784. [CrossRef]
37. Power, E.A.; Thirunamachandran, T. Circular dichroism: A general theory based on quantum electrodynamics. *J. Chem. Phys.* **1974**, *60*, 3695–3701. [CrossRef]
38. Bialynicki-Birula, I.; Bialynicka-Birula, Z. The role of the Riemann-Silberstein vector in classical and quantum theories of electromagnetism. *J. Phys. A Math. Gen.* **2013**, *46*, 053001. [CrossRef]
39. Fernandez-Corbaton, I.; Molina-Terriza, G. Role of duality symmetry in transformation optics. *Phys. Rev. B* **2013**, *88*, 085111. [CrossRef]
40. Bliokh, K.Y.; Kivshar, Y.S.; Nori, F. Magnetoelectric effects in local light-matter interactions. *Phys. Rev. Lett.* **2014**, *113*, 033601. [CrossRef] [PubMed]
41. Bialynicki-Birula, I. On the wave function of the photon. *Acta Phys. Pol.-Ser. A Gen. Phys.* **1994**, *86*, 97–116. [CrossRef]
42. Andrews, D.L. Photon-based and classical descriptions in nanophotonics: A review. *J. Nanophoton.* **2014**, *8*, 081599. [CrossRef]
43. Lipkin, D.M. Existence of a new conservation law in electromagnetic theory. *J. Math. Phys.* **1964**, *5*, 696–700. [CrossRef]
44. Fushchich, W.; Nikitin, A. The complete sets of conservation laws for the electromagnetic field. *J. Phys. A Math. Gen.* **1992**, *25*, L231–L233. [CrossRef]
45. Ibragimov, N.H. Symmetries, Lagrangian and Conservation Laws for the Maxwell Equations. *Acta Appl. Math.* **2009**, *105*, 157–187. [CrossRef]
46. Philbin, T.G. Lipkin's conservation law, Noether's theorem, and the relation to optical helicity. *Phys. Rev. A* **2013**, *87*, 043843. [CrossRef]
47. Bliokh, K.Y.; Nori, F. Characterizing optical chirality. *Phys. Rev. A* **2011**, *83*, 021803. [CrossRef]
48. Coles, M.M.; Andrews, D.L. Chirality and angular momentum in optical radiation. *Phys. Rev. A* **2012**, *85*, 063810. [CrossRef]
49. Anco, S.C.; Pohjanpelto, J. Classification of local conservation laws of Maxwell's equations. *Acta Appl. Math.* **2001**, *69*, 285–327. [CrossRef]
50. Alpeggiani, F.; Bliokh, K.; Nori, F.; Kuipers, L. Electromagnetic helicity in complex media. *arXiv*, 2018. [CrossRef] [PubMed]
51. Nieto-Vesperinas, M. Optical theorem for the conservation of electromagnetic helicity: Significance for molecular energy transfer and enantiomeric discrimination by circular dichroism. *Phys. Rev. A* **2015**, *92*, 023813. [CrossRef]
52. Andrews, D.L.; Allcock, P. A quantum electrodynamical foundation for molecular photonics. In *Modern Nonlinear Optics, Part 1*; Evans, M.W., Ed.; Wiley: New York, NY, USA, 2001; Volume 119, pp. 603–675.
53. Power, E.A.; Zienau, S. Coulomb gauge in non-relativistic quantum electrodynamics and the shape of spectral lines. *Philos. Trans. R. Soc. A* **1959**, *251*, 427–454. [CrossRef]
54. Woolley, R. Charged particles, gauge invariance, and molecular electrodynamics. *Int. J. Quant. Chem.* **1999**, *74*, 531–545. [CrossRef]
55. Andrews, D.L.; Jones, G.A.; Salam, A.; Woolley, R.G. Perspective: Quantum Hamiltonians for optical interactions. *J. Chem. Phys.* **2018**, *148*, 040901. [CrossRef] [PubMed]
56. Proskurin, I.; Ovchinnikov, A.S.; Nosov, P.; Kishine, J.-I. Optical chirality in gyrotropic media: Symmetry approach. *New J. Phys.* **2017**, *19*, 063021. [CrossRef]

57. Bayer, M.; Gutbrod, T.; Reithmaier, J.P.; Forchel, A.; Reinecke, T.L.; Knipp, P.A.; Dremin, A.A.; Kulakovskii, V.D. Optical Modes in Photonic Molecules. *Phys. Rev. Lett.* **1998**, *81*, 2582–2585. [CrossRef]
58. Sakai, K.; Yamamoto, T.; Sasaki, K. Nanofocusing of structured light for quadrupolar light-matter interactions. *Sci. Rep.* **2018**, *8*, 7746. [CrossRef] [PubMed]
59. McAtee, P.D.; Lakhtakia, A. Experimental and theoretical investigation of the co-occurrence of linear and circular dichroisms for oblique incidence of light on chiral sculptured thin films. *J. Opt. Soc. Am. A* **2018**, *35*, 1131–1139. [CrossRef]
60. Woolley, R.G. Is there a quantum definition of a molecule? *J. Math. Chem.* **1998**, *23*, 3–12. [CrossRef]
61. Salam, A. *Molecular Quantum Electrodynamics. Long-Range Intermolecular Interactions*; Wiley: Hoboken, NJ, USA, 2010.
62. Fiutak, J. The multipole expansion in quantum theory. *Can. J. Phys.* **1963**, *41*, 12–20. [CrossRef]
63. Buhmann, S.Y.; Safari, H.; Scheel, S.; Salam, A. Body-assisted dispersion potentials of diamagnetic atoms. *Phys. Rev. A* **2013**, *87*, 012507. [CrossRef]
64. Forbes, K.A.; Bradshaw, D.S.; Andrews, D.L. Identifying diamagnetic interactions in scattering and nonlinear optics. *Phys. Rev. A* **2016**, *94*, 033837. [CrossRef]
65. Forbes, K.A. Role of magnetic and diamagnetic interactions in molecular optics and scattering. *Phys. Rev. A* **2018**, *97*, 053832. [CrossRef]
66. Zangwill, A. *Modern Electrodynamics*; Cambridge University Press: Cambridge, UK, 2013.
67. Barcellona, P.; Passante, R.; Rizzuto, L.; Buhmann, S.Y. Dynamical Casimir-Polder interaction between a chiral molecule and a surface. *Phys. Rev. A* **2016**, *93*, 032508. [CrossRef]
68. Barron, L.D.; Buckingham, A.D. Rayleigh and Raman optical activity. *Annu. Rev. Phys. Chem.* **1975**, *26*, 381–396. [CrossRef]
69. Andrews, D.L. Rayleigh and Raman optical-activity—An analysis of the dependence on scattering angle. *J. Chem. Phys.* **1980**, *72*, 4141–4144. [CrossRef]
70. Andrews, D.L. Symmetry characterization in molecular multiphoton spectroscopy. *Spectrochim. Acta Part A* **1990**, *46*, 871–885. [CrossRef]
71. Ghosh, D.C. A Quest for the Origin of Barrier to the Internal Rotation of Hydrogen Peroxide (H_2O_2) and Fluorine Peroxide (F_2O_2). *Int. J. Mol. Sci.* **2006**, *7*, 289–319. [CrossRef]
72. Grishanin, B.; Zadkov, V. Photoinduced chirality of hydrogen peroxide molecules. *J. Exp. Theor. Phys.* **1999**, *89*, 669–676. [CrossRef]
73. Sersic, I.; van de Haar, M.A.; Arango, F.B.; Koenderink, A.F. Ubiquity of optical activity in planar metamaterial scatterers. *Phys. Rev. Lett.* **2012**, *108*, 223903. [CrossRef] [PubMed]
74. Lodahl, P.; Mahmoodian, S.; Stobbe, S.; Rauschenbeutel, A.; Schneeweiss, P.; Volz, J.; Pichler, H.; Zoller, P. Chiral quantum optics. *Nature* **2017**, *541*, 473–480. [CrossRef] [PubMed]
75. Fedotov, V.A.; Schwanecke, A.S.; Zheludev, N.I.; Khardikov, V.V.; Prosvirnin, S.L. Asymmetric transmission of light and enantiomerically sensitive plasmon resonance in planar chiral nanostructures. *Nano Lett.* **2007**, *7*, 1996–1999. [CrossRef]
76. Terhune, R.; Maker, P.; Savage, C. Measurements of nonlinear light scattering. *Phys. Rev. Lett.* **1965**, *14*, 681. [CrossRef]
77. Wallace, R. Diagrammatic perturbation theory of multiphoton transitions. *Mol. Phys.* **1966**, *11*, 457–470. [CrossRef]
78. Jenkins, R.D.; Andrews, D.L.; Dávila Romero, L.C. A new diagrammatic methodology for non-relativistic quantum electrodynamics. *J. Phys. B At. Mol. Opt. Phys.* **2002**, *35*, 445–468. [CrossRef]
79. Bradshaw, D.S.; Andrews, D.L. Quantum channels in nonlinear optical processes. *J. Nonlinear Opt. Phys. Mat.* **2009**, *18*, 285–299. [CrossRef]
80. Andrews, D.L.; Thirunamachandran, T. Hyper-Raman scattering by chiral molecules. *J. Chem. Phys.* **1979**, *70*, 1027–1030. [CrossRef]
81. Williams, M.D.; Ford, J.S.; Andrews, D.L. Hyper-Rayleigh scattering in centrosymmetric systems. *J. Chem. Phys.* **2015**, *143*, 124301. [CrossRef] [PubMed]
82. Stokes, A. Quantum optical dipole radiation fields. *Eur. J. Phys.* **2016**, *37*, 034001. [CrossRef]
83. Bradshaw, D.S.; Andrews, D.L. Interparticle interactions: Energy potentials, energy transfer, and nanoscale mechanical motion in response to optical radiation. *J. Phys. Chem. A* **2013**, *117*, 75–82. [CrossRef] [PubMed]

84. Bradshaw, D.S.; Andrews, D.L. Chiral discrimination in optical trapping and manipulation. *New J. Phys.* **2014**, *16*, 103021. [CrossRef]
85. Bradshaw, D.S.; Andrews, D.L. Manipulating particles with light: Radiation and gradient forces. *Eur. J. Phys.* **2017**, *38*, 034008. [CrossRef]
86. Butler, P.H. *Point Group Symmetry Applications: Methods and Tables*; Springer Science & Business Media: New York, NY, USA, 2012.
87. Kim, S.K. *Group Theoretical Methods and Applications to Molecules and Crystals*; Cambridge University Press: Cambridge, UK, 1999.
88. Grinter, R. Photon angular momentum: Selection rules and multipolar transition moments. *J. Phys. B At. Mol. Opt. Phys.* **2008**, *41*, 095001. [CrossRef]
89. Grinter, R. Characterizing emitted electromagnetic radiation: The vector-spherical-harmonic/Bessel-function description. *J. Phys. B At. Mol. Opt. Phys.* **2014**, *47*, 075004. [CrossRef]
90. Stone, A.J. Properties of Cartesian-spherical transformation coefficients. *J. Phys. B At. Mol. Opt. Phys.* **1976**, *9*, 485. [CrossRef]
91. Grinter, R.; Jones, G.A. Interpreting angular momentum transfer between electromagnetic multipoles using vector spherical harmonics. *Opt. Lett.* **2018**, *43*, 367–370. [CrossRef] [PubMed]
92. Jones, G.A.; Grinter, R. The plane- and spherical-wave descriptions of electromagnetic radiation: A comparison and discussion of their relative merits. *Eur. J. Phys.* **2018**, *39*, 5. [CrossRef]
93. Coope, J.A.R.; Snider, R.F.; McCourt, F.R. Irreducible Cartesian tensors. *J. Chem. Phys.* **1965**, *43*, 2269–2275. [CrossRef]
94. Coope, J.A.R.; Snider, R.F. Irreducible Cartesian Tensors. II. General Formulation. *J. Math. Phys.* **1970**, *11*, 1003–1017. [CrossRef]
95. Jerphagnon, J. Invariants of the third-rank Cartesian tensor: Optical nonlinear susceptibilities. *Phys. Rev. B* **1970**, *2*, 1091–1098. [CrossRef]
96. Jerphagnon, J.; Chemla, D.; Bonneville, R. The description of the physical properties of condensed matter using irreducible tensors. *Adv. Phys.* **1978**, *27*, 609–650. [CrossRef]
97. Andrews, D.L.; Ghoul, W.A. Irreducible fourth-rank Cartesian tensors. *Phys. Rev. A* **1982**, *25*, 2647–2657. [CrossRef]
98. Gutiérrez-Cuevas, R.; Alonso, M.A. Complete confined bases for beam propagation in Cartesian coordinates. *J. Opt. Soc. Am. A* **2017**, *34*, 1697–1702. [CrossRef] [PubMed]
99. Stone, A.J. Transformation between cartesian and spherical tensors. *Mol. Phys.* **1975**, *29*, 1461–1471. [CrossRef]
100. Thyssen, J.; Schwerdtfeger, P.; Bender, M.; Nazarewicz, W.; Semmes, P.B. Quadrupole and hexadecapole couplings for 127 I in Li 127 I. *Phys. Rev. A* **2001**, *63*, 022505. [CrossRef]
101. Bancewicz, T.; Ożgo, Z. Irreducible spherical representation of some fourth-rank tensors. *J. Comput. Methods Sci. Eng.* **2010**, *10*, 129–138.
102. Andrews, D.L.; Thirunamachandran, T. The hyper-Raman effect: A new approach to vibrational mode classification and assignment of spectral-lines. *J. Chem. Phys.* **1978**, *68*, 2941–2951. [CrossRef]
103. Andrews, D.L.; Ghoul, W.A. Polarization studies in multi-photon absorption-spectroscopy. *J. Chem. Phys.* **1981**, *75*, 530–538. [CrossRef]
104. Andrews, D.L. The theory of double-beam three-photon absorption. *J. Chem. Phys.* **1982**, *77*, 2831–2835. [CrossRef]
105. Andrews, D.L. The theory of double-beam three-photon absorption. II. Polarization-ratio analysis. *J. Chem. Phys.* **1983**, *78*, 1731–1734. [CrossRef]
106. Andrews, D.L.; Wilkes, P.J. Irreducible tensors and selection-rules for three-frequency absorption. *J. Chem. Phys.* **1985**, *83*, 2009–2014. [CrossRef]
107. Kielich, S.; Bancewicz, T. Symmetric and non-symmetric hyper-Raman scattering: Its polarization states and angular dependences. *J. Raman Spectrosc.* **1990**, *21*, 791–796. [CrossRef]
108. Mikhailov, V. Addition of Arbitrary Number of Identical Angular Momenta. *J. Phys. A Math. Gen.* **1977**, *10*, 147–153. [CrossRef]
109. Rashid, M.A. Addition of arbitrary number of identical angular momenta. *J. Phys. A Math. Gen.* **1977**, *10*, L135–L137. [CrossRef]

110. Allcock, P.; Andrews, D.L. Six-wave mixing: Secular resonances in a higher-order mechanism for second-harmonic generation. *J. Phys. B At. Mol. Opt. Phys.* **1997**, *30*, 3731–3742. [CrossRef]
111. Lin, S.J.; Hands, I.D.; Andrews, D.L.; Meech, S.R. Optically induced second harmonic generation by six-wave mixing: A novel probe of solute orientational dynamics. *J. Phys. Chem. A* **1999**, *103*, 3830–3836. [CrossRef]
112. Lin, S.J.; Hands, I.D.; Andrews, D.L.; Meech, S.R. Phase matching and optical geometry considerations in ultrafast non-degenerate six-wave-mixing experiments. *Opt. Commun.* **2000**, *174*, 285–290. [CrossRef]
113. Moll, K.D.; Homoelle, D.; Gaeta, A.L.; Boyd, R.W. Conical Harmonic Generation in Isotropic Materials. *Phys. Rev. Lett.* **2002**, *88*, 153901. [CrossRef] [PubMed]
114. Coles, M.M.; Williams, M.D.; Andrews, D.L. Second harmonic generation in isotropic media: Six-wave mixing of optical vortices. *Opt. Express* **2013**, *21*, 12783–12789. [CrossRef] [PubMed]
115. Christie, J.; Lockwood, D. Selection Rules for Three-and Four-Photon Raman Interactions. *J. Chem. Phys.* **1971**, *54*, 1141–1154. [CrossRef]
116. Zyss, J. Molecular engineering implications of rotational invariance in quadratic nonlinear optics: From dipolar to octupolar molecules and materials. *J. Chem. Phys.* **1993**, *98*, 6583–6599. [CrossRef]
117. Kleinman, D.A. Nonlinear dielectric polarization in optical media. *Phys. Rev.* **1962**, *126*, 1977–1979. [CrossRef]
118. Wagnière, G. Theoretical investigation of Kleinman symmetry in molecules. *Appl. Phys. B* **1986**, *41*, 169–172. [CrossRef]
119. Ford, J.S.; Andrews, D.L. Molecular Tensor Analysis of Third-Harmonic Scattering in Liquids. *J. Phys. Chem. A* **2018**, *122*, 563–573. [CrossRef] [PubMed]
120. McClain, W.M. Excited state symmetry assignment through polarized two-photon absorption studies of fluids. *J. Chem. Phys.* **1971**, *55*, 2789–2796. [CrossRef]
121. McClain, W.M. Polarization dependence of three-photon phenomena for randomly oriented molecules. *J. Chem. Phys.* **1972**, *57*, 2264. [CrossRef]
122. Bancewicz, T. Excess hyperpolarizabilities: The irreducible tensor approach. *J. Math. Chem.* **2012**, *50*, 1570–1581. [CrossRef]
123. Scholes, G.D.; Andrews, D.L. Damping and higher multipole effects in the quantum electrodynamical model for electronic energy transfer in the condensed phase. *J. Chem. Phys.* **1997**, *107*, 5374–5384. [CrossRef]
124. Andrews, D.L. Optical angular momentum: Multipole transitions and photonics. *Phys. Rev. A* **2010**, *81*, 033825. [CrossRef]
125. Andrews, D.L. On the conveyance of angular momentum in electronic energy transfer. *Phys. Chem. Chem. Phys.* **2010**, *12*, 7409–7417. [CrossRef] [PubMed]
126. Smith, S.N.A.; Andrews, D.L. Three-dimensional ensemble averages for tensorial interactions in partially oriented, multi-particle systems. *J. Phys. A Math. Gen.* **2011**, *44*, 395001. [CrossRef]
127. Jeffreys, H. On isotropic tensors. *Math. Proc. Camb. Philos. Soc.* **1973**, *73*, 173–176. [CrossRef]
128. Andrews, D.L.; Blake, N.P. Three-dimensional rotational averages in radiation molecule interactions: An irreducible Cartesian tensor formulation. *J. Phys. A Math. Gen.* **1989**, *22*, 49–60. [CrossRef]
129. Andrews, D.L.; Ghoul, W.A. Eighth rank isotropic tensors and rotational averages. *J. Phys. A Math. Gen.* **1981**, *14*, 1281–1290. [CrossRef]
130. Molina-Terriza, G.; Torres, J.P.; Torner, L. Management of the angular momentum of light: Preparation of photons in multidimensional vector states of angular momentum. *Phys. Rev. Lett.* **2002**, *8801*, 013601. [CrossRef] [PubMed]
131. Babiker, M.; Bennett, C.R.; Andrews, D.L.; Dávila Romero, L.C. Orbital angular momentum exchange in the interaction of twisted light with molecules. *Phys. Rev. Lett.* **2002**, *89*, 143601. [CrossRef] [PubMed]
132. Andrews, D.L.; Harlow, M.J. Phased and Boltzmann-weighted rotational averages. *Phys. Rev. A* **1984**, *29*, 2796–2806. [CrossRef]
133. Wagnière, G. The evaluation of three-dimensional rotational averages. *J. Chem. Phys.* **1982**, *76*, 473–480. [CrossRef]
134. Friese, D.H.; Beerepoot, M.T.P.; Ruud, K. Rotational averaging of multiphoton absorption cross sections. *J. Chem. Phys.* **2014**, *141*, 204103. [CrossRef] [PubMed]
135. Forbes, K.A.; Andrews, D.L. Optical orbital angular momentum: Twisted light and chirality. *Opt. Lett.* **2018**, *43*, 435–438. [CrossRef] [PubMed]
136. Andrews, D.L.; Sherborne, B.S. A symmetry analysis of electric-field-induced spectra. *Chem. Phys.* **1984**, *88*, 1–5. [CrossRef]

137. Coles, M.M.; Leeder, J.M.; Andrews, D.L. Static and dynamic modifications to photon absorption: The effects of surrounding chromophores. *Chem. Phys. Lett.* **2014**, *595–596*, 151–155. [CrossRef]
138. Andrews, D.L.; Bittner, A.M. Influence of a magnetic-field on line-intensities in the optical-spectra of free molecules. *J. Chem. Soc. Faraday Trans.* **1991**, *87*, 513–516. [CrossRef]
139. Wagnière, G.; Meier, A. The influence of a static magnetic field on the absorption coefficient of a chiral molecule. *Chem. Phys. Lett.* **1982**, *93*, 78–81. [CrossRef]
140. Train, C.; Rikken, G.; Verdaguer, M. Non-Centrosymmetric Molecular Magnets. In *Molecular Magnetic Materials: Concepts and Applications*; Sieklucka, B., Pincowicz, D., Eds.; Wiley-VCH: Weinheim, Germany, 2016; pp. 301–322.
141. Raupach, E.; Rikken, G.L.J.A.; Train, C.; Malézieux, B. Modelling of magneto-chiral enantioselective photochemistry. *Chem. Phys.* **2000**, *261*, 373–380. [CrossRef]
142. Banerjee-Ghosh, K.; Ben Dor, O.; Tassinari, F.; Capua, E.; Yochelis, S.; Capua, A.; Yang, S.-H.; Parkin, S.S.P.; Sarkar, S.; Kronik, L.; et al. Separation of enantiomers by their enantiospecific interaction with achiral magnetic substrates. *Science* **2018**. [CrossRef] [PubMed]

© 2018 by the author. Licensee MDPI, Basel, Switzerland. This article is an open access article distributed under the terms and conditions of the Creative Commons Attribution (CC BY) license (http://creativecommons.org/licenses/by/4.0/).

Review

Quasi-Lie Brackets and the Breaking of Time-Translation Symmetry for Quantum Systems Embedded in Classical Baths

Alessandro Sergi [1,2,3,*], **Gabriel Hanna** [4], **Roberto Grimaudo** [3,5] **and Antonino Messina** [3,6]

1. Dipartimento di Scienze Matematiche e Informatiche, Scienze Fisiche e Scienze della Terra, Università degli Studi di Messina, Contrada Papardo, 98166 Messina, Italy
2. Institute of Systems Science, Durban University of Technology, P.O. Box 1334, Durban 4000, South Africa
3. Istituto Nazionale di Fisica Nucleare, Sez. di Catania,95123 Catania, Italy; roberto.grimaudo01@unipa.it (R.G.); antonino.messina@unipa.it (A.M.)
4. Department of Chemistry, University of Alberta, 11227 Saskatchewan Drive Edmonton, Edmonton, AB T6G 2G2, Canada; gabriel.hanna@ualberta.ca
5. Dipartimento di Fisica e Chimica dell'Universitá di Palermo, Via Archirafi 36, I-90123 Palermo, Italy
6. Dipartimento di Matematica e Informatica, Università degli Studi di Palermo, Via Archirafi 34, I-90123 Palermo, Italy
* Correspondence: asergi@unime.it

Received: 13 September 2018; Accepted: 12 October 2018; Published: 16 October 2018

Abstract: Many open quantum systems encountered in both natural and synthetic situations are embedded in classical-like baths. Often, the bath degrees of freedom may be represented in terms of canonically conjugate coordinates, but in some cases they may require a non-canonical or non-Hamiltonian representation. Herein, we review an approach to the dynamics and statistical mechanics of quantum subsystems embedded in either non-canonical or non-Hamiltonian classical-like baths which is based on operator-valued quasi-probability functions. These functions typically evolve through the action of quasi-Lie brackets and their associated Quantum-Classical Liouville Equations, or through quasi-Lie brackets augmented by dissipative terms. Quasi-Lie brackets possess the unique feature that, while conserving the energy (which the Noether theorem links to time-translation symmetry), they violate the time-translation symmetry of their algebra. This fact can be heuristically understood in terms of the dynamics of the open quantum subsystem. We then describe an example in which a quantum subsystem is embedded in a bath of classical spins, which are described by non-canonical coordinates. In this case, it has been shown that an off-diagonal open-bath geometric phase enters into the propagation of the quantum-classical dynamics. Next, we discuss how non-Hamiltonian dynamics may be employed to generate the constant-temperature evolution of phase space degrees of freedom coupled to the quantum subsystem. Constant-temperature dynamics may be generated by either a classical Langevin stochastic process or a Nosé–Hoover deterministic thermostat. These two approaches are not equivalent but have different advantages and drawbacks. In all cases, the calculation of the operator-valued quasi-probability function allows one to compute time-dependent statistical averages of observables. This may be accomplished in practice using a hybrid Molecular Dynamics/Monte Carlo algorithms, which we outline herein.

Keywords: quasi-lie brackets; quantum-classical Liouville equation; hybrid quantum-classical systems; breaking of time-translation symmetry; classical spin dynamics; Langevin dynamics; Nosé–Hoover dynamics

1. Introduction

A growing community of physicists is interested in both monitoring and controlling the time evolution of small numbers of quantum degrees of freedom (DOF) that are embedded in noisy and uncontrollable environments [1–3]. A specific case of such a system is encountered when the environment is classical-like in nature. This situation is one of fundamental importance because, ultimately, we and our experimental tools behave classically, at least from a coarse-grained perspective. In recent years, we have also witnessed a rising interest in nano-mechanical, opto-mechanical and other types of hybrid quantum systems [4–26]. Such systems often exhibit an interplay between classical and quantum effects, allowing them to be modeled by means of hybrid quantum-classical methods.

It has been known for a long time, that the dynamics and statistical mechanics of a quantum subsystem coupled to classical-like DOF can be formulated in terms of operator-valued quasi-probability functions in phase space [27–32]. For example, the dynamics of nano-mechanical oscillators has been previously described by one of the authors in terms of operator-valued quasi-probability functions [33]. Such functions evolve through quasi-Lie brackets [34–43], which can also be augmented by dissipative terms when the energy is not conserved [44,45]. When the bath is described by canonically conjugate variables (and only in this case), a hybrid quantum-classical formalism may be derived. Starting from a fully quantum representation of the subsystem and bath DOF, one can perform a partial Wigner transform [46] (over the bath DOF) and then take its semiclassical limit [47]. The resulting equation of motion is commonly known as the quantum-classical Liouville equation (QCLE) [48–60]. The QCLE has been used to study a wide variety of problems [61–75] and a number of in-depth reviews on the basic formulation of the theory exist [76–89]. The mathematical structure underlying the QCLE is dictated by a quasi-Lie bracket [42,43,90,91]. Quasi Lie brackets are known within the community of classical molecular dynamics simulators as non-Hamiltonian brackets [92–94]. Mathematicians have also studied very similar structures known as almost Poisson brackets or quasi-Lie algebras [95–99]. It is interesting to note that the quasi-Lie (or non-Hamiltonian) structure of the QCLE [30,31,34–43] has both favorable and unfavorable aspects associated with it. Because the antisymmetry of the quasi-Lie bracket ensures energy conservation, one is able to verify the stability of numerical integration algorithms. However, because the quasi-Lie algebra is not invariant under time translation, the initially classical DOF acquire a quantum character as time flows, implying that one never has a true dynamical theory of quantum and classical DOF but only an approximated dynamics of a full quantum system [100]. This is somewhat paradoxical because energy conservation is linked to time-translation symmetry through the Noether theorem; nevertheless, quasi-Lie brackets break the time-translation symmetry of the algebra (which can be seen as a signature of the effect of the classical bath on the quantum subsystem).

This review deals with situations where the bath DOF are described in terms of non-canonical coordinates [101,102] or non-Hamiltonian coordinates [92–94], and situations where dissipation must be taken explicitly into account [44,45]. In all these cases, we will see that the operator-valued probability functions will develop new functional dependences and novel definitions of the quasi-Lie brackets will have to be introduced. In particular, we will first describe the case of a classical spin bath [90,91], as an example of a bath described by non-canonical coordinates [101,102]. It has been shown that for such a bath an off-diagonal [103] open-path [104–106] geometric phase [107–109] enters into the propagation of the quantum-classical dynamics. We will then describe the case of a non-Hamiltonian bath, which arises when the bath coordinates coupled to the quantum subsystem are also coupled to a large bath (which does not directly interact with the quantum subsystem and whose detailed dynamics is not of interest). In such cases, the secondary bath acts as a thermal reservoir and can be described either by means of stochastic processes [110] (e.g., Langevin dynamics [45]), or by means of non-Hamiltonian fictitious coordinates acting as deterministic thermostats (e.g., the Nosé–Hoover thermostat [111,112]). Both Langevin and Nosé–Hoover deterministic time evolutions are examples of non-Hamiltonian dynamics. However, only Nosé–Hoover dynamics is defined solely in terms of a quasi-Lie bracket [42,43]. Instead, explicit dissipative dynamics requires that diffusive terms be added to the bracket.

The quantum-classical equations of motion herein discussed can be implemented in silico using a variety of simulation algorithms [78,113–123]. We will sketch out one such integration algorithm, which unfolds the quantum-classical dynamics of the operator-valued quasi-probability function in terms of piecewise-deterministic trajectories evolving on the adiabatic energy surfaces of the system under study [78,113].

The structure of this review is as follows. In Section 2, we illustrate the algebraic approach used to formulate the dynamics of a quantum subsystem embedded in a classical-like environment with canonically conjugate coordinates. In Section 3, we show how this formalism can be generalized to the case of a bath described by non-canonical variables, namely a collection of classical spins. Here, we will also show how an off-diagonal open-path geometric phase enters into the time evolution of the operator-valued quasi-probability function of the system. In Section 4, we show how the formalism allows us to also treat stochastic classical-like baths undergoing Langevin dynamics. Finally, in Section 5, we shed light on the quasi-Lie algebra established by the quantum-classical brackets and show how their antisymmetric structure is exploited to achieve thermal control of the bath DOF by means of deterministic thermostats such as the Nosé–Hoover and Nosé–Hoover chain thermostats. Our conclusions and perspectives are given in Section 6.

2. Quasi-Lie Brackets and Hybrid Quantum-Classical Systems

Classical and quantum dynamics share the same algebraic structure [124,125], which is realized by means of Poisson brackets in the classical case and commutators in the quantum theory. Poisson brackets have a symplectic structure that is easily represented in matrix form [102,126]. Both Poisson brackets and commutators define Lie algebras. In terms of commutators, a Lie algebra possesses the following properties:

$$[\hat{\chi}_1, \hat{\chi}_2] = -[\hat{\chi}_2, \hat{\chi}_2], \quad (1)$$

$$[\hat{\chi}_1\hat{\chi}_2, \hat{\chi}_3] = \hat{\chi}_1[\hat{\chi}_2, \hat{\chi}_3] + [\hat{\chi}_1, \hat{\chi}_3]\hat{\chi}_2, \quad (2)$$

$$[c, \hat{\chi}_j] = 0, \quad (3)$$

where c is a so-called c-number and $\hat{\chi}_j, j = 1, 2, 3$ are quantum operators. In order to have a Lie algebra, together with Equations (1)–(3), the Jacobi relation must also hold

$$\mathcal{J} = [\hat{\chi}_1, [\hat{\chi}_2, \hat{\chi}_3]] + [\hat{\chi}_3, [\hat{\chi}_1, \hat{\chi}_2]] + [\hat{\chi}_2, [\hat{\chi}_3, \hat{\chi}_1]] = 0. \quad (4)$$

The time-translation invariance of the commutator algebra follows from the Jacobi relation, which therefore states an integrability condition. If $\hat{H}$ is not explicitly time-dependent, the antisymmetry of the commutator (1), arising from the antisymmetry of the symplectic matrix $\mathcal{B}$, ensures that the energy is a constant of motion: $d\hat{H}/dt = i\hat{\mathcal{L}}\hat{H} = 0$. Energy conservation under time-translation is a fundamental property shared by the algebra of Poisson brackets and the algebra of commutators that is in agreement with Noether theorem.

Now, let us consider a hybrid quantum-classical system, in which the quantum subsystem, described by a few canonically conjugate operators $(\hat{q}, \hat{p}) = \hat{x}$ is embedded in a classical bath with many DOF, described by many canonically conjugate phase space coordinates, $X = (Q, P)$. We will assume that the Hamiltonian of this hybrid system has the form

$$\begin{aligned} \hat{H}_W(X) &= \frac{P^2}{2M} + \frac{\hat{p}^2}{2m} + V_W(\hat{q}, Q) \\ &= \frac{P^2}{2M} + \hat{h}_W(Q), \end{aligned} \quad (5)$$

where m and M are the masses of the subsystem and bath DOF, respectively, and V_W is the potential energy describing the interactions among the subsystem DOF, among the bath DOF, and between these two sets of DOF. The last equality on the right-hand side of Equation (5) defines the adiabatic Hamiltonian $\hat{h}_W(Q)$ of the system. It has been known for many years that the statistical mechanics of such hybrid quantum-classical systems may be formulated in terms of an operator-valued quasi-probability function $\hat{W}(X,t)$ [27–32]. Specifically, the statistical average of hybrid quantum-classical operators, representing a dynamical property of the system, may be calculated according to

$$\langle \hat{\chi} \rangle(t) = \text{Tr}' \int dX \hat{W}(X,t) \hat{\chi}_W(X) , \qquad (6)$$

where Tr' denotes the partial trace involving a complete set of states of the quantum subsystem.

The operator-valued quasi-probability function in phase space evolves according to

$$\frac{\partial}{\partial t} \hat{W}(X,t) = -\frac{i}{\hbar} \begin{bmatrix} \hat{H}_W & \hat{W}(X,t) \end{bmatrix} \mathcal{D} \begin{bmatrix} \hat{H}_W \\ \hat{W}(X,t) \end{bmatrix} = -\frac{i}{\hbar} [\hat{H}_W, \hat{W}(X,T)]_\mathcal{D} = -i\hat{\mathcal{L}}^\mathcal{D} \hat{\chi} , \qquad (7)$$

where $\mathcal{D}$ is an antisymmetric matrix super-operator defined by

$$\mathcal{D} = \begin{bmatrix} 0 & 1 - \frac{\overleftarrow{\nabla} \mathcal{B} \overrightarrow{\nabla}}{2i\hbar^{-1}} \\ -\left(1 - \frac{\overleftarrow{\nabla} \mathcal{B} \overrightarrow{\nabla}}{2i\hbar^{-1}}\right) & 0 \end{bmatrix} , \qquad (8)$$

with $\nabla = (\partial/\partial Q, \partial/\partial P) = \partial/\partial X$, and

$$\overleftarrow{\nabla} \mathcal{B} \overrightarrow{\nabla} = \sum_{I,J=1}^{2N} \overleftarrow{\nabla}_I \mathcal{B}_{IJ} \overrightarrow{\nabla}_J \qquad (9)$$

denotes the Poisson bracket operator. The last equality on the right-hand side of Equation (7) defines the quantum-classical Liouville operator $i\hat{\mathcal{L}}^\mathcal{D}$. Equation (7) is the QCLE [48–60] of the system.

The QCLE in Equation (7) is founded upon a quasi-Lie bracket, which we may write explicitly as

$$[\hat{\chi}_1(X), \hat{\chi}_2(X)]_\mathcal{D} = \begin{bmatrix} \hat{\chi}_1(X) & \hat{\chi}_2(X) \end{bmatrix} \mathcal{D} \begin{bmatrix} \hat{\chi}_1(X) \\ \hat{\chi}_2(X) \end{bmatrix} , \qquad (10)$$

where $\mathcal{D}$ is the antisymmetric matrix operator defined in Equation (8). However, in contrast to the Lie brackets of quantum and classical mechanics, the quasi-Lie bracket defined in Equation (10) violates the Jacobi relation (4):

$$\mathcal{J}_\mathcal{D} = [\hat{\chi}_1(X), [\hat{\chi}_2(X), \hat{\chi}_3(X)]_\mathcal{D}]_\mathcal{D} + [\hat{\chi}_3(X), [\hat{\chi}_1(X), \hat{\chi}_2(X)]_\mathcal{D}]_\mathcal{D} + [\hat{\chi}_2(X), [\hat{\chi}_3(X), \hat{\chi}_1(X)]_\mathcal{D}]_\mathcal{D} \neq 0 . \qquad (11)$$

The failure of the Jacobi implies that the algebra of quasi-Lie brackets is not invariant under time-translation. For example, it can be generally proven that

$$e^{it\mathcal{L}^\mathcal{D}} [\hat{\chi}_1(X,0), \hat{\chi}_2(X)] \neq \left[e^{it\mathcal{L}^\mathcal{D}} \hat{\chi}_1(X), e^{it\mathcal{L}^\mathcal{D}} \hat{\chi}_2(X) \right] . \qquad (12)$$

On the other hand, the quasi-Lie bracket conserves the energy $e^{it\mathcal{L}^\mathcal{D}} \hat{H}_W(X) = \hat{H}_W(X)$. Hence, the dynamics generated by the QCLE displays energy conservation and lack of time-translation invariance of the bracket algebra. The situation is surprising because one does not expect a broken time-translation invariance symmetry in an isolated system. However, while a total hybrid quantum-classical system is closed from the point of view of energy conservation, the quasi-Lie bracket describes the irreversible transfer of quantum information from the subsystem to the classical DOF, which acquire a quantum character as the time flows. In this sense, one can heuristically argue that the lack of time-translation invariance or the algebra is a mere consequence of the open dynamics of the quantum subsystem.

2.1. Derivation of the QCLE through a Partial Wigner Transform

When the bath DOF are described by canonically conjugate variables (and only in this case), the hybrid quantum can be derived by performing a partial Wigner transform of the quantum Liouville equation (QLE) over the bath DOF and taking a semiclassical limit of the resulting equations. To this end, let us consider the fully quantum counterpart to the Hamiltonian in Equation (5):

$$\hat{H} = \frac{\hat{P}^2}{2M} + \frac{\hat{p}^2}{2m} + V(\hat{q}, \hat{Q}) \,. \tag{13}$$

The quantum statistical state of the system is described by the density matrix (or statistical operator) $\hat{\rho}(t)$. The time dependence of the density matrix is dictated by the QLE:

$$\frac{d}{dt}\hat{\rho}(t) = -\frac{i}{\hbar}[\hat{H}, \hat{\rho}(t)] = -\frac{i}{\hbar} \begin{bmatrix} \hat{\rho} & \hat{H} \end{bmatrix} \mathcal{B} \begin{bmatrix} \hat{\rho} \\ \hat{H} \end{bmatrix}, \tag{14}$$

where $[...,...]$ denotes the commutator, and $\mathcal{B}$ is the symplectic matrix [102,126]:

$$\mathcal{B} = \begin{bmatrix} 0 & 1 \\ -1 & 0 \end{bmatrix}. \tag{15}$$

The average of an operator $\hat{\chi}$ defined on the same Hilbert space of the system is calculated by

$$\langle \hat{\chi} \rangle(t) = \text{Tr}\,(\hat{\rho}(t)\hat{\chi}) \,, \tag{16}$$

where Tr denotes the trace operation. Now, in order to derive a classical-like description of the bath, one introduces the partial Wigner transform of the density matrix $\hat{\rho}$ over the $\hat{X}$'s:

$$\hat{W}(X,t) = \frac{1}{2\pi\hbar}\int dZ e^{iP\cdot Z/\hbar}\langle Q - \frac{Z}{2}|\hat{\rho}(t)|Q + \frac{Z}{2}\rangle \,. \tag{17}$$

The symbol $\hat{W}$ denotes an operator-valued Wigner function (also known as the partially-Wigner transformed density matrix), which is both an operator in the Hilbert space of the $\hat{q}$'s and a function of the bath coordinates X. The partial Wigner transform of an arbitrary operator $\hat{\chi}$ is analogously given by

$$\hat{\chi}_W(X) = \int dZ e^{iP\cdot Z/\hbar}\langle Q - \frac{Z}{2}|\hat{\chi}|Q + \frac{Z}{2}\rangle \,. \tag{18}$$

Taking the partial Wigner transform of Equation (16) leads to the expression for the average of $\hat{\chi}$ given in Equation (6). The partial Wigner transform of the Hamiltonian in Equation (13) is given in Equation (5).

Upon taking the partial Wigner transform of the QLE, Equation (14), and truncating the resulting equation after first order in $\hbar$, one arrives at the QCLE

$$\begin{aligned}\frac{\partial}{\partial t}\hat{W}(X,t) &= -\frac{i}{\hbar}[\hat{H}_W, \hat{W}(X,t)] + \frac{1}{2}\hat{H}_W\overrightarrow{\nabla}\mathcal{B}\overrightarrow{\nabla}\hat{W}(X,t) - \frac{1}{2}\hat{W}(X,t)\overleftarrow{\nabla}\mathcal{B}\overrightarrow{\nabla}\hat{H}_W \\ &= -i\mathcal{L}\hat{W}(X,t) \,,\end{aligned} \tag{19}$$

where the last equality defines the quantum Liouville operator $i\mathcal{L} = (i/\hbar)[\hat{H}_W, \cdot] - (1/2)(\hat{H}_W \overleftarrow{\nabla} \mathcal{B} \overrightarrow{\nabla} \cdot) + (1/2)(\cdot \overleftarrow{\nabla} \mathcal{B} \overrightarrow{\nabla} \hat{H}_W)$. To arrive at Equation (19), we have used the partial Wigner transform of a product of operators,

$$(\hat{\chi}_1 \hat{\chi}_2)_W (X) = \hat{\chi}_{1,W}(X) e^{\frac{i\hbar}{2} \overleftarrow{\nabla} \mathcal{B} \overrightarrow{\nabla}} \hat{\chi}_{2,W}(X), \tag{20}$$

and truncated the exponential after first order in $\hbar$, i.e.,

$$e^{\frac{i\hbar}{2} \overleftarrow{\nabla} \mathcal{B} \overrightarrow{\nabla}} \approx 1 + \frac{i\hbar}{2} \overleftarrow{\nabla} \mathcal{B} \overrightarrow{\nabla}. \tag{21}$$

It should be noted that Equation (21) is exact for Hamiltonians with quadratic bath terms and bilinear coupling between the $\hat{x}$ and X DOF. In Ref. [47], it is shown how the linear expansion can be performed in terms of the parameter $\mu = \sqrt{m/M}$, which is small in cases where the bath DOF are much more massive than those of the subsystem. Equation (19) is exactly equivalent to Equation (7).

2.2. Integration Algorithm

A number of algorithms, which depend on the basis representation, exist for approximately solving the QCLE [50,51,54–56,60,78,113–123]. Herein, we illustrate the so-called Sequential Short-Time Propagation (SSTP) algorithm [78,113], which offers a good compromise between accuracy and simplicity of implementation. The SSTP algorithm is based on the representation of the QCLE in the adiabatic basis, which is defined by the eigenvalue equation

$$\hat{h}_W |\alpha; Q\rangle = E_\alpha(Q) |\alpha; Q\rangle. \tag{22}$$

The representation of the QCLE in the adiabatic basis is sketched in Appendix A. In the adiabatic basis, the QCLE is given by Equation (A1) and the quantum-classical Liouville super-operator matrix elements are given in Equation (A4).

To derive the SSTP algorithm, we divide the time interval t into n equal small steps $\tau = t/n$. If one is able to calculate the propagation over a single τ, the dynamics over the whole interval can be reconstructed by sequential iteration of the procedure. Let us then consider the quantum-classical propagator over a small step τ for the matrix elements of the operator-valued quasi-probability function $\hat{W}(X)$ in the adiabatic basis. Such a propagator is written as

$$\left(e^{-i\tau\mathcal{L}}\right)_{\alpha\alpha',\beta\beta'} \approx \delta_{\alpha\beta} \delta_{\alpha'\beta'} e^{-i \int_0^\tau ds \, \omega_{\alpha\alpha'}(s)} e^{-i\tau L_{\alpha\alpha'}} \left(1 + \tau \mathcal{T}_{\alpha\alpha',\beta\beta'}\right). \tag{23}$$

On the right-hand side of Equation (23), we have introduced $\omega_{\alpha\alpha'}$, the Bohr frequency defined in Equation (A3), $iL_{\alpha\alpha'}$ is a classical-like Liouville operator, defined in Equation (A5), and $\mathcal{T}_{\alpha\alpha',\beta\beta'}$ is the transition operator defined in Equation (A7). The SSTP dynamics of the matrix elements of $\hat{W}(X,t)$ is given by

$$W_{\alpha\alpha'}(X,\tau) = \sum_{\beta\beta'} \delta_{\alpha\beta} \delta_{\alpha'\beta'} e^{-i \int_0^\tau ds \, \omega_{\alpha\alpha'}(s)} e^{-i\tau L_{\alpha\alpha'}} \left(1 + \tau \mathcal{T}_{\alpha\alpha',\beta\beta'}\right) W_{\beta\beta'}(X). \tag{24}$$

When τ is infinitesimal, the right-hand side of Equations (23) and (24), become essentially equal to the left-hand side, as can be seen from the Dyson identity [113].

The transition operator is purely off-diagonal. Its action generates quantum transitions in the subsystems and changes the bath momenta accordingly. Upon setting the transition operator to zero, we obtain an adiabatic expression for the propagator. If the non-adiabatic effects are not too strong, they may be treated in a perturbative fashion by sampling the action of the transition operator in a stochastic fashion. Typically, researchers have used [43,62,64–75,78,82,85,87–91,113,118–122] the following expressions for the probabilities of making a transition (jump) and not-making a transition, respectively:

$$\mathcal{P}_J = \frac{|\tau \frac{P}{M} \cdot d_{\alpha\beta}|}{1 + |\tau \frac{P}{M} \cdot d_{\alpha\beta}|}, \qquad (25)$$

$$\mathcal{Q}_{NO-J} = \frac{1}{1 + |\tau \frac{P}{M} \cdot d_{\alpha\beta}|}. \qquad (26)$$

Another important technical ingredient of the algorithm is the approximation of the transition operator in Equation (A7) with its momentum-jump form:

$$\mathcal{T}^{MJ}_{\alpha\alpha',\beta\beta'} = \delta_{\alpha'\beta'} \frac{P}{M} \cdot d_{\alpha\beta} e^{(E_\alpha - E_\beta) M \partial / \partial (P \cdot \hat{d}_{\alpha\beta})^2} + \delta_{\alpha\beta} \frac{P}{M} \cdot d^*_{\alpha'\beta'} e^{(E'_\alpha - E'_\beta) M \partial / \partial (P \cdot \hat{d}^*_{\alpha'\beta'})^2}, \qquad (27)$$

where $\hat{d}_{\alpha\beta}$ is the normalized coupling vector. Within the momentum-jump approximation [77,78], the action of the transition operator on the bath momenta can be easily obtained in closed form:

$$e^{(E_\alpha - E_\beta) M \partial / \partial (P \cdot \hat{d}_{\alpha\beta})^2} P = P - P\left(P \cdot \hat{d}_{\alpha\beta}\right) + \hat{d}_{\alpha\beta} \sqrt{\left(P \cdot \hat{d}_{\alpha\beta}\right)^2 + M\left(E_\alpha - E_\beta\right)}. \qquad (28)$$

Considering Equations (6) and (24), together with its SSTP implementation just described, one can see that the solution of the QCLE can be obtained from an ensemble of classical-like trajectories, where each trajectory (whose initial conditions arise from a Monte Carlo sampling [127] of the X's), involves deterministic evolution segments on a given adiabatic energy surfaces interspersed with stochastic quantum transitions, caused by the momentum-jump operator in Equation (27).

The SSTP algorithm [78,113] maps the calculation of averages through the QCLE (19) onto a stochastic process. It is a hybrid Molecular Dynamics/Monte Carlo procedure suffering from two main problems. The first is given by the momentum-jump approximation, which is not valid in general. One can avoid this approximation by devising different integration schemes, but usually at the expense of other approximations [123]. The second problem is not just associated with the SSTP algorithm, but it is common to all Monte Carlo approaches to the calculation of quantum averages: the *infamous* sign-problem. The sign-problem is one of the major unsolved problems in the physics of quantum systems. Within the SSTP algorithm, it manifests itself both through the oscillating phase factors associated with the propagation on mean-energy surfaces and through the accumulation of fluctuating weights associated with the Monte Carlo sampling of the quantum transitions. In practice, upon analyzing the results obtained by means of this algorithm [43,62,64–75,78,82,85,87–91,113,115–122], we can conclude that the more quantum is the character of the bath the greater is the error in the calculation of the averages.

The mapping of the calculation of averages via the SSTP algorithm onto a stochastic process is reminiscent of the approach to open quantum system dynamics provided by the Stochastic Liouville Equation (SLE) [128–131]. However, in contrast to the SLE, the QCLE is a deterministic equation that explicitly takes into account all the DOF of the system without approximating the memory of the total hybrid quantum-classical system. The stochastic process only enters through the specific hybrid Molecular Dynamics/Monte Carlo implementation provided by the SSTP algorithm. Indeed, a recently proposed scheme of integration [123] does not involve any stochastic process whatsoever.

3. Classical Spin Baths

Contrary to what some books in quantum mechanics state (in the authors's knowledge, an exception is Schulman's book [132]), the concept of spin can be defined in an entirely classical way [132–136]. In practice, spinors provide a more fundamental representation of the rotation group than that given by tensors [132–136]. Hence, one can think of a collection, e.g., a bath, of DOF comprising classical spinors (or, for brevity, spins): a classical spin-bath. An example of a classical spin baths is given by the Classical Heisenberg Model [137], whose Hamiltonian is

$$H_{CHS} = \sum_{a=x,y,z} \sum_{I,J}^{N} S_a^I C_{IJ}^a S_a^J, \qquad (29)$$

where $\mathbf{S}^I$ are N classical vectors obeying the constraint

$$\left(S_x^I\right)^2 + \left(S_y^I\right)^2 + \left(S_z^I\right)^2 = 1, \qquad (30)$$

for $I = 1, ..., N$, and the C_{IJ}^a are coupling constants. However, since the generalization to baths with many spins is straightforward, in the following, we will illustrate the theory using a bath comprising a single classical spin. Consider a classical spin vector $\mathbf{S}$, with components S_a, $a = x, y, z$, and Hamiltonian $H^S(\mathbf{S})$. Let us define the spin gradient as $\nabla^\mathbf{S} = \partial/\partial \mathbf{S}$, which in terms of the spin components is written as $\nabla_a^S = \partial/\partial S_a$, with $a = x, y, z$. The equations of motion of the spin are then written as

$$\dot{\mathbf{S}} = \mathcal{B}^\mathbf{S} \nabla^\mathbf{S} H_S, \qquad (31)$$

where

$$\mathcal{B}^\mathbf{S} = \begin{bmatrix} 0 & S_z & -S_y \\ -S_z & 0 & S_x \\ S_y & -S_x & 0 \end{bmatrix}. \qquad (32)$$

One can also adopt the compact form $\mathcal{B}_{ab}^\mathbf{S} = \sum_{c=x,y,z} \epsilon_{abc} S_c$ and $a, b = x, y, z$ of the antisymmetric matrix $\mathcal{B}^\mathbf{S}$, where ϵ_{abc} is the Levi–Civita pseudo-tensor. The Casimir $C_2 = \mathbf{S} \cdot \mathbf{S}$ is preserved by the equations of motion (31), independently of the form of the spin Hamiltonian $H^S(\mathbf{S})$. In addition, the dynamics has a zero phase space compressibility $\kappa^S = \nabla \mathbf{S} \cdot \dot{\mathbf{S}} = 0$. The classical phase space flow of the spin is defined through the non-canonical bracket

$$\sum_{a,b} A(\mathbf{S}) \overleftarrow{\nabla}_a^S \mathcal{B}_{ab}^S \overrightarrow{\nabla}_b^S B(\mathbf{S}) = A(\mathbf{S}) \overleftarrow{\nabla}^S \mathcal{B}^S \overrightarrow{\nabla}^S B(\mathbf{S}), \qquad (33)$$

where $A = A(\mathbf{S})$ and $B = B(\mathbf{S})$ are arbitrary functions of the spin DOF.

Consider now the hybrid quantum-classical Hamiltonian of a quantum subsystem coupled to the classical spin

$$\begin{aligned} \hat{\mathcal{H}}(\mathbf{S}) &= \hat{H}(\{\hat{x}\}) + V_C(\{\hat{x}\}, \mathbf{S}) + H^S(\mathbf{S}) \\ &= \hat{h}_\mathbf{S}(\mathbf{S}) + H^S(\mathbf{S}), \end{aligned} \qquad (34)$$

describing a quantum subsystem in terms of the Hamiltonian $\hat{H}(\{\hat{x}\})$, depending on the operators $\{\hat{x}\}$, $V(\{\hat{x}\}, \mathbf{S})$ is the subsystem-spin interaction potential, and the second line of the equation defines the adiabatic Hamiltonian $\hat{h}_\mathbf{S}$. The quantum-classical dynamics of the operator-valued quasi-probability function (defined in the spinor space of the total system), $\hat{W}^\mathbf{S}(\mathbf{S}, t)$, is dictated by the spin-bath QCLE [90,91]

$$\begin{aligned} \frac{\partial}{\partial t} \hat{W}^\mathbf{S}(\mathbf{S}, t) &= -\frac{i}{\hbar} \left[\hat{\mathcal{H}}(\mathbf{S}) \quad \hat{W}^\mathbf{S}(\mathbf{S}, t) \right] \mathcal{D}^\mathbf{S} \left[\begin{array}{c} \hat{\mathcal{H}}(\mathbf{S}) \\ \hat{W}^\mathbf{S}(\mathbf{S}, t) \end{array} \right] \\ &= -\frac{i}{\hbar} [\hat{\mathcal{H}}(\mathbf{S}), \hat{W}^\mathbf{S}(\mathbf{S}, t)]_{\mathcal{D}^\mathbf{S}}, \end{aligned} \qquad (35)$$

where

$$\mathcal{D}^\mathbf{S} = \begin{bmatrix} 0 & 1 + \frac{i\hbar}{2} \overleftarrow{\nabla} \mathcal{B}^S \overrightarrow{\nabla} \\ -1 - \frac{i\hbar}{2} \overleftarrow{\nabla} \mathcal{B}^S \overrightarrow{\nabla} & 0 \end{bmatrix}. \qquad (36)$$

We next set out to represent Equation (35) in the adiabatic basis $|\alpha; \mathbf{S}\rangle$ defined by the eigenvalue equation

$$\hat{h}_\mathbf{S}(\mathbf{S})|\alpha; \mathbf{S}\rangle = E_\alpha(\mathbf{S})|\alpha; \mathbf{S}\rangle . \qquad (37)$$

It should be noted that, in contrast to the case of canonically conjugate phase space coordinates which depends only on the positions Q and not on the conjugate momenta P, this adiabatic basis depends on all the non-canonical spin coordinates $\mathbf{S}$. In this basis, Equation (35) becomes

$$\begin{aligned}\frac{\partial}{\partial t}\hat{W}^\mathbf{S}_{\alpha\alpha'} = & -i\omega_{\alpha\alpha'}W^\mathbf{S}_{\alpha\alpha'} - H^\mathbf{S}\overleftarrow{\nabla}^\mathbf{S}\boldsymbol{\mathcal{B}}^\mathbf{S}\langle\alpha|\overrightarrow{\nabla}^\mathbf{S}\hat{W}^\mathbf{S}|\alpha'\rangle \\ & +\frac{1}{2}\langle\alpha|\hat{h}_\mathbf{S}\overleftarrow{\nabla}^\mathbf{S}\boldsymbol{\mathcal{B}}^\mathbf{S}\overrightarrow{\nabla}^\mathbf{S}\hat{W}^\mathbf{S}|\alpha'\rangle - \frac{1}{2}\langle\alpha|\hat{W}^\mathbf{S}\overleftarrow{\nabla}^\mathbf{S}\boldsymbol{\mathcal{B}}^\mathbf{S}\overrightarrow{\nabla}^\mathbf{S}\hat{h}_\mathbf{S}|\alpha'\rangle ,\end{aligned} \qquad (38)$$

where $\omega_{\alpha\alpha'} = E_\alpha(\mathbf{S}) - E_{\alpha'}(\mathbf{S})/\hbar$ is the Bohr frequency. Defining the spin coupling vector

$$d^\mathbf{S}_{\alpha\alpha'} = \langle\alpha; \mathbf{S}|\overrightarrow{\nabla}^\mathbf{S}|\alpha'; \mathbf{S}\rangle, \qquad (39)$$

one finds the two identities

$$\langle\alpha; \mathbf{S}|\left(\overrightarrow{\nabla}^\mathbf{S}\hat{W}^\mathbf{S}(\mathbf{S})\right)|\alpha'; \mathbf{S}\rangle = \overrightarrow{\nabla}^\mathbf{S}W^\mathbf{S}_{\alpha\alpha'}(\mathbf{S}) + \sum_\beta d^\mathbf{S}_{\alpha\beta}W^\mathbf{S}_{\beta\alpha'}(\mathbf{S}) - \sum_{\beta'}W^\mathbf{S}_{\alpha\beta'}(\mathbf{S})d^\mathbf{S}_{\beta'\alpha'} \qquad (40)$$

$$\langle\alpha; \mathbf{S}|\left(\overrightarrow{\nabla}^\mathbf{S}\hat{h}_\mathbf{S}(\mathbf{S})\right)|\alpha'; \mathbf{S}\rangle = \overrightarrow{\nabla}^\mathbf{S}h^{\alpha\alpha'}_\mathbf{S} - \Delta E_{\alpha\alpha'}d^\mathbf{S}_{\alpha\alpha'} \qquad (41)$$

where $\Delta E_{\alpha\alpha'} = E_\alpha - E_{\alpha'}$. Using Equations (40) and (41), the spin-bath QCLE may be rewritten as

$$\frac{\partial}{\partial t}W^\mathbf{S}_{\alpha\alpha'}(\mathbf{S}, t) = -\sum_{\beta\beta'}\left(i\omega_{\alpha\alpha'}\delta_{\alpha\beta}\delta_{\alpha\alpha'} + iL_{\alpha\alpha'}\delta_{\alpha\beta}\delta_{\alpha\alpha'} + \mathcal{T}^\mathbf{S}_{\alpha\alpha',\beta\beta'} + \mathcal{S}_{\alpha\alpha',\beta\beta'}\right)W^\mathbf{S}_{\beta\beta'}(\mathbf{S}, t) , \qquad (42)$$

where we have defined the classical-like spin-Liouville operator

$$\begin{aligned}iL_{\alpha\alpha'} & = H_\mathbf{S}\overleftarrow{\nabla}^\mathbf{S}\boldsymbol{\mathcal{B}}^\mathbf{S}\overrightarrow{\nabla}^\mathbf{S} + \frac{1}{2}E_{\alpha'}\overleftarrow{\nabla}^\mathbf{S}\boldsymbol{\mathcal{B}}^\mathbf{S}\overrightarrow{\nabla}^\mathbf{S} + \frac{1}{2}E_\alpha\overleftarrow{\nabla}^\mathbf{S}\boldsymbol{\mathcal{B}}^\mathbf{S}\overrightarrow{\nabla}^\mathbf{S} \\ & = \left(\boldsymbol{\mathcal{B}}^\mathbf{S}\overrightarrow{\nabla}^\mathbf{S}H^\mathbf{S}_{\alpha\alpha'}\right)\cdot\overrightarrow{\nabla}^\mathbf{S} ,\end{aligned} \qquad (43)$$

with the average adiabatic Hamiltonian

$$H^\mathbf{S}_{\alpha\alpha'} = H_\mathbf{S} + \frac{1}{2}(E_\alpha + E_{\alpha'}) . \qquad (44)$$

The transition operator for the spin bath is given by

$$\begin{aligned}\mathcal{T}^\mathbf{S}_{\alpha\alpha',\beta\beta'} = & \; d^\mathbf{S}_{\alpha\beta}\cdot\left(\boldsymbol{\mathcal{B}}^\mathbf{S}\overrightarrow{\nabla}^\mathbf{S}H_\mathbf{S}\right)\delta_{\beta'\alpha'} + \frac{1}{2}\Delta E_{\alpha\beta}d^\mathbf{S}_{\alpha\beta}\cdot\left(\boldsymbol{\mathcal{B}}^\mathbf{S}\overrightarrow{\nabla}^\mathbf{S}\right)\delta_{\alpha'\beta'} \\ & +d^{\mathbf{S}*}_{\alpha'\beta'}\cdot\left(\boldsymbol{\mathcal{B}}^\mathbf{S}\overrightarrow{\nabla}^\mathbf{S}H_\mathbf{S}\right)\delta_{\alpha\beta} + \frac{1}{2}\Delta E_{\alpha'\beta'}d^{\mathbf{S}*}_{\alpha'\beta'}\cdot\left(\boldsymbol{\mathcal{B}}^\mathbf{S}\overrightarrow{\nabla}^\mathbf{S}\right)\delta_{\alpha\beta} .\end{aligned} \qquad (45)$$

The limit $d^\mathbf{S}_{\alpha\alpha} \to 0$ of the spin transition operator in Equation (45) provides the form of the standard transition operator for canonical conjugate coordinates, given in Equation (A7). Finally, because of the spin nature of the bath, one finds a higher order transition operator (which does not appear in the case of canonical conjugate bath coordinates):

$$\begin{aligned}\mathcal{S}_{\alpha\alpha',\beta\beta'} = & \; \frac{1}{2}\Delta E_{\alpha\sigma}d^\mathbf{S}_{\alpha\sigma}\boldsymbol{\mathcal{B}}^\mathbf{S}d^\mathbf{S}_{\sigma\beta}\delta_{\alpha'\beta'} + \frac{1}{2}\Delta E_{\alpha\beta}d^\mathbf{S}_{\alpha\beta}\boldsymbol{\mathcal{B}}^\mathbf{S}d^{\mathbf{S}*}_{\alpha'\beta'} \\ & +\frac{1}{2}\Delta E_{\alpha'\sigma'}d^{\mathbf{S}*}_{\alpha'\sigma'}\boldsymbol{\mathcal{B}}^\mathbf{S}d^{\mathbf{S}*}_{\sigma'\beta'}\delta_{\alpha\beta} + \frac{1}{2}\Delta E_{\alpha'\beta'}d^{\mathbf{S}*}_{\alpha'\beta'}\boldsymbol{\mathcal{B}}^\mathbf{S}d^\mathbf{S}_{\alpha\beta} \\ & -\frac{1}{2}(E_\alpha + E_{\alpha'})\overleftarrow{\nabla}^\mathbf{S}\boldsymbol{\mathcal{B}}^\mathbf{S}\cdot d^\mathbf{S}_{\alpha\beta}\delta_{\alpha'\beta'} - \frac{1}{2}(E_\alpha + E_{\alpha'})\overleftarrow{\nabla}^\mathbf{S}\boldsymbol{\mathcal{B}}^\mathbf{S}\cdot d^{\mathbf{S}*}_{\alpha'\beta'}\delta_{\alpha\beta} .\end{aligned} \qquad (46)$$

The adiabatic limit of the spin-bath QCLE in (42) can be taken by setting to zero the off-diagonal elements of $d_{\alpha\alpha'}$, which appear in the operators in Equations (45) and (46). This is physically reasonable whenever the coupling between the different adiabatic energy surfaces is negligible. One obtains

$$\begin{aligned}\mathcal{T}^{S,ad}_{\alpha\alpha',\beta\beta'} &= (d^S_{\alpha\alpha} + d^{S*}_{\alpha'\alpha'})\mathcal{B}^S\vec{\nabla}^S H_S \delta_{\alpha\beta}\delta_{\beta'\alpha'} \\ &= -i(\phi^S_{\alpha\alpha} - \phi^S_{\alpha'\alpha'})\mathcal{B}^S\vec{\nabla}^S \delta_{\alpha\beta}\delta_{\beta'\alpha'}\,.\end{aligned} \qquad (47)$$

The geometric phase

$$\phi^S_{\alpha\alpha} = -i d^S_{\alpha\alpha} \qquad (48)$$

has been introduced exploiting the purely imaginary character of $d^S_{\alpha\alpha}$. Similarly, the higher order transition operator becomes

$$\mathcal{S}^{ad}_{\alpha\alpha',\beta\beta'} = -\frac{i}{2}\sum_{I,J}\left(\phi^S_{\alpha\alpha} - \phi^S_{\alpha'\alpha'}\right)\mathcal{B}^S\vec{\nabla}^S (E_\alpha + E_{\alpha'})\delta_{\alpha\alpha}\delta_{\alpha'\alpha'} \qquad (49)$$

Putting everything together, the adiabatic approximation of the spin-bath QCLE may be written as

$$\frac{\partial}{\partial t}W^S_{\alpha\alpha'}(\mathbf{S},t) = \left[-i\omega_{\alpha\alpha'} - i\left(\phi^S_{\alpha\alpha} - \phi^S_{\alpha'\alpha'}\right)\mathcal{B}\vec{\nabla}^S H^{\alpha\alpha'}_S - H^{\alpha\alpha'}_S \overleftarrow{\nabla}^S \mathcal{B}\vec{\nabla}^S\right]W^S_{\alpha\alpha'}(\mathbf{S},t). \qquad (50)$$

In Equation (50), the phase $\omega_{\alpha\alpha'}$ has a dynamical nature while the phase $\phi^S_{\alpha\alpha}$ is of a geometric origin and it can be considered an instance of the famous Berry phase [107–109]. Interestingly, Equation (35) predicts that the geometric phase $\phi^S_{\alpha\alpha}$ can be non-zero also for open paths of the classical spins of the bath (open-path Berry phases were discussed in Ref. [104]). Moreover, the phase factor $\phi^S_{\alpha\alpha} - \phi^S_{\alpha'\alpha'}$ is purely off-diagonal (off-diagonal Berry phases for environments described by canonically conjugate variables were discussed in Refs. [103,105,106]). It is worth mentioning that the geometric phase $\phi^S_{\alpha\alpha}$ is predicted also for non-adiabatic dynamics.

When the total Hamiltonian is time-independent, as the one in Equation (34), the adiabatic evolution of the matrix elements of the spin-bath operator-valued quasi-probability function, given by Equation (50), can be rewritten as

$$\frac{\partial}{\partial t}W^S_{\alpha\alpha'}(\mathbf{S},t) = \left[-i\omega_{\alpha\alpha'} - \left(\langle\alpha,S|\frac{d}{dt}|\alpha,S\rangle - \langle\alpha',S|\frac{d}{dt}|\alpha',S\rangle\right) - H^{\alpha\alpha'}_S \overleftarrow{\nabla}^S \mathcal{B}\vec{\nabla}^S\right]W^S_{\alpha\alpha'}(\mathbf{S},t). \qquad (51)$$

Using the Dyson identity, one can obtain the following form for $\hat{W}^S(\mathbf{S},t)$ in terms of the adiabatic propagator:

$$\begin{aligned}W^S_{\alpha\alpha'}(\mathbf{S},t) =\ & \exp\left[-i\int_{t_0}^t dt'\omega_{\alpha\alpha'}(t')\right]\exp\left[-\int_{t_0}^t dt'\left(\langle\alpha,S|\frac{d}{dt'}|\alpha,S\rangle - \langle\alpha',S|\frac{d}{dt'}|\alpha',S\rangle\right)\right] \\ & \times \exp\left[-(t-t_0)H^{\alpha\alpha'}_S \overleftarrow{\nabla}^S \mathcal{B}\vec{\nabla}^S\right]W^S_{\alpha\alpha'}(\mathbf{S},t_0)\,.\end{aligned} \qquad (52)$$

Equation (52) provides a convenient starting point for devising numerical integration schemes based on the SSTP propagation scheme [113].

In Ref. [91], the following model Hamiltonian was considered:

$$\hat{H}(\mathbf{S}) = -\Omega\hat{\sigma}_x - c_1 b\hat{\sigma}_z - \mu\mathbf{S}\cdot\boldsymbol{\sigma} - c_2 b S_z + \frac{S_z^2}{2} \qquad (53)$$

$$\hat{H}(\mathbf{S}) = \hat{h}_S(\mathbf{S}) - c_2 b S_z + \frac{S_z^2}{2}, \qquad (54)$$

where Ω, c_1, and c_2 are real parameters, b is the z component of the magnetic field $\mathbf{B} = (0,0,b)$, while $\boldsymbol{\sigma} = (\sigma_x, \sigma_y, \sigma_z)$ is a vector having the Pauli matrices σ_x, σ_y, and σ_z as components. The SSTP algorithm was applied to Equation (52) and the action of the classical like Liouville operator

$H_{\alpha\alpha'}^S \overleftarrow{\nabla}^S \mathcal{B}^S \overrightarrow{\nabla}^S$ was evaluated using time reversible integration algorithms based on the symmetric break-up of the Liouville propagator [138–140].

4. Stochastic Classical Baths

Consider a quantum-classical system comprising a quantum subsystem and a classical environment whose classical phase space coordinates are partitioned into two sets: one set $X = (Q, P)$ interacts directly with the quantum subsystem while the second set $X' = (Q', P')$ interacts only with the coordinates X (and therefore is not directly coupled to the quantum subsystem). We assume that the detailed dynamics of the coordinates X' is not interesting: their function is just that of working as a thermal bath, leading to dissipative dynamics [44].

An equation of motion for the hybrid quantum-classical system composed of the quantum subsystem and the classical DOF X only has been derived using projection operator methods [44]. It takes the form,

$$\frac{\partial}{\partial t}\hat{W}(X,t) = -\frac{i}{\hbar}\left[\hat{H}_W \ \hat{W}(X,t)\right]\mathcal{D}\left[\begin{array}{c}\hat{H}_W \\ \hat{W}(X,t)\end{array}\right]$$
$$+ \ \zeta\overrightarrow{\nabla}_P\left(\frac{P}{M} + k_BT\overrightarrow{\nabla}_P\right)\hat{W}(X,t), = -i\hat{\mathcal{L}}^D\hat{W}(X,t), \quad (55)$$

where $\nabla_P = \partial/\partial P$, ζ is the friction constant, k_B is the Boltzmann constant, and T is the temperature of the bath. The Hamiltonian in Equation (55) is defined in Equation (5). However, in the present case, we must interpret $V_W(\hat{q}, Q)$ as the potential of mean force arising from the average over the primed bath variables Q'. The Liouville operator $i\hat{\mathcal{L}}^D$, defined on the right-hand side of Equation (55), determines the dissipative dynamics of the system. This Fokker–Planck-like operator and the potential of mean force make the dissipative quantum-classical Liouville operator in Equation (55) different from that describing an isolated quantum-classical system [47]. In particular, the term $\zeta\overrightarrow{\nabla}_P\left[(P/M) + k_BT\overrightarrow{\nabla}_P\right]$ directly breaks the time-translation symmetry leading to diffusive motion and energy dissipation.

The dissipative Liouville operator can be written in the adiabatic basis as

$$i\hat{\mathcal{L}}^D_{\alpha\alpha'\beta\beta'} = \left(i\omega_{\alpha\alpha'}(R) + iL^K_{\alpha\alpha'}\right)\delta_{\alpha\beta}\delta_{\alpha'\beta'} + \mathcal{T}_{\alpha\alpha'\beta\beta'}, \quad (56)$$

where we have defined the Kramers operator as

$$iL^K_{\alpha\alpha'} = \left[\frac{P}{M}\overrightarrow{\nabla}_Q + \frac{1}{2}\left(F^{\alpha}_W + F^{\alpha'}_W\right)\overrightarrow{\nabla}_P - \zeta\overrightarrow{\nabla}_P\left(\frac{P}{M} + k_BT\overrightarrow{\nabla}_P\right)\right]. \quad (57)$$

The quantum-classical average of any operator or dynamical variable $\hat{\chi}(X)$ can be written as

$$\langle\hat{\chi}\rangle(t) = \sum_{\alpha\alpha'\beta\beta'}\int dX \chi_{\alpha'\alpha}(X)\exp[-i\mathcal{L}^D_{\alpha\alpha'\beta\beta'}t]W^{\beta\beta'}(X)$$
$$= \sum_{\alpha\alpha'\beta\beta'}\int dX W^{\beta\beta'}(X)\exp[i\mathcal{L}^{DB}_{\beta'\beta\alpha'\alpha}t]\chi_{\alpha'\alpha}(R,P), \quad (58)$$

where $i\mathcal{L}^{DB}_{\beta'\beta\alpha'\alpha}$ is the backward operator, defined as

$$i\hat{\mathcal{L}}^{DB}_{\alpha\alpha'\beta\beta'} = \left(i\omega_{\alpha\alpha'}(R) + iL^{KB}_{\alpha\alpha'}\right)\delta_{\alpha\beta}\delta_{\alpha'\beta'} + \mathcal{T}_{\alpha\alpha'\beta\beta'} \quad (59)$$

The backward Kramers $iL^{KB}_{\alpha\alpha'}$ operator is written as

$$iL^{KB}_{\alpha\alpha} = \left[\frac{P}{M}\overrightarrow{\nabla}_Q + \frac{1}{2}\left(F^{\alpha}_W + F^{\alpha'}_W\right)\overrightarrow{\nabla}_P - \zeta\left(\frac{P}{M} - k_BT\overrightarrow{\nabla}_P\right)\overrightarrow{\nabla}_P\right]\delta_{\alpha\beta}\delta_{\alpha'\beta'}. \quad (60)$$

According to the classical theory of random processes [110], the time evolution under the backward Kramers operator $i\mathcal{L}^{KB}_{\alpha\alpha'\beta\beta'}$ can be unfolded it via an average over realizations of stochastic Langevin trajectories. In such a picture, the classical trajectory segments obey the Langevin equations of motion,

$$\dot{Q} = \frac{P}{M}, \tag{61}$$

$$\dot{P} = -\frac{\zeta}{M}P + \frac{1}{2}\left(F_W W^\alpha + F_W^{\alpha'}\right) + \mathcal{R}(t), \tag{62}$$

where $\mathcal{R}(t)$ is a Gaussian white noise process with the properties,

$$\langle \mathcal{R}(t) \rangle = 0, \tag{63}$$

$$\langle \mathcal{R}(t)\mathcal{R}(t') \rangle = 2k_B T \zeta \delta(t-t'). \tag{64}$$

To Equations (61) and (62), one can associate a time-dependent Langevin–Liouville operator

$$iL^L_{\alpha\alpha'}(t) = \frac{P}{M}\vec{\nabla}_Q + \left(-\frac{\zeta}{M}P + \frac{1}{2}(F_W^\alpha + F_W^\alpha) + \mathcal{R}(t)\right)\vec{\nabla}_P, \tag{65}$$

and a time-ordered propagator

$$\mathcal{U}^L_{\alpha\alpha'}(t,0) = \mathcal{T}\exp\left[\int_0^t dt' iL^L_{\alpha\alpha'}(t')\right]. \tag{66}$$

In order to generate the stochastic Langevin trajectories, we can use a total time-dependent Langevin–Liouville super-operator

$$i\mathcal{L}^L_{\alpha\alpha'\beta\beta'}(t) = \left(i\omega_{\alpha\alpha'}(Q) + iL^L_{\alpha\alpha'}(t)\right)\delta_{\alpha\beta}\delta_{\alpha'\beta'} + \mathcal{T}_{\alpha\alpha'\beta\beta'} \tag{67}$$

and the associated propagator

$$\mathcal{U}^L_{\alpha\alpha'\beta\beta'}(t,0) = \mathcal{T}\exp\left[\int_0^t dt' i\mathcal{L}^L_{\alpha\alpha'\beta\beta'}(t')\right]. \tag{68}$$

Within such a Langevin picture, the quantum-classical average of any operator $\hat{\chi}(X)$ can be calculated as

$$\langle \hat{\chi} \rangle(t) = \sum_{\alpha\alpha'\beta\beta'} \int dX W^{\beta\beta'}(Q)\overline{\mathcal{U}^L_{\beta\beta'\alpha\alpha'}(t)\chi_{\alpha'\alpha}(Q)} \tag{69}$$

where the over-line denotes an average over an ensemble of stochastic Langevin trajectories. Since they are independent from each other, the order in which the average over phase space and the average over the stochastic Langevin process are performed can be permuted. Hence, one can write

$$\langle \hat{\chi}(X,t) \rangle = \sum_{\alpha\alpha'\beta\beta'} \int dRdP W^{\beta\beta'}(X)\overline{\mathcal{U}^L_{\beta\beta'\alpha\alpha'}(t)\chi'_{\alpha'\alpha}(XP)}. \tag{70}$$

Equation (70) allows one to calculate averages in a quantum-classical dissipative system as phase space weighted averages over many Langevin trajectories.

In Ref. [45], a quantum subsystem with two energy levels interacting with a dissipative classical quartic oscillator was considered. The Hamiltonian of the hybrid quantum-classical system reads

$$\hat{H}_W(X) = \frac{P^2}{2M} + V_q(Q) - \hbar\Omega\hat{\sigma}_x - \hbar\gamma_0 Q\hat{\sigma}_z, \tag{71}$$

where $V_q(Q) = \frac{a}{4}R^4 - \frac{b}{2}R^2$, Ω, a, b, and γ_0 are real parameters, M is the mass of the quartic oscillator, and $\hat{\sigma}_x$ and $\hat{\sigma}_z$ are Pauli matrices.

The calculation of quantum-classical averages using the dynamics defined by the time-dependent Langevin–Liouville propagator $\mathcal{U}^L_{ss'}(t)$ in Equation (68) is no more complicated than that for deterministic quantum-classical dynamics. The momentum-jump approximation [77,78] and a simple generalization of the SSTP algorithm [78,113] to the time dependent propagator were used in Ref. [45]. The explicitly time-dependent propagator $\mathcal{U}^L_{ss'}(t)$ must be defined as a time ordered product. A simple way to achieve that is to employ the decomposition scheme devised by Suzuki [141]. Details of the numerical procedures are found in Ref. [45].

5. Non-Hamiltonian Dynamics in Thermal Baths

By exploiting the antisymmetric structure of the quantum-classical commutator, arising from the matrix operator $\mathcal{D}$ given in Equation (8), one can impose the thermodynamic constraints of constant temperature on the classical-like DOF [42,43]. Following Refs. [92–94], constant-temperature dynamics for the classical bath coordinates, as defined through the non-Hamiltonian Nosé–Hoover equations of motion, can be introduced by modifying the matrix $\mathcal{B}$ and augmenting in a minimal way the dimension of the phase space bath. The classical Nosé–Hoover thermostat is briefly discussed in Appendix B.

As in the classical case, the Nosé variables are

$$X^N \equiv (Q, Q_\eta, P, P_\eta), \tag{72}$$

where Q_η and P_η are the Nosé coordinate and momentum. The Nosé quantum-classical Hamiltonian is obtained by adding the Nosé kinetic energy $P_\eta^2/2M_\eta$ and potential energy Nk_BTQ_η to $\hat{H}_W$ in Equation (5)

$$H^N = \frac{P^2}{2M} + \frac{P_\eta^2}{2M_\eta} + Nk_BTQ_\eta + \hat{h}_W(Q), \tag{73}$$

where M_η is the Nosé inertial parameter, k_B is the Boltzmann constant, T is the constant temperature, and N is the number of Q coordinates. Using the matrix $\mathcal{B}^N$ in Equation (A13), the classical phase space quasi-Hamiltonian bracket of two variables A_1 and A_2 can be defined as

$$A_1 \overleftarrow{\nabla}^N \mathcal{B}^N \overrightarrow{\nabla}^N A_2 = \sum_{I,J=1}^{2(N+1)} A_1 \overleftarrow{\nabla}^N_I \mathcal{B}^N_{IJ} \overrightarrow{\nabla}^N_J A_2. \tag{74}$$

The explicit form of the matrix operator, which defines the quantum-classical bracket and the law of motion through Equation (19), is then given by

$$\mathcal{D}^N = \begin{bmatrix} 0 & 1 - \frac{\overleftarrow{\nabla}^N \mathcal{B}^N \overrightarrow{\nabla}^N}{2i\hbar^{-1}} \\ -\left(1 - \frac{\overleftarrow{\nabla}^N \mathcal{B}^N \overrightarrow{\nabla}^N}{2i\hbar^{-1}}\right) & 0 \end{bmatrix}. \tag{75}$$

The Nosé–Hoover QCLE for the operator-valued quasi-probability function $\hat{W}^N(X^N, t)$ is given by

$$\begin{aligned}\frac{d}{dt}\hat{W}^N(X^N, t) &= -i\mathcal{L}^N W^N(X^N, t) - \kappa^N(X^N)W^N(X^N, t) \\ &= -\frac{i}{\hbar}\begin{bmatrix} \hat{H}^N & \hat{W}^N(X^N, t) \end{bmatrix} \cdot \mathcal{D}^N \cdot \begin{bmatrix} \hat{H}^N \\ \hat{W}^N(X^N, t) \end{bmatrix} - \kappa^N(X^N)W^N(X^N, t).\end{aligned} \tag{76}$$

The presence of the term $-\kappa^N(X^N)\hat{W}^N(X^N,t)$ in the left-hand side of Equation (76) derives from the passage from the Heisenberg to the Schrödinger picture, as it is explained in Appendix C.

Upon considering the term in the right-hand side of (76), one obtains

$$\hat{H}^N\overleftarrow{\nabla}^N B^N\overrightarrow{\nabla}^N\hat{\chi}(X^N,t) - \hat{\chi}(X^N,t)\overleftarrow{\nabla}^N B^N\overrightarrow{\nabla}^N\hat{H}_N = \frac{\partial \hat{V}}{\partial Q}\frac{\partial \hat{\chi}(X^N,t)}{\partial P} + \frac{\partial \hat{\chi}(X^N,t)}{\partial P}\frac{\partial \hat{V}}{\partial Q}$$
$$- 2F_{Q_\eta}\frac{\partial \hat{\chi}(X^N,t)}{\partial P_\eta} - 2\frac{P}{M}\frac{\partial \hat{\chi}(X^N,t)}{\partial Q} \quad (77)$$
$$- 2\frac{P_\eta}{M_\eta}\frac{\partial \hat{\chi}(X^N,t)}{\partial Q_\eta} + 2\frac{P_\eta}{M_\eta}P\frac{\partial \hat{\chi}(X^N,t)}{\partial P},$$

where $F_{Q_\eta} = \frac{P^2}{M} - Nk_BT$. Finally, using the above result, the Nosé–Hoover QCLE reads

$$\frac{d}{dt}\hat{W}^N(X^N,t) = -\frac{i}{\hbar}\left(H^N\hat{W}^N(X^N,t) - \hat{\chi}(X^N,t)H^N\right) + \frac{1}{2}\left(\frac{\partial \hat{W}^N(X^N,t)}{\partial P}\frac{\partial \hat{V}}{\partial Q} + \frac{\partial \hat{V}}{\partial Q}\frac{\partial \hat{\chi}(X^N,t)}{\partial P}\right)$$
$$-\frac{P}{M}\frac{\partial \hat{W}^N(X^N,t)}{\partial Q} - \frac{P_\eta}{M_\eta}\frac{\partial \hat{\chi}(X^N,t)}{\partial Q_\eta} + \frac{P_\eta}{M_\eta}P\frac{\partial \hat{\chi}(X^N,t)}{\partial P} - F_{Q_\eta}\frac{\partial \hat{W}^N(X^N,t)}{\partial P_\eta}. \quad (78)$$

In the adiabatic states defined in Equation (22), Equation (78) reads

$$\frac{d}{dt}\hat{W}^N_{\alpha\alpha'}(X^N,t) = -\sum_{\beta\beta'} i\mathcal{L}^N_{\alpha\alpha',\beta\beta'}\hat{W}^N_{\beta\beta'}(X^N,t), \quad (79)$$

where

$$i\mathcal{L}^N_{\alpha\alpha',\beta\beta'} = i\omega_{\alpha\alpha'}\delta_{\alpha\beta}\delta_{\alpha'\beta'} + \delta_{\alpha\beta}\delta_{\alpha'\beta'}i\hat{L}^N_{\alpha\alpha'} + \mathcal{T}_{\alpha\alpha',\beta\beta'}. \quad (80)$$

We have used the definition of the Bohr frequency $\omega_{\alpha\alpha'}$ in Equation (A3) and of the transition operator $\mathcal{T}_{\alpha\alpha',\beta\beta'}$ in Equation (A7) in Appendix A. We have introduced a classical-like Nosé–Liouville operator

$$i\hat{L}^N_{\alpha\alpha'} = \frac{P}{M}\frac{\partial}{\partial Q} + \frac{1}{2}\left(F^\alpha + F^{\alpha'}\right)$$
$$-P\frac{P_\eta}{M_\eta}\frac{\partial}{\partial P} + \frac{P_\eta}{M_\eta}\frac{\partial}{\partial Q_\eta} + F_{Q_\eta}\frac{\partial}{\partial P_\eta}\frac{\partial}{\partial P}. \quad (81)$$

The existence of the stationary operator-valued Nosé quasi-probability function $\hat{W}^{N,e}(X^N)$ is discussed in Appendix C.

Nosé–Hoover Chain Thermal Baths

The Nosé–Hoover thermostat suffers from lack of ergodic dynamics when the bath has high frequencies of motion. The Nosè–Hoover chain [142] is a more general non-Hamiltonian thermostat that solves the ergodicity problems suffered by the standard Nosé–Hoover thermostat in the case of stiff variables. The Nosè–Hoover chain thermostat can also be formulated in a quantum-classical framework with minimal changes with respect to what is shown in Section 5. To this end, considering for simplicity a chain of just two thermostat coordinates, one can define the classical phase space point as

$$X^{\text{NHC}} = (R, Q_{\eta_1}, Q_{\eta_2}, P, P_{\eta_1}, P_{\eta_2}), \quad (82)$$

$$\hat{H}^{\text{NHC}} = \frac{\hat{p}^2}{2m} + \frac{P^2}{2M} + \frac{P^2_{\eta_1}}{2M_{\eta_1}} + \frac{P^2_{\eta_2}}{2M_{\eta_2}} \quad (83)$$
$$+ \hat{V}(\hat{q}, R) + Nk_BTQ_{\eta_1} + k_BTQ_{\eta_2},$$

where M_{η_1} and M_{η_2} are the inertial parameters of the thermostat variables. As shown in Ref. [92,93], one can define an antisymmetric matrix

$$\mathcal{B}^{\text{NHC}} = \begin{bmatrix} 0 & 0 & 0 & 1 & 0 & 0 \\ 0 & 0 & 0 & 0 & 1 & 0 \\ 0 & 0 & 0 & 0 & 0 & 1 \\ -1 & 0 & 0 & 0 & -P & 0 \\ 0 & -1 & 0 & P & 0 & -P_{\eta_1} \\ 0 & 0 & -1 & 0 & P_{\eta_1} & 0 \end{bmatrix}. \tag{84}$$

The matrix $\mathcal{B}^{\text{NHC}}$ can be used to define the quasi-Hamiltonian bracket according to Equation (9). The Nosé–Hoover chain classical equations of motion in phase space [92] are then given by

$$\dot{X} = -X^{\text{NHC}} \overleftarrow{\nabla}^{\text{NHC}} \mathcal{B}^{\text{NHC}} \overrightarrow{\nabla}^{\text{NHC}} \hat{H}^{\text{NHC}}. \tag{85}$$

Quantum-classical dynamics is then introduced using the matrix super-operator

$$\mathcal{D}^{\text{NHC}} = \begin{bmatrix} 0 & 1 - \frac{\overleftarrow{\nabla}^{\text{NHC}} \mathcal{B}^{\text{NHC}} \overrightarrow{\nabla}^{\text{NHC}}}{2i\hbar^{-1}} \\ -\left(1 - \frac{\overleftarrow{\nabla}^{\text{NHC}} \mathcal{B}^{\text{NHC}} \overrightarrow{\nabla}^{\text{NHC}}}{2i\hbar^{-1}}\right) & 0 \end{bmatrix}. \tag{86}$$

The quantum-classical equations of motion can then be written as

$$\frac{d\hat{\chi}}{dt} = \frac{i}{\hbar} \begin{bmatrix} \hat{H}^{\text{NHC}} & \hat{\chi} \end{bmatrix} \cdot \mathcal{D}^{\text{NHC}} \cdot \begin{bmatrix} \hat{H}^{\text{NHC}} \\ \hat{\chi} \end{bmatrix}. \tag{87}$$

The equations of motion can be represented using the adiabatic basis obtaining the Liouville super-operator

$$i\mathcal{L}^{\text{NHC}}_{\alpha\alpha',\beta\beta'} = (i\omega_{\alpha\alpha'} + iL^{\text{NHC}}_{\alpha\alpha'})\delta_{\alpha\beta}\delta_{\alpha'\beta'} - \mathcal{T}_{\alpha\alpha',\beta\beta'}, \tag{88}$$

where

$$iL^{\text{NHC}}_{\alpha\alpha'} = \frac{P}{M}\frac{\partial}{\partial R} + \frac{1}{2}(F^\alpha + F^{\alpha'})\frac{\partial}{\partial P} + \sum_{k=1}^{2}\left(\frac{P_{\eta_k}}{M_{\eta_k}}\frac{\partial}{\partial Q_{\eta_k}} + F_{Q_{\eta_k}}\frac{\partial}{\partial P_{\eta_k}}\right) - \frac{P_{\eta_2}}{M_{\eta_2}}P_{\eta_1}\frac{\partial}{\partial P_{\eta_1}}, \tag{89}$$

with $F_{Q_{\eta_2}} = (P^2_{\eta_1}/M_{\eta_1}) - k_B T$. The proof of the existence of stationary density matrix in the case of Nosé–Hoover chains follows the same logic of the simpler Nosé–Hoover case. In the adiabatic basis, the density matrix stationary up to order bar has the same form as that given in Equations (A50) and (A52). One has just to replace Equation (A50) for the order zero term with

$$W_{\alpha\alpha'}^{\text{NHC},e,(0)} = \frac{1}{Z^{\text{NHC}}} e^{-\beta\left[\frac{P^2}{2M} + E_\alpha(R) + \sum_{k=1}^{2}\left(\frac{P^2_{\eta_k}}{2M_{\eta_k}}\right) + Nk_BTQ_{\eta_1} + k_BTQ_{\eta_2}\right]} \tag{90}$$

with an obvious definition of Z^{NHC}.

6. Conclusions and Perspectives

In this review, we discussed how to mathematically describe the dynamics and statistical mechanics of quantum subsystems embedded in classical baths. The formalism is founded on an operator-valued quasi-probability function evolving through a QCLE defined in terms of a quasi-Lie bracket. It is worth emphasizing that the QCLE is a fully deterministic equation that takes into account explicitly *all* the DOF of the system, i.e., it describes the quantum and classical DOF of the total hybrid system. Hence, the QCLE generates a unitary dynamics, conserving both the system's probability and energy. However, the time-translation invariance of the quasi-Lie bracket algebra is broken. This situation is surprising: one does not expect a broken time-translation invariance symmetry in an isolated system when all its degrees of freedom are taken into account. This can be seen as a signature of the effect of the classical bath on the quantum subsystem, and of the back-reaction of the subsystem onto the bath. In other words, the total hybrid system is closed from the point of view of energy and probability conservation but, because of the above mentioned back-reaction, it is also open: the quasi-Lie bracket describes the irreversible transfer of quantum information onto the classical DOF. We also reviewed how the hybrid quantum-classical theory can be derived from a partial Wigner transform and a semiclassical limit of the QLE only in the case when the bath is described by canonically conjugate coordinates. After this, we discussed how to treat quantum subsystems embedded in both non-canonical and non-Hamiltonian bath. In all cases, the mathematical object representing the state of the system is an operator-valued quasi-probability function that depend on the coordinates of the bath and whose equation of motion depends on the specific case under study. It is explained how classical spin baths are described in terms of non-canonical coordinates and how this fact leads to the appearance of an off-diagonal open-path geometric phase in the dynamics of the operator-valued quasi-probability function of the system. We then discussed how the effect of thermal baths can be implemented by means of a stochastic, quantum-classical Langevin dynamics and by means of a deterministic, non-Hamiltonian Nosé–Hoover thermostatted dynamics. The formulation of the dynamics in both the spin and Nosé–Hoover case was achieved by generalizing the quasi-Lie bracket of the canonical case.

The formalisms were presented in such a way to shed light on practical implementation via computer simulation algorithms. The particular class of algorithms upon which we focused is based on the unfolding of the evolution of the operator-valued quasi-probability function in terms of piecewise-deterministic trajectories evolving on the adiabatic energy surfaces of the system. These methods scales favorably in terms of bath DOF but, to date, have been limited to relatively short time intervals and Markovian systems. When the dynamics is non-Markovian, the memory function, i.e., the autocorrelation function of the random force [3,110], cannot be approximated by a delta function. The memory function of the bath can be expected to become more and more different from a delta function as the quantum character of the bath becomes more pronounced (for example, at low temperature) and as the subsystem-bath coupling grows in strength.

The QCLE discussed herein constitutes an approach to open quantum system dynamics (in the case of hybrid quantum-classical systems) that is both distinct and complementary to that given by master equations [3,110]. Within the QCLE approach, the degrees of freedom of the bath are not integrated out of the dynamics but are explicitly taken into account at every time step. Hence, there is no memory function to be approximated and bath properties can be calculated with the same ease with which subsystem properties are computed. The limitations of the QCLE approach are mostly numerical in character and arise in the SSTP algorithm, herein discussed, from the momentum-jump approximation and the accumulation of fluctuating statistical weights associated with the Monte Carlo sampling of the quantum transitions of the subsystem.

The QCLE-based approach to quantum dynamics in classical baths has proven to be successful in modeling a variety of quantum processes in the condensed phase. Nevertheless, the currently algorithms also present significant challenges, necessitating the need for further improvements and developments. In light of the above, we hope that this review will attract the attention of a broad community of researchers and spur further work along this direction. In addition to further algorithm developments, we are interested in broadening the scope of applications studied by this approach. For example, based on preliminary results, we believe that this approach can be successfully applied to studying the interplay between quantum and classical fluctuations in hybrid nanoscale devices.

Author Contributions: A.S. is responsible for the conceptualization and the original draft preparation; G.H., R.G., and A.M. are responsible for writing - review & editing; A.M. is responsible for the supervision of the project.

Funding: A.S. and R.G. acknowledge support by research funds in memory of Francesca Palumbo, difc 3100050001d08+, University of Palermo.

Conflicts of Interest: The authors declare no conflict of interest.

Abbreviations

The following abbreviations are used in this manuscript:

DOF   Degrees of Freedom
QCLE Quantum-Classical Liouville Equation
QLE   Quantum Liouville Equation
SSTP Sequential Short-Time Propagation

Appendix A. Representation in the Adiabatic Basis

In the adiabatic basis, Equation (19) reads

$$\frac{d}{dt}W_{\alpha\alpha'}(X,t) = -\sum_{\beta\beta'} i\mathcal{L}_{\alpha\alpha',\beta\beta'} W_{\beta\beta'}(X,t) , \tag{A1}$$

where

$$W_{\alpha\alpha'}(X,t) = \langle \alpha; Q|\hat{W}(X,t)|\alpha'; Q\rangle \tag{A2}$$

are the matrix elements of the density matrix. Upon defining the Bohr frequency as

$$\omega_{\alpha\alpha'} = \frac{E_\alpha - E_{\alpha'}}{\hbar} , \tag{A3}$$

the Liouville super-operator may be written as

$$i\mathcal{L}_{\alpha\alpha',\beta\beta'} = i\omega_{\alpha\alpha'}\delta_{\alpha\beta}\delta_{\alpha'\beta'} + \delta_{\alpha\beta}\delta_{\alpha'\beta'}iL_{\alpha\alpha'} + \mathcal{T}_{\alpha\alpha',\beta\beta'} . \tag{A4}$$

We have also introduced a classical-like Liouville operator

$$iL_{\alpha\alpha'} = \frac{P}{M}\frac{\partial}{\partial Q} + \frac{1}{2}\left(F_W^\alpha + F_W^{\alpha'}\right)\frac{\partial}{\partial P} , \tag{A5}$$

where

$$F_W^\alpha = -\frac{\partial E_\alpha}{\partial Q} \tag{A6}$$

is the Hellmann–Feynman force.

In Equation (A4), the transition operator $\mathcal{T}_{\alpha\alpha',\beta\beta'}$ is defined as

$$\mathcal{T}_{\alpha\alpha',\beta\beta'} = \delta_{\alpha'\beta'}\frac{P}{M}\cdot d_{\alpha\beta}\left(1 + \frac{1}{2}S_{\alpha\beta}\cdot\frac{\partial}{\partial P}\right) + \delta_{\alpha\beta}\frac{P}{M}\cdot d_{\alpha'\beta'}^*\left(1 + \frac{1}{2}S_{\alpha'\beta'}^*\cdot\frac{\partial}{\partial P}\right) . \tag{A7}$$

In turn, the transition operator is defined in terms of the shift vector

$$S_{\alpha\alpha'} = \frac{(E_\alpha - E_{\alpha'})}{\frac{P}{M}\cdot d_{\alpha\alpha'}}d_{\alpha\beta} \tag{A8}$$

and of the coupling vector

$$d_{\alpha\alpha'} = \langle \alpha; Q | \frac{\partial}{\partial Q} | \alpha'; Q \rangle . \tag{A9}$$

Appendix B. The Nosè–Hoover Thermostat

The Nosè–Hoover thermostat was originally formulated in Refs. [111,112]. Herein, we follow Refs. [92–94]. The Hamiltonian of the subsystem with phase space coordinates (R, P) is:

$$H^B = \frac{P^2}{2M} + V(R) , \tag{A10}$$

where $V(R)$ is the potential energy. One can introduce an extended system comprised by the coordinates of the original subsystem augmented with the additional variables Q_η and conjugate momentum P_η. The dimension of such an extended phase space is obviously $2N + 2$, which is computationally tractable whenever N is computationally tractable. As a consequence, the phase space point of the extended system is

$$X^N = \begin{bmatrix} R \\ Q_\eta \\ P \\ P_\eta \end{bmatrix} , \tag{A11}$$

while the energy reads:

$$H^N = H^B + 3Nk_B T Q_\eta + \frac{P_\eta^2}{2M_\eta} , \tag{A12}$$

where M_η is a fictitious mass associated with the additional degree of freedom, k_B is Boltzmann constant, and T the bath constant temperature. In order to define time evolution, we abandon the Hamiltonian structure of the theory. To this end, using the general formalism of Refs. [92–94], we introduce the antisymmetric matrix:

$$\mathcal{B}^N = \begin{bmatrix} 0 & 0 & 1 & 0 \\ 0 & 0 & 0 & 1 \\ -1 & 0 & 0 & -P \\ 0 & -1 & P & 0 \end{bmatrix} , \tag{A13}$$

so that Nosé's equations of motion can be written as

$$\dot{X}_K^N = \sum_{I,J=1}^{2(N+1)} X_K^N \overleftarrow{\nabla}_I^N \mathcal{B}_{IJ}^N \overrightarrow{\nabla}_J^N H^N = \sum_{J=1}^{2N} B_{KJ}^N \overrightarrow{\nabla}_J^N H^N , \tag{A14}$$

where the first equality on the right-hand side of Equations (A14) introduces the Nosé bracket, while the extended phase space gradient is denoted as $\nabla_J^N = \partial / \partial X_J^N$. We remark here that the Nosé bracket does not satisfy the Jacobi relation [92–94], and thus defines a quasi-Hamiltonian algebra. The Liouville equation for the Nosé distribution function is

$$\begin{aligned} \frac{\partial}{\partial t} W^N(X^N, t) &= -\sum_{K=1}^{2(N+1)} \nabla_K^N \left(\dot{X}_K^N W^N(X^N, t) \right) \\ &= -\left(\sum_{K=1}^{2(N+1)} \dot{X}_K \overrightarrow{\nabla}_K^N - \kappa^N \right) W^N(X^N, t) = 0 , \end{aligned} \tag{A15}$$

where the compressibility of the phase space reads:

$$\kappa^N = \sum_{k=1}^{2(N+1)} \nabla_K^N \dot{X}_k = \sum_{k,j=1}^{2(N+1)} B_{KJ}^N \overleftarrow{\nabla}_K^N \overrightarrow{\nabla}_J^N H^N . \tag{A16}$$

As implied by Equation (A16), Nosé's phase space flow has a non-zero compressibility (however, this does not always occur for a quasi-Hamiltonian dynamics). In terms of the Nosé bracket, the equilibrium Liouville equation for Nosé distribution function reads:

$$W^N(X^N) \overleftarrow{\nabla}^N \mathcal{B}^N \overrightarrow{\nabla}^N H^N = -\kappa^N W^N(X^N) . \tag{A17}$$

By direct substitution, one can verify that the solution of Equation (A17) is:

$$W^N(X^N) \propto \exp[-w] \delta(E - H^N) , \tag{A18}$$

where w is defined by the equation $dw/dt = \kappa^N$. Equations (A14) can be written explicitly in the form:

$$\dot{R} = \frac{P}{M}, \tag{A19}$$

$$\dot{P} = -\frac{\partial V}{\partial R} - P\frac{P_\eta}{M_\eta}, \tag{A20}$$

$$\dot{Q}_\eta = \frac{P_\eta}{M_\eta}, \tag{A21}$$

$$\dot{P}_\eta = \frac{P^2}{M} - Nk_B T . \tag{A22}$$

In order to write explicitly the Nosé distribution function, it is useful to introduce the following extended phase space function:

$$H^T = H^B + \frac{P_\eta^2}{2M_\eta} . \tag{A23}$$

Using the equations of motion, one finds

$$\frac{dH^T}{dt} = -Nk_B T \frac{P_\eta}{M_\eta} , \tag{A24}$$

which is related to the compressibility by

$$\kappa^N = -N\frac{P_\eta}{M_\eta} = \beta\frac{dH^T}{dt} . \tag{A25}$$

At this point, we have all the ingredients that are needed to prove that extended phase space averages of functions of the subsystem coordinates (R, P) can be written as canonical averages. We start by considering

$$\langle A(R,P) \rangle_N \propto \int dX^N e^{-\int \kappa^N dt} \delta(E - H^N) A(R,P)$$

$$= \int dR dP dQ_\eta dP_\eta e^{-\beta \int \frac{dH^T}{dt} dt} \delta(E - H^N) A(R,P) \tag{A26}$$

$$= \int dR dP dQ_\eta dP_\eta e^{-\beta H^T} \delta(E - H^N) A(R,P) .$$

The integral

$$\int dQ_\eta \delta(E - H^N) \tag{A27}$$

is calculated by using the identity

$$\delta(f(Q_\eta)) = \sum_{\{Q_{\eta_0}\}} \frac{\delta(Q_\eta - Q_{\eta_0})}{\frac{df}{dQ_\eta}(Q_{\eta_0})}, \tag{A28}$$

where the sum runs over the zeros Q_{η_0} of $f(Q_\eta)$. Upon identifying $f(Q_\eta) = E - H^N$, one gets $Q_{\eta_0} = H^T - E/N$ and

$$\delta(f(Q_\eta)) = \frac{\delta\left(Q_\eta - \beta(\mathcal{H}_T - E)/N\right)}{3Nk_BT} \tag{A29}$$

with the above results, the integral over Q_η becomes a trivial Gaussian integral over P_η:

$$\int dP_\eta e^{-\beta \frac{P_\eta^2}{2M_\eta}} = \sqrt{\pi M_\eta k_B T}. \tag{A30}$$

Finally, one obtains:

$$\langle A(R,P)\rangle_N \propto \int dR dP e^{-\beta H^B} A(R,P) \equiv \langle A(R,P)\rangle_{\text{can}}. \tag{A31}$$

Hence, averages in the canonical ensemble can be calculated by letting the trajectories evolve according to Nosé's dynamics.

The quasi-Hamiltonian Nosè dynamics is a well-established tool of molecular dynamics simulations. In practice, it is adopted whenever one wants to calculate dynamical properties at constant temperature and/or study phase transitions. Discussions and pointers to the relevant literature on the subject can be found in Ref. [127].

Appendix C. Stationary Operator-Valued Nosé Quasi-Probability Function

The quantum average of any operator $\hat{W}^N(X^N)$, in a dynamics where the temperature of the X degrees of freedom is controlled by the Nosè–Hoover thermostat can be calculated as

$$\langle \hat{\chi}(X^N, t)\rangle = \text{Tr}' \int dX^N \, \hat{W}^N(X^N, t)\hat{\chi}(X^N). \tag{A32}$$

The action of $\exp\left(i\mathcal{L}^N t\right)$ can be transferred from $\hat{\chi}(X^N)$ to $\hat{W}^N(X^N)$ by using the cyclic invariance of the trace and integrating by parts the terms coming from the classical brackets. One can write

$$i\mathcal{L}^N = \frac{i}{\hbar}\left[\hat{H}^N, \ldots\right] - \frac{1}{2}\hat{H}^N \overleftarrow{\vec{\nabla}}^N \boldsymbol{\mathcal{B}} \overrightarrow{\vec{\nabla}}^N - \overleftarrow{\vec{\nabla}}^N \boldsymbol{\mathcal{B}} \overrightarrow{\vec{\nabla}}^N \hat{H}^N\}. \tag{A33}$$

The action of $i\mathcal{L}^N$ on an arbitrary operator $\hat{\chi}(X^N)$ is defined by

$$i\mathcal{L}^N \hat{\chi} = \frac{i}{\hbar}[\hat{H}^N, \hat{\chi}] - \frac{1}{2}\hat{H}^N \overleftarrow{\vec{\nabla}}^N \boldsymbol{\mathcal{B}} \overrightarrow{\vec{\nabla}}^N \hat{\chi} - \hat{\chi} \overleftarrow{\vec{\nabla}}^N \boldsymbol{\mathcal{B}} \overrightarrow{\vec{\nabla}}^N \hat{H}^N \tag{A34}$$

when integrating by parts the right-hand side, one obtains a term proportional to the compressibility $\kappa^N = \vec{\nabla}^N \boldsymbol{\mathcal{B}} \vec{\nabla}^N \hat{H}^N$. As a result, the quantum Liouville operator, partially depending on phase space variables, is non-Hermitian

$$\left(i\hat{\mathcal{L}}^N\right)^\dagger = -i\hat{\mathcal{L}}^N - \kappa^N. \tag{A35}$$

The average value can then be written as

$$\langle \hat{\chi} \rangle = \text{Tr}' \int dX \, \hat{\chi}(X^N) \exp\left[-(i\mathcal{L}^N + \kappa^N)t\right] \hat{W}^N(X^N) \,. \tag{A36}$$

The operator-valued Nosè quasi-probability function evolves under the equation:

$$\frac{\partial}{\partial t}\hat{W}^N(X^N,t) = -\frac{i}{\hbar}\left[\hat{H}^N, \hat{W}^N(X^N,t)\right] + \frac{1}{2}\left(H^N \overleftarrow{\nabla}^N \mathcal{B}^N \overrightarrow{\nabla}^N \hat{W}^N(X^N,t) - \hat{W}^N(X^N,t)\overleftarrow{\nabla}^N \mathcal{B}^N \overrightarrow{\nabla}^N \hat{H}^N\right) \\ -\kappa^N(X)\hat{W}_N(X,t) \,. \tag{A37}$$

The stationary operator-valued Nosè quasi-probability function $\hat{W}^{N,e}$ is defined by

$$(i\mathcal{L}^N + \kappa^N)\hat{W}^{N,e} = 0 \,. \tag{A38}$$

To find the explicit expression, one can follow Ref. [41]: the density matrix is expanded in powers of $\hbar$

$$\hat{W}^{N,e} = \sum_{k=0}^{\infty} \hbar^n \hat{W}^{N,e,(k)} \tag{A39}$$

and an explicit solution in the adiabatic basis is searched for. On such a basis, the Nosè–Liouville operator is expressed by Equation (80) and the Nosè Hamiltonian is given by

$$\begin{aligned} H_N^\alpha &= \frac{P^2}{2M} + \frac{P_\eta^2}{2M_\eta} + Nk_BTQ_\eta + E_\alpha(R) \\ &= H_\alpha^P(R,P) + \frac{P_\eta^2}{2M_\eta} + Nk_BTQ_\eta \,. \end{aligned} \tag{A40}$$

One obtains an infinite set of equations corresponding to the various power of $\hbar$

$$iH^N_{\alpha\alpha'}W^{N,e,(0)}_{\alpha\alpha'} = 0 \tag{A41}$$

$$iH^N_{\alpha\alpha'}W^{N,e,(k+1)}_{\alpha\alpha'} = (iL^N_{\alpha\alpha'} + \kappa^N)W^{N,e,(k)}_{\alpha\alpha'} + \sum_{\beta\beta'} \mathcal{T}_{\alpha\alpha',\beta\beta'}W^{N,e,(k)}_{\beta\beta'} \quad (k \geq 1) \,. \tag{A42}$$

In order to ensure that a solution can be found by recursion, one must discuss the solution of Equation (A42) when calculating the diagonal elements $W^{(n)\alpha\alpha}_{Ne}$ in terms of the off-diagonal ones $W^{(n)\alpha\alpha'}_{Ne}$. To this end, using $W^{N,e,(k)}_{\alpha\alpha'} = (W^{N,e,(k)}_{\alpha'\alpha})^*$, $\mathcal{T}_{\alpha\alpha,\beta\beta'} = \mathcal{T}^*_{\alpha\alpha,\beta'\beta}$ and the fact that $\mathcal{T}_{\alpha\alpha,\beta\beta} = 0$ when a real basis is chosen, it is useful to re-write Equation (A42) in the form

$$(iL^N_{\alpha\alpha} + \kappa^N)W^{Ne,(k)}_{\alpha\alpha} = \sum_{\beta > \beta'} 2\text{Re}\left(\mathcal{T}_{\alpha\alpha,\beta\beta'}W^{N,e(k)}_{\beta\beta'}\right) \,. \tag{A43}$$

One has [92] $(-iL^N_{\alpha\alpha} - \kappa^N)^\dagger = iL^N_{\alpha\alpha}$. The right-hand side of this equation is expressed by means of the generalized bracket in Equation (74): H^α_N and any general function $f(H^\alpha_N)$ are constants of motion under the action of $iL^N_{\alpha\alpha}$. The phase space compressibility κ^N associated with the generalized bracket in the case of Nosè dynamics is

$$\begin{aligned} \kappa^N_\alpha &= -\beta\frac{d}{dt}\left(\frac{P^2}{2M} + \frac{P_\eta^2}{2M_\eta} + E_\alpha(R)\right) \\ &= -\beta N\frac{P_\eta}{M_\eta} = -\beta N\frac{d}{dt}H^T_\alpha \,, \end{aligned} \tag{A44}$$

where N is the number of classical momenta P in the Hamiltonian.

To ensure that a solution to Equation (A43) exists, one must invoke the theorem of Fredholm alternative, requiring that the right-hand side of Equation (A43) is orthogonal to the null space of $(iL_{\alpha\alpha}^N)^\dagger = -iL_{\alpha\alpha}^N - \kappa^N$ [143]. The null-space of this operator is defined by the equation $(iL_{\alpha\alpha}^N + \kappa^N)G_\alpha(X) = 0$, with $G_\alpha(X) = f(H_\alpha^N)\exp(-w_\alpha^N)$. Hence, the condition to be satisfied is

$$\int dX^N e^{-w_\alpha} \sum_{\beta > \beta'} 2\text{Re}\left(\mathcal{T}_{\alpha\alpha,\beta\beta'} W_{\beta\beta'}^{N,e,(k)}\right) f(H_\alpha^N) = 0 . \qquad (A45)$$

The fact that $2\exp(-w_\alpha)\text{Re}\left(\mathcal{T}_{\alpha\alpha,\beta\beta'} W_{\beta\beta'}^{N,e,(k)}\right)$ and $f(H_N^\alpha)$ are, respectively, an odd and an even function of P guarantees the validity of Equation (A45).

The formal solution of Equation (A43) can then be written as

$$W_{\alpha\alpha}^{N,e,(k)} = (iL_{\alpha\alpha}^N + \kappa^N)^{-1} \sum_{\beta > \beta'} 2\text{Re}\left(\mathcal{T}_{\alpha\alpha,\beta\beta'} W_{\beta\beta'}^{N,e,(k)}\right) , \qquad (A46)$$

and the formal solution of Equation (A42) for $\alpha \neq \alpha'$ as

$$W_{\alpha\alpha'}^{N,e,(n+1)} = \frac{i}{E_{\alpha\alpha'}}(iL_{\alpha\alpha'}^N + \kappa^N)W_{\alpha\alpha'}^{N,e,(k)} - \frac{i}{H_{\alpha\alpha'}^N}\sum_{\beta\beta'}\mathcal{T}_{\alpha\alpha',\beta\beta'} W_{\beta\beta'}^{N,e,(k)} . \qquad (A47)$$

Equations (A46) and (A47) allow one to calculate $W_{\alpha\alpha'}^{N,e}$ to all orders in $\hbar$ once $W_{\alpha\alpha'}^{N,e,(0)}$ is given. This order zero term is obtained by the solution of $(iL_{\alpha\alpha}^N + \kappa^N)W_{\alpha\alpha}^{N,e,(0)} = 0$. All higher order terms are obtained by the action of $H_{\alpha\alpha'}^N$, the imaginary unit i and $\mathcal{T}_{\alpha\alpha'\beta\beta'}$ (involving factors of $d_{\alpha\alpha'}$, P and derivatives with respect to P. Hence, one can conclude that functional dependence of $W_{Ne}^{(0)\alpha\alpha}$ on the Nosè variables Q_η and P_η is preserved in higher order terms $W_{\alpha\alpha'}^{N,e,(n)}$. One can find a stationary solution to order $\hbar$ by considering the first two equations of the set given by Equations (A41) and (A42):

$$\left[\hat{H}^N, \hat{W}^{N,e,(0)}\right] = 0 \quad (\text{for } k = 0) , \qquad (A48)$$

$$i\left[\hat{H}^N, \hat{W}^{N,e,(1)}\right] = +\frac{1}{2}\left(\hat{H}^N \overleftarrow{\nabla} \mathcal{B}^N \overrightarrow{\nabla} \hat{W}^{N,e,(0)} + \hat{W}^{N,e,(0)} \overleftarrow{\nabla} \mathcal{B}^N \overrightarrow{\nabla} \hat{H}^N\right) \quad (\text{for } k = 1) . \qquad (A49)$$

For the $\hbar^0$ term, one can make the ansatz

$$\hat{W}_{\alpha\beta}^{N,e,(0)} = \frac{1}{Z^N} e^{w_\alpha^N} \delta\left(\mathcal{E}_\alpha - H_\alpha^N\right) \delta_{\alpha\beta} , \qquad (A50)$$

where Z^N is

$$Z^N = \sum_\alpha \int d\mathcal{M}\, \delta\left(\mathcal{E}_\alpha - H_\alpha^N\right) \qquad (A51)$$

and obtain

$$\hat{W}_{\alpha\alpha'}^{N,e,(1)} = -i\frac{P}{M}d_{\alpha\alpha'}\hat{W}_{\alpha\alpha}^{N,e,(0)}\left[\frac{1 - e^{-\beta(E_{\alpha'} - E_\alpha)}}{E_\alpha - E_{\alpha'}} + \frac{\beta}{2}\left(1 + e^{-\beta(E_{\alpha'} - E_\alpha)}\right)\right] \qquad (A52)$$

for the $\hbar$ term.

Equations (A50) and (A52) give the explicit form of the stationary solution of the Nosè-Liouville equation up to order $\mathcal{O}(\hbar)$. One can now prove that, when calculating averages of quantum-classical operators depending only on physical phase space variables, $\mathcal{G}_\alpha(R,P)$, the canonical form of the stationary density is obtained. It can be noted that it will suffice to prove this result for the

$\hbar^0$ term since, as discussed before, the differences with the standard case are contained therein. Indeed, when calculating

$$\langle \mathcal{G}_\alpha(R,P) \rangle \propto = \sum_\alpha \int dX^N e^{-w_\alpha^N} \mathcal{G}_\alpha(R,P) \delta(\mathcal{E}_\alpha - H_\alpha^T - Nk_B T Q_\eta) , \quad \text{(A53)}$$

considering the integral of the delta function over Nosè variables, one has

$$\int dP_\eta dQ_\eta \, e^{-N\eta} \delta(\mathcal{E}_\alpha - H_\alpha^T - Nk_B T Q_\eta) = \text{const} \times \exp[-\beta H_\alpha^T] , \quad \text{(A54)}$$

where the property $\delta(f(s)) = [df/ds]_{s=s_0}^{-1} \delta(s - s_0)$ has been used (s_0 is the zero of $f(s)$).

References

1. Caldeira, A.O. *An Introduction to Macroscopic Quantum Phenomena and Quantum Dissipation*; Cambridge University Press: Cambridge, UK, 2014.
2. Weiss, U. *Quantum Dissipative Systems*; World Scientific: Singapore, 1999.
3. Breuer, H.-P.; Petruccione, F. *The Theory of Open Quantum Systems*; Clarendon Press: Oxford, UK, 2002.
4. Carr, S.M.; Lawrence, W.E.; Wybourne, M.N. Accessibility of quantum effects in mesomechanical systems. *Phys. Rev. B* **2001**, *64*, 220101. [CrossRef]
5. Armour, A.D.; Blencowe, M.P.; Schwab, K.C. Entanglement and Decoherence of a Micromechanical Resonator via Coupling to a Cooper-Pair Box. *Phys. Rev. Lett.* **2002**, *88*, 148301. [CrossRef] [PubMed]
6. Irish, E.K. Schwab, K. Quantum measurement of a coupled nanomechanical resonator–Cooper-pair box system. *Phys. Rev. B* **2003**, *68*, 155311. [CrossRef]
7. Blencowe, M. Quantum electromechanical systems. *Phys. Rep.* **2004** *395*, 159–222. [CrossRef]
8. Katz, I.; Retzker, A.; Straub, R.; Lifshitz, R. Signatures for a Classical to Quantum Transition of a Driven Nonlinear Nanomechanical Resonator. *Phys. Rev. Lett.* **2007**, *99*, 040404. [CrossRef] [PubMed]
9. Claudon, J.; Zazunov, A.; Hekking, F.W.J.; Buisson, O. Rabi-like oscillations of an anharmonic oscillator: Classical versus quantum interpretation. *Phys. Rev. B* **2008**, *78*, 184503. [CrossRef]
10. Tiwari, R.P.; Stroud, D. Numerical study of energy loss by a nanomechanical oscillator coupled to a Cooper-pair box. *Phys. Rev. B* **2008**, *77*, 214520. [CrossRef]
11. Katz, I.; Lifshitz, R.; Retzker, A.; Straub, R. Classical to quantum transition of a driven nonlinear nanomechanical resonator. *New J. Phys.* **2008**, *10*, 125023. [CrossRef]
12. Chakraborty, A. Buckled nano rod—A two state system: Quantum effects on its dynamics. *Mol. Phys.* **2009**, *107*, 1777–1786. [CrossRef]
13. Guo, L.-Z.; Zheng, Z.-G.; Li, X.-Q. Quantum dynamics of mesoscopic driven Duffing oscillators. *EPL* **2010**, *90*, 10011. [CrossRef]
14. Galve, F. Propagation properties and limitations on the attainable entanglement in a driven harmonic chain. *Phys. Rev. A* **2011**, *84*, 012318. [CrossRef]
15. Brouard, S.; Alonso, D.; Sokolovski, D. Decoherence of a quantum harmonic oscillator monitored by a Bose-Einstein condensate. *Phys. Rev. A* **2011**, *84*, 012114. [CrossRef]
16. Brown, K.R.; Ospelkaus, C.; Colombe, Y.; Wilson, A.C.; Leibfried, D.; Wineland, D.J. Coupled quantized mechanical oscillators. *Nature* **2011**, *471*, 196–199. [CrossRef] [PubMed]
17. Rips, S.; Kiffner, M.; Wilson-Rae, I.; Hartmann, M.J. Steady-state negative Wigner functions of nonlinear nanomechanical oscillators. *New J. Phys.* **2012**, *14*, 023042. [CrossRef]
18. Chakraborty, A. Buckled nano rod—A two state system and quantum effects on its dynamics using system plus reservoir model. *Mol. Phys.* **2011**, *109*, 517–526. [CrossRef]
19. Metelmann, A.; Brandes, T. Adiabaticity in semiclassical nanoelectromechanical systems. *Phys. Rev. B* **2011**, *84*, 155455. [CrossRef]
20. Eom, K.; Park, H.S.; Yoon, D.S.; Kwon, T. Nanomechanical resonators and their applications in biological/chemical detection: Nanomechanics principles. *Phys. Rep.* **2011**, *503*, 115–163. [CrossRef]

21. Li, Q.; Kapulkin, A.; Anderson, D.; Tan, S.M.; Pattanayak, A.K. Experimental signatures of the quantum–classical transition in a nanomechanical oscillator modeled as a damped-driven double-well problem. *Phys. Scr.* **2012**, *151*, 014055. [CrossRef]
22. Poot, M.; van der Zant, H.S.J. Mechanical systems in the quantum regime. *Phys. Rep.* **2012**, *511*, 273–335. [CrossRef]
23. Xiang, Z.; Ashhab, S.; You, J.Q.; Nori, F. Hybrid quantum circuits: Superconducting circuits interacting with other quantum systems. *Rev. Mod. Phys.* **2013**, *85*, 623–653. [CrossRef]
24. Imboden, M.; Mohanty, P. Dissipation in nanoelectromechanical systems. *Phys. Rep.* **2014**, *534*, 89–146. [CrossRef]
25. Zeng, W.; Nie, W.; Li, L.; Chen, A. Ground-state cooling of a mechanical oscillator in a hybrid optomechanical system including an atomic ensemble. *Sci. Rep.* **2017**, *7*, 17258. [CrossRef] [PubMed]
26. Gu, X.; Kockum, A.F.; Miranowicz, A.; Liu, Y.; Nori, F. Microwave photonics with superconducting quantum circuits. *Phys. Rep.* **2017**, *718*, 1–102. [CrossRef]
27. Silin, V.P. The Kinetics of Paramagnetic Phenomena. *Zh. Teor. Eksp. Fiz.* **1956**, *30*, 421–422.
28. Rukhazade, A.A.; Silin, V.P. On the magnetic susceptibility of a relativistic electron gas. *Soviet Phys. JETP* **1960**, *11*, 463–464.
29. Balescu, R. A Covariant Formulation of Relativistic Quantum Statistical Mechanics. I. Phase Space Description of a Relativistic Quantum Plasma. *Acta Phys. Aust.* **1968**, *28*, 336.
30. Zhang, W.Y.; Balescu, R. Statistical Mechanics of a spin-polarized plasma. *J. Plasma Phys.* **1988**, *40*, 199–213. [CrossRef]
31. Balescu, R.; Zhang, W.Y. Kinetic equation, spin hydrodynamics and collisional depolarization rate in a spin polarized plasma. *J. Plasma Phys.* **1988**, *40*, 215–234. [CrossRef]
32. Osborn, T.A.; Kondrat'eva, M.F.; Tabisz, G.C.; McQuarrie, B.R. Mixed Weyl symbol calculus and spectral line shape theory. *J. Phys. A Math. Gen.* **1999**, *32*, 4149. [CrossRef]
33. Beck, G.M.; Sergi, A. Quantum dynamics of a nano-rod under compression. *Phys. Lett. A* **2013**, *377*, 1047–1051. [CrossRef]
34. Aleksandrov, I.V.; Naturforsch, Z. The Statistical Dynamics of a System Consisting of a Classical and a Quantum Subsystem. *Z. Naturforsch. A* **1981**, *36* 902–908. [CrossRef]
35. Gerasimenko, V.I. Dynamical equations of quantum-classical systems. *Theor. Math. Phys.* **1982**, *50*, 49–55. [CrossRef]
36. Boucher, W.; Traschen, J. Semiclassical physics and quantum fluctuations. *Phys. Rev. D* **1988**, *37*, 3522. [CrossRef]
37. Petrina, D.Y.; Gerasimenko, V.I.; Enolskii, V.Z. Equations of motion of one class of quantum-classical systems. *Sov. Phys. Dokl.* **1990**, *35*, 925.
38. Martens, C.C.; Fang, J.Y. Semiclassical-Limit Molecular Dynamics on Multiple Electronic Surfaces. *J. Chem. Phys.* **1996**, *106*, 4918–4930. [CrossRef]
39. Prezhdo, O.V.; Kisil, V.V. Mixing quantum and classical mechanics. *Phys. Rev. A* **1997**, *56*, 162. [CrossRef]
40. Donoso, A.; Martens, C.C. Simulation of Coherent Nonadiabatic Dynamics Using Classical Trajectories. *J. Phys. Chem. A* **1998**, *102*, 4291–4300. [CrossRef]
41. Nielsen, S.; Kapral, R.; Ciccotti, G. Statistical mechanics of quantum-classical systems. *J. Chem. Phys.* **2001**, *115*, 5805–5815. [CrossRef]
42. Sergi, A. Non-Hamiltonian Commutators in Quantum Mechanics. *Phys. Rev. E* **2005**, *72*, 066125. [CrossRef] [PubMed]
43. Sergi, A. Deterministic constant-temperature dynamics for dissipative quantum systems. *J. Phys. A Math. Theor.* **2007**, *40*, F347. [CrossRef]
44. Kapral, R. Quantum-Classical Dynamics in a Classical Bath. *J. Phys. Chem. A* **2001**, *105*, 2885–2889. [CrossRef]
45. Sergi, A.; Kapral, R. Nonadiabatic Reaction Rates for Dissipative Quantum-Classical Systems. *J. Chem. Phys.* **2003**, *119*, 12776–12783. [CrossRef]
46. Wigner, E. On the Quantum Correction For Thermodynamic Equilibrium. *Phys. Rev.* **1932**, *40*, 749. [CrossRef]
47. Kapral, R.; Ciccotti, G. Mixed quantum-classical dynamics. *J. Chem. Phys.* **1999**, *110*, 8919–8929. [CrossRef]
48. Donoso, A.; Martens, C.C. Semiclassical multistate Liouville dynamics in the adiabatic representation. *J. Chem. Phys.* **2000**, *112*, 3980–3989. [CrossRef]

49. Donoso, A.; Kohen, D.; Martens, C.C. Simulation of nonadiabatic wave packet interferometry using classical trajectories. *J. Chem. Phys.* **2000**, *112*, 7345–7354. [CrossRef]
50. Wan, C.-C.; Schofield, J. Exact and asymptotic solutions of the mixed quantum-classical Liouville equation. *J. Chem. Phys.* **2000**, *112*, 4447–4459. [CrossRef]
51. Wan, C.-C.; Schofield, J. Mixed quantum-classical molecular dynamics: Aspects of the multithreads algorithm. *J. Chem. Phys.* **2000**, *113*, 7047–7054. [CrossRef]
52. Santer, M.; Manthe, U.; Stock, G. Quantum-classical Liouville description of multidimensional nonadiabatic molecular dynamics. *J. Chem. Phys.* **2001**, *114*, 2001–2012. [CrossRef]
53. Horenko, I.; Schmidt, B.; Schütte, C. A theoretical model for molecules interacting with intense laser pulses: The Floquet-based quantum-classical Liouville equation. *J. Chem. Phys.* **2001**, *115*, 5733–5743. [CrossRef]
54. Wan, C.-C.; Schofield, J. Solutions of mixed quantum-classical dynamics in multiple dimensions using classical trajectories. *J. Chem. Phys.* **2002**, *116*, 494–506. [CrossRef]
55. Horenko, I.; Salzmann, C.; Schmidt, B.; Schütte, C. Quantum-classical Liouville approach to molecular dynamics: Surface hopping Gaussian phase-space packets. *J. Chem. Phys.* **2002**, *117*, 11075–11088. [CrossRef]
56. Horenko, I.; Schmidt, B.; Schütte, C. Multidimensional classical Liouville dynamics with quantum initial conditions. *J. Chem. Phys.* **2002**, *117*, 4643–4650. [CrossRef]
57. Sergi, A.; Kapral, R. Quantum-Classical Dynamics of Nonadiabatic Chemical Reactions. *J. Chem. Phys.* **2003**, *118*, 8566–8575. [CrossRef]
58. Horenko, I.; Weiser, M.; Schmidt, B.; Schütte, C. Fully adaptive propagation of the quantum-classical Liouville equation. *J. Chem. Phys.* **2004**, *120*, 8913–8923. [CrossRef] [PubMed]
59. Riga, J.M.; Martens, C.C. Simulation of environmental effects on coherent quantum dynamics in many-body systems. *J. Chem. Phys.* **2004**, *120*, 6863–6873. [CrossRef] [PubMed]
60. Roman, E.; Martens, C.C. Semiclassical Liouville method for the simulation of electronic transitions: Single ensemble formulation. *J. Chem. Phys.* **2004**, *121*, 11572–11580. [CrossRef] [PubMed]
61. Thorndyke, B.; Micha, D.A. Photodissociation dynamics from quantum-classical density matrix calculations. *Chem. Phys. Lett.* **2005**, *403*, 280–286. [CrossRef]
62. Hanna, G.; Kapral, R. Quantum-classical Liouville dynamics of nonadiabatic proton transfer. *J. Chem. Phys.* **2005**, *122*, 244505. [CrossRef] [PubMed]
63. Riga, J.M.; Fredj, E.; Martens, C.C. Simulation of vibrational dephasing of I2 in solid Kr using the semiclassical Liouville method. *J. Chem. Phys.* **2006**, *124*, 064506. [CrossRef] [PubMed]
64. Kim, H.; Hanna, G.; Kapral, R. Analysis of kinetic isotope effects for nonadiabatic reactions. *J. Chem. Phys.* **2006**, *125*, 084509. [CrossRef] [PubMed]
65. Hanna, G.; Geva, E. Vibrational energy relaxation of a hydrogen-bonded complex dissolved in a polar liquid via the mixed quantum-classical Liouville methods. *J. Phys. Chem. B* **2008**, *112*, 4048–4058. [CrossRef] [PubMed]
66. Sergi, A.; Sinayskiy, I.; Petruccione, F. Numerical and Analytical Approach to the Quantum Dynamics of Two Coupled Spins in Bosonic Baths. *Phys. Rev. A* **2009**, *80*, 012108. [CrossRef]
67. Hanna, G.; Geva, E. Multi-dimensional spectra via the mixed quantum-classical Liouville method: Signatures of nonequilibrium dynamics. *J. Phys. Chem. B* **2009**, *113*, 9278–9288. [CrossRef] [PubMed]
68. Rekik, N.; Hsieh, C.-Y.; Freedman, H.; Hanna, G. A mixed quantum-classical Liouville study of the population dynamics in a model photo-induced condensed phase electron transfer reaction. *J. Chem. Phys.* **2013**, *138*, 144106. [CrossRef] [PubMed]
69. Martinez, F.; Rekik, N.; Hanna, G. Simulation of nonlinear optical signals via approximate solutions of the quantum-classical Liouville equation: Application to the pump-probe spectroscopy of a condensed phase electron transfer reaction. *Chem. Phys. Lett.* **2013**, *573*, 77–83. [CrossRef]
70. Shakib, F.; Hanna, G. An analysis of model proton-coupled electron transfer reactions via the mixed quantum-classical Liouville approach. *J. Chem. Phys.* **2014**, *141*, 044122. [CrossRef] [PubMed]
71. Uken, D.A.; Sergi, A. Quantum dynamics of a plasmonic metamolecule with a time-dependent driving. *Theor. Chem. Acc.* **2015**, *134*, 141. [CrossRef]
72. Shakib, F.; Hanna, G. New insights into the nonadiabatic state population dynamics of model proton-coupled electron transfer reactions from the mixed quantum-classical Liouville approach. *J. Chem. Phys.* **2016**, *144*, 024110. [CrossRef] [PubMed]

73. Martinez, F.; Hanna, G. Mixed quantum-classical simulations of transient absorption pump-probe signals for a photo-induced electron transfer reaction coupled to an inner-sphere vibrational mode. *J. Phys. Chem. A* **2016**, *120*, 3196–3205. [CrossRef] [PubMed]
74. Li, M.; Freedman, H.; Dell'Angelo, D.; Hanna, G. A model platform for rapid, robust, directed, and long-range vibrational energy transport: Insights from a mixed quantum-classical study of a 1D molecular chain. *AIP Conf. Proc.* **2017**, *1906*, 030007.
75. Freedman, H.; Hanna, G. Mixed quantum-classical Liouville simulation of vibrational energy transfer in a model alpha-helix at 300 K. *Chem. Phys.* **2016**, *477*, 74–88. [CrossRef]
76. Micha, D.A.; Thorndyke, B. Dissipative dynamics in many-atom systems: A density matrix treatment. *Int. J. Quant. Chem.* **2002**, *90*, 759–771. [CrossRef]
77. Kapral, R.; Ciccotti, G. A Statistical Mechanical Theory of Quantum Dynamics in Classical Environments. In *Bridging Time Scales: Molecular Simulations for the Next Decade*; Nielaba, P., Mareschal, M., Ciccotti, G., Eds.; Springer: Berlin, Germay, 2002; p. 445.
78. Sergi, A.; MacKernan, D.; Ciccotti, G.; Kapral, R. Simulating Quantum Dynamics in Classical Environments. *Theor. Chem. Acc.* **2003**, *110*, 49–58. [CrossRef]
79. Micha, D.A.; Thorndyke, B. The Quantum-Classical Density Operator for Electronically Excited Molecular Systems. *Adv. Quantum Chem.* **2004**, *47*, 293–314.
80. Sergi, A.; Kapral, R. Nonadiabatic Chemical Reactions. *Comp. Phys. Commun.* **2005**, *169*, 400–403. [CrossRef]
81. Kapral, R.; Sergi, A. Dynamics of Condensed Phase Proton and Electron Transfer Processes. In *Handbook of Theoretical and Computational Nanotechnology*; Rieth M., Schommers, W., Eds.; American Scientific Publishers: Valencia, CA, USA, 2005; Chapter 92.
82. Kapral, R. Progress in the Theory of Mixed Quantum-Classical Dynamics. *Ann. Rev. Phys. Chem.* **2006**, *57*, 129–157. [CrossRef] [PubMed]
83. Ciccotti, G.; Coker, D.; Kapral, R. Quantum Statistical Mechanics with Trajectories. In *Quantum Dynamics of Complex Molecular Systems*; Micha, D.A., Burghardt, I., Eds.; Springer: Berlin, Germany, 2007; p. 275.
84. Micha, D.; Leathers, A.; Thorndyke, B. Density Matrix Treatment of Electronically Excited Molecular Systems: Applications to Gaseous and Adsorbate Dynamics. In *Quantum Dynamics of Complex Molecular Systems*; Springer: Berlin, Germany, 2007; Volume 83, p. 165.
85. Grunwald, R.; Kelly, A.; Kapral, R. Quantum Dynamics in Almost Classical Environments. In *Energy Transfer Dynamics in Biomaterial Systems*; Burghardt, I., May, V., Micha, D.A., Bittner, E.R., Eds.; Springer: Berlin, Germany, 2009; p. 383.
86. Bonella, S.; Coker, D.; MacKernan, D.; Kapral, R.; Ciccotti, G. Trajectory Based Simulations of Quantum-Classical Systems. In *Energy Transfer Dynamics in Biomaterial Systems*; Burghardt, I., May, V., Micha, D.A., Bittner, E.R., Eds.; Springer: Berlin, Germany, 2009; p. 415.
87. Hanna, G.; Kapral, R. Quantum-Classical Liouville Dynamics of Condensed Phase Quantum Processes. In *Reaction Rate Constant Computations: Theories and Applications*; Han, K., Chu, T., Eds.; Royal Society of Chemistry: London, UK, 2013; p. 233.
88. Kapral, R. Quantum Dynamics in Open Quantum-Classical Systems. *J. Phys. Condens. Matter* **2015**, *27*, 073201. [CrossRef] [PubMed]
89. Kapral, R. Surface hopping from the perspective of quantum-classical Liouville dynamics. *Chem. Phys.* **2016**, *481*, 77–83. [CrossRef]
90. Sergi, A. Alessandro Sergi, Communication: Quantum dynamics in classical spin baths. *J. Chem. Phys.* **2013**, *139*, 031101. [CrossRef] [PubMed]
91. Sergi, A. Computer Simulation of Quantum Dynamics in a Classical Spin Environment. *Theor. Chem. Acc.* **2014**, *133*, 1495. [CrossRef]
92. Sergi, A.; Ferrario, M. Non-Hamiltonian Equations of Motion with a Conserved Energy. *Phys. Rev. E* **2001**, *64*, 056125. [CrossRef] [PubMed]
93. Sergi, A. Non-Hamiltonian Equilibrium Statistical Mechanics. *Phys. Rev. E* **2003**, *67*, 021101. [CrossRef] [PubMed]
94. Sergi, A.; Giaquinta, P.V. On the geometry and entropy of non-Hamiltonian phase space. *J. Stat. Mech. Theory Exp.* **2007**, *2*, P02013. [CrossRef]
95. Grabowski, J.; Urbafiski, E. Algebroids—General differential calculi on vector bundles. *J. Geom. Phys.* **1999**, *31*, 111–141. [CrossRef]

96. Grabowska, K.; Urbanski, P.; Grabowski, J. Geometrical mechanics on algebroids. *Int. J. Geom. Meth. Mod. Phys.* **2006**, *3*, 559–575. [CrossRef]
97. Grabowski, J.; Urbański, E. Lie algebroids and poisson-nijenhuis structures. *Rep. Math. Phys.* **1997**, *40*, 195–208. [CrossRef]
98. De Leon, M.; Marrero, J.C.; de Diego, D.M. Linear almost poisson structures and hamilton-jacobi equation. Applications to nonholonomic mechanics. *J. Geom. Mech.* **2010**, *2*, 159. [CrossRef]
99. Grabowska, K.; Grabowski, J. Variational calculus with constraints on general algebroids. *J. Phys. A Math. Theor.* **2008**, *41*, 175204. [CrossRef]
100. Caro, J.; Salcedo, L.L. Impediments to mixing classical and quantum dynamics. *Phys. Rev. A* **1999**, *60*, 842–852. [CrossRef]
101. Sergi, A. Variational Principle and phase space measure in non-canonical coordinates. *Atti della Accademia Peloritana dei Pericolanti Classe di Scienze Fisiche, Matematiche e Naturali* **2005**, *83*, C1A0501003.
102. McCauley, J.L. *Classical Mechanics*; Cambridge University Press: Cambridge, UK, 1998.
103. Manini, N.; Pistolesi, F. Off-diagonal Geometric phases. *Phys. Rev. Lett.* **2000**, *85*, 3067–3071. [CrossRef] [PubMed]
104. Pati, A.K. Adiabatic Berry phase and Hannay angle for open paths. *Ann. Phys.* **1998**, *270*, 178–197. [CrossRef]
105. Filipp, S.; Sjöqvist, E. Off-diagonal generalization of the mixed-state geometric phase. *Phys. Rev. A* **2003**, *68*, 042112. [CrossRef]
106. Englman, R.; Yahalom, A.; Baer, M. The open path phase for degenerate and non-degenerate systems and its relation to the wave function and its modulus. *Eur. Phys. J. D* **2000**, *8*, 1. [CrossRef]
107. Berry, M.V. Quantal Phase Factors Accompanying Adiabatic Changes. *Proc. R. Soc. Lond. Ser. A* **1984**, *392*, 45–57. [CrossRef]
108. Shapere, A.; Wilczek, F. (Eds.) *Geometric Phases in Physics*; World Scientific: Singapore, 1989.
109. Mead, C.A. The geometric phase in molecular systems. *Rev. Mod. Phys.* **1992**, *64*, 51. [CrossRef]
110. Gardiner, C.W. *Handbook of Stochastic Methods*; Springer: New York, NY, USA, 2002.
111. Nosé, S. A molecular dynamics method for simulations in the canonical ensemble. *Mol. Phys.* **1984**, *52*, 255–268. [CrossRef]
112. Hoover, W.G. Canonical dynamics: Equilibrium phase-space distributions. *Phys. Rev. A* **1985**, *31*, 1695. [CrossRef]
113. MacKernan, D.; Kapral, R.; Ciccotti, G. Sequential short-time propagation of quantum classical dynamics. *J. Phys. Condens. Matter* **2002**, *14*, 9069. [CrossRef]
114. Kernan, D.M.; Ciccotti, G.; Kapral, R. Trotter-Based Simulation of Quantum-Classical Dynamics. *J. Phys. Chem. B* **2008**, *112*, 424–432. [CrossRef] [PubMed]
115. Sergi, A.; Petruccione, F. Sampling Quantum Dynamics at Long Time. *Phys. Rev. E* **2010**, *81*, 032101. [CrossRef] [PubMed]
116. Uken, D.A.; Sergi, A.; Petruccione, F. Stochastic Simulation of Nonadiabatic Dynamics at Long Time. *Phys. Scr.* **2011**, *143*, 014024. [CrossRef]
117. Uken, D.A.; Sergi, A.; Petruccione, F. Filtering Schemes in the Quantum-Classical Liouville Approach to Non-adiabatic Dynamics. *Phys. Rev. E* **2013**, *88*, 033301. [CrossRef] [PubMed]
118. Martinez, F.; Hanna, G. Assessment of approximate solutions of the quantum-classical Liouville equation for dynamics simulations of quantum subsystems embedded in classical environments. *Mol. Simul.* **2014**, *41*, 107–122. [CrossRef]
119. Dell'Angelo, D.; Hanna, G. Self-consistent filtering scheme for efficient calculations of observables via the mixed quantum-classical Liouville approach. *J. Chem. Theory Comput.* **2016**, *12*, 477–485. [CrossRef] [PubMed]
120. Dell'Angelo, D.; Hanna, G. Using multi-state transition filtering to improve the accuracy of expectation values via mixed quantum-classical Liouville dynamics. *AIP Conf. Proc.* **2016**, *1790*, 020009.
121. Dell'Angelo, D.; Hanna, G. On the performance of multi-state transition filtering in mixed quantum-classical Liouville surface-hopping simulations: Beyond two-and three-state quantum Subsystems. *Theor. Chem. Acc.* **2018**, *137*, 15. [CrossRef]
122. Dell'Angelo, D.; Hanna, G. Importance of eigenvector sign consistency in computations of expectation values via mixed quantum-classical surface-hopping dynamics. *Theor. Chem. Acc.* **2017**, *136*, 75. [CrossRef]

123. Liu, J.; Hanna, G. Efficient and deterministic propagation of mixed quantum-classical Liouville dynamics. *J. Phys. Chem. Lett.* **2018**, *9*, 3928–3933. [CrossRef] [PubMed]
124. Dirac, P.A.M. *Lessons in Quantum Mechanics*; Dover: New York, NY, USA, 2001.
125. Balescu, R. *Equilibrium and Non-Equilibrium Statistical Mechanics*; Wiley: New York, NY, USA, 1975.
126. Goldstein, H. *Classical Mechanics*; Addison-Wesley: London, UK, 1980.
127. Frenkel, D.; Smit, B. *Understanding Molecular Simulation*; Academic Press: San Diego, CA, USA, 1996.
128. Abergel, D.; Palmer, A.G., III. On the Use of the Stochastic Liouville Equation in Nuclear Magnetic Resonance: Application to $R_{1\rho}$ Relaxation in the Presence of Exchange. *Concepts Magn. Reson.* **2003**, *19*, 134–148. [CrossRef]
129. La Cour Jansen, T.; Zhuang, W.; Mukamel, S. Stochastic Liouville equation simulation of multidimensional vibrational line shapes of trialanine. *J. Chem. Phys.* **2004**, *121*, 10577–10598. [CrossRef] [PubMed]
130. La Cour Jansen, T.; Hayashi, T.; Zhuang, W.; Mukamel, S. Stochastic Liouville equations for hydrogen-bonding fluctuations and their signatures in two-dimensional vibrational spectroscopy of water. *J. Chem. Phys.* **2005**, *123*, 114504. [CrossRef]
131. Sanda, F.; Zhuang, W.; Jansen, T.L.; Hayashi, T.; Mukamel, S. Signatures of Chemical Exchange in 2D Vibrational Spectroscopy; Simulations Based on the Stochastic Liouville Equations. In *Ultrafast Phenomena XV*; Springer Series in Chemical Physics; Corkum, P., Jonas, D.M., Miller, R.J.D., Weiner, A.M., Eds.; Springer: Berlin, Germany, 2007; Volume 88, pp. 401–403.
132. Schulman, L.S. *Techniques and Applications of Path Integration*; Dover Publications Inc.: New York, NY, USA, 2005.
133. Cartan, È. *The Theory of Spinors*; Dover Publications Inc.: New York, NY, USA, 1981.
134. Hladik, J. *Spinors in Physics*; Springer: Berlin, Germany, 1999.
135. Carmeli, M. *Classical Fields*; John Wiley & Sons: New York, NY, USA, 1982.
136. Barut, A.O. *Electrodynamics and Classical Theory of Fields and Particles*; Dover Publications Inc.: New York, NY, USA, 1980.
137. Huang, K. *Statistical Mechanics*; John Wiley & Sons: Singapore, 1987.
138. Tuckerman, M.; Martyna, G.J.; Berne, B.J. Reversible multiple time scale molecular dynamics. *J. Chem. Phys.* **1992**, *97*, 1990–2001. [CrossRef]
139. Martyna, G.J.; Tuckerman, M.; Tobias, D.J.; Klein, M.L. Explicit reversible integrators for extended systems dynamics. *Mol. Phys.* **1996**, *87*, 1117–1157. [CrossRef]
140. Sergi, A.; Ferrario, M.; Costa, D. Reversible integrators for basic extended system molecular dynamics. *Mol. Phys.* **1999**, *97*, 825–832. [CrossRef]
141. Suzuki, M. General Decomposition Theory of Ordered Exponentials. *Proc. Jpn. Acad. Ser. B* **1993**, *69*, 161–166. [CrossRef]
142. Martyna, G.J.; Klein, M.L.; Tuckerman, M. Nosè–Hoover chains: The canonical ensemble via continuous dynamics. *J. Chem. Phys.* **1992**, *92*, 2635–2643. [CrossRef]
143. Courant, R.; Hilbert, D. *Methods of Mathematical Physics I*; Interscience: New York, NY, USA, 1953.

© 2018 by the authors. Licensee MDPI, Basel, Switzerland. This article is an open access article distributed under the terms and conditions of the Creative Commons Attribution (CC BY) license (http://creativecommons.org/licenses/by/4.0/).

Article

Ultrafast Dynamics of High-Harmonic Generation in Terms of Complex Floquet Spectral Analysis

Hidemasa Yamane [†] and Satoshi Tanaka *,[†]

Department of Physical Science, Osaka Prefecture University, Gakuen-cho 1-1, Sakai 599-8531, Japan; s_h.yamane@p.s.osakafu-u.ac.jp
* Correspondence: stanaka@p.s.osakafu-u.ac.jp; Tel.: +81-72-254-9710
† These authors contributed equally to this work.

Received: 3 July 2018; Accepted: 30 July 2018; Published: 1 August 2018

Abstract: We studied the high-harmonic generation (HHG) of a two-level-system (TLS) driven by an intense monochromatic phase-locked laser based on complex spectral analysis with the Floquet method. In contrast with phenomenological approaches, this analysis deals with the whole process as a coherent quantum process based on microscopic dynamics. We have obtained the time-frequency resolved spectrum of spontaneous HHG single-photon emission from an excited TLS driven by a laser field. Characteristic spectral features of the HHG, such as the plateau and cutoff, are reproduced by the present model. Because the emitted high-harmonic photon is represented as a superposition of different frequencies, the Fano profile appears in the long-time spectrum as a result of the quantum interference of the emitted photon. We reveal that the condition of the quantum interference depends on the initial phase of the driving laser field. We have also clarified that the change in spectral features from the short-time regime to the long-time regime is attributed to the interference between the interference from the Floquet resonance states and the dressed radiation field.

Keywords: high-harmonic generation; fano effect; quantum interference; Floquet method; complex spectral analysis

---

1. Introduction

The advent of ultrafast strong light sources has opened up a new era of optical science, called attosecond physics [1,2]. Recent advances in the manipulation techniques of a laser light pulse, such as carrier-envelope-phase (CEP) control, has enabled us to induce coherent electron motion with sub-femtosecond precision. The well-controlled coherent electron motion is a source of the high-harmonic pulse in the XUV region to the X-ray region [3–10]. These attosecond light pulses are now the most powerful tool to explore the real-time dynamics of electronic motions in atoms and molecules on the attosecond time scale [11–16].

Recently, HHG has been experimentally observed also from solids, such as in semiconductors [17–21], topological phase materials [22], low-dimensional materials [23], thin films [24], and amorphous solids [25]. Some have used two-color light beams, such as near infrared and far infrared, to clearly distinguish the different electronic excitations, which are sometimes called high-order sideband generation (HSG) [26–29]. Even though coherent electronic motion in solids is different from that of atoms and molecules, characteristic features of the HHG spectrum, such as the plateau and cutoff, are common to those from atoms and molecules. These experiments indicate that the fundamental mechanism of the HHG photon emission is the same, whether from atoms, molecules, or solids [30]. Therefore, it is essential to clarify the microscopic mechanism that determines how the quantum coherence of an electron induced by the driving field is transferred to a high-harmonic photon through nonlinear interaction between the electron and the driving field. Developing the microscopic theory of HHG, one has to keep in mind that HHG photon emission is a spontaneous photon emission because the radiation field with the high-harmonic frequency is initially in the quantum vacuum [31,32]. For a precise argument, it is necessary to include the entire HHG process as a coherent quantum process, including the free radiation field as an environment.

Most conventional theories, however, have employed phenomenological treatments that may terminate the quantum coherence in the HHG process: for example, the HHG emission spectrum is calculated classically based on Maxwell's equation [2,6,33–35], or the Markov approximation is used in a master equation with a phenomenological parameter [18,27,30,36]. In fact, how to interpret the dissipative spontaneous photon emission within the framework of quantum mechanics has been a debate since the early days of the theory [37–40], as an irreversible decay process contradicts the time-reversible quantum dynamics. As a solution to the problem, a new formalism, i.e., *complex spectral analysis*, has been explored over the last two decades, so that the Hamiltonian can take complex eigenvalues by expanding the vector space to the *extended Hilbert space* [41–48].

In this contribution, we apply complex spectral analysis to study the HHG, using the Floquet method to take into account a non-perturbative interaction between the electron and the driving field. The total system under consideration consists of not only the strongly coupled matter and driving laser field, but also the free radiation field with a continuous spectrum. We solve the complex eigenvalue problem of the Floquet Hamiltonian of the total system in the extended Hilbert space. The time evolution of the quantum state is then described by the eigenstate expansion of the total system, and thereby the quantum coherence is retained. We study the HHG of a two-level system (TLS) driven by an intense monochromatic phase-locked laser and obtain the analytical expression for the time-frequency resolved spectral amplitude for a HHG single-photon observation.

We show that the calculated HHG spectrum exhibits the characteristic features of HHG from solids. We reveal that the quantum interference of the Floquet resonance states causes a Fano-type dip structure in the HHG spectrum. Moreover, we show that quantum interference between the Floquet resonance states and the dressed field states are responsible for the temporal change in the HHG spectrum from the adiabatic regime to the stationary regime. Because the superposition of the photon states with different frequencies depends on the initial phase of the driving laser, we can quantum mechanically control the HHG photon emission by changing the phase of the laser.

2. Model

We consider spontaneous emission from a TLS excited by a delta-function pulse. The excited TLS is driven by a monochromatic phase-locked laser field with amplitude $\mathcal{E}_0$ and frequency $\omega/2$ with a phase θ relative to the excitation delta-pulse as shown in Figure 1. Starting from the minimal coupling Hamiltonian under the dipole approximation, the total Hamiltonian, composed of the electronic system and the radiation field, is represented by [31].

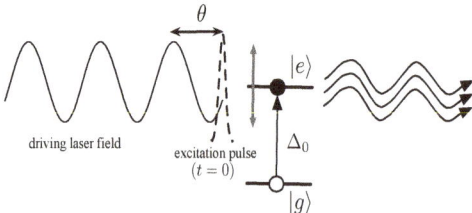

Figure 1. High-harmonic generation of a driven TLS.

$$\hat{H}(t) = E_1|1\rangle\langle 1| + (E_1 + \Delta_0)|2\rangle\langle 2| + \mathcal{E}(t)\left(|1\rangle\langle 2| + |2\rangle\langle 1|\right) \\ + \int \omega_k \hat{a}_k^\dagger \hat{a}_k dk + \lambda \int \mathcal{C}_k \left(|1\rangle\langle 2| + |2\rangle\langle 1|\right)(\hat{a}_k + \hat{a}_k^\dagger)dk \,, \qquad (1)$$

where $|1\rangle$ and $|2\rangle$ represent the ground and excited states with energies E_1 and E_2, respectively, and $\Delta_0 \equiv E_2 - E_1$. We classically describe the monochromatic phase-locked laser field as

$$\mathcal{E}(t) \equiv \mathcal{E}_0 \cos\left(\frac{\omega t + \theta}{2}\right). \qquad (2)$$

Conversely, the scattered bosonic radiation field is dealt with quantum mechanically: $\hat{a}_k$ ($\hat{a}_k^\dagger$) represents the free radiation field with energy $\omega_k = c|k|$, and λ is a dimensionless coupling constant, where we consider $c = 1$. The coupling coefficient C_k is given by $C_k = \sqrt{\omega_k}$ except for a constant factor [31].

As shown in Appendix A, under the condition of

$$2\frac{\mathcal{E}_0^2}{\omega} \gg \Delta_0 \gg |\mathcal{E}_0|, \qquad (3)$$

the Hamiltonian $\hat{H}(t)$ can be written in terms of the adiabatic basis of $|g\rangle$ and $|e\rangle$ given in Equation (A12). Under the rotating wave approximation, the Hamiltonian is given by

$$\hat{H}_{\rm ad}(t) = E_0|g\rangle\langle g| + (E_e + A\cos(\omega t + \theta))|e\rangle\langle e| + \int \omega_k \hat{a}_k^\dagger \hat{a}_k dk + \lambda \int C_k \left(|e\rangle\langle g|\hat{a}_k + |g\rangle\langle e|\hat{a}_k^\dagger\right) dk, \qquad (4)$$

where we have defined the renormalized amplitude as

$$A \equiv \frac{2\mathcal{E}_0^2}{\Delta_0}. \qquad (5)$$

In this paper, we consider a one-dimensional system for simplicity, which does not influence our main results. Hereafter, we simply write $\hat{H}_{\rm ad}(t)$ as $\hat{H}(t)$.

Because the number of elementary excitations does not change in $\hat{H}(t)$, the evolution of the state is closed in a single-excitation subspace of the dressed atom states of $|e\rangle \otimes |0_k\rangle \equiv |d\rangle$ and $|g\rangle \otimes |1_k\rangle \equiv |k\rangle$ [49]. Then, the Hamiltonian $\hat{H}(t)$ in this subspace is represented by

$$\hat{H}(t) = (\Delta_0 + A\cos(\omega t + \theta))|d\rangle\langle d| + \int \omega_k |k\rangle\langle k| dk + \lambda \int C_k(|d\rangle\langle k| + |k\rangle\langle d|) dk, \qquad (6)$$

where the energy difference is defined by $\Delta_0 \equiv E_e - E_0$, and we take E_0 as the origin of energy. In terms of the renormalized amplitude, the intensity of the driving field is given by $|A\cos(\omega t + \theta)|^2$, so that the maximum intensity is A^2.

Because the Hamiltonian is time-periodic, $\hat{H}(t + T) = \hat{H}(t)$ with $T \equiv 2\pi/\omega$, the Floquet theorem may be applied: the wave vector can be written as

$$|\Psi(t)\rangle = \sum_\xi c_\xi e^{-iz_\xi t}|\Phi_\xi(t)\rangle, \qquad (7)$$

with a periodic Floquet eigenfunction $|\Phi_\xi(t+T)\rangle = |\Phi_\xi(t)\rangle$ with the Floquet quasi-energy z_ξ [50–52]. The composite space $\mathcal{F} \equiv \mathcal{R} \otimes \mathcal{T}$ is made up of the configuration space $\mathcal{R}$ and the space $\mathcal{T}$ of periodic functions in time with period T [50]. The conjugate basis set $\{|\kappa_n\rangle\}$ to the time basis set $\{|t\rangle\}$ is constructed as

$$|\kappa_n\rangle \equiv \frac{1}{T}\int_0^T dt\, e^{i\kappa_n t}|t\rangle, \qquad (8)$$

where $\kappa_n \equiv n\omega = 2\pi n/T$ $(n=0,1,\cdots)$ [53]. It is well known that the Floquet eigenstate possesses mode-translational symmetry [50]

$$|\Phi_\zeta^{(n)}(t)\rangle = \exp[i\kappa_n t]|\Phi_\zeta^{(0)}(t)\rangle \, , \quad z_\zeta^{(n)} = z_\zeta^{(0)} + n\omega \, , \tag{9}$$

where (n) denotes the Floquet mode index and ζ classifies a state within a Floquet mode space of (n).

In terms of the conjugate basis set, the Floquet Hamiltonian is represented by

$$\begin{aligned}\hat{H}_F = &\sum_{n=-\infty}^{\infty} (\Delta_0 + n\omega)|d,\kappa_n\rangle\!\rangle\langle\!\langle d,\kappa_n| + \sum_{n=-\infty}^{\infty} \frac{A}{2}\left[e^{i\theta}|d,\kappa_{n+1}\rangle\!\rangle\langle\!\langle d,\kappa_n| + e^{-i\theta}|d,\kappa_n\rangle\!\rangle\langle\!\langle d,\kappa_{n+1}|\right] \\ &+ \sum_{n=-\infty}^{\infty}\int(\omega_k + n\omega)|k,\kappa_n\rangle\!\rangle\langle\!\langle k,\kappa_n|dk + \lambda\sum_{n=-\infty}^{\infty}\int \mathcal{C}_k[|k,\kappa_n\rangle\!\rangle\langle\!\langle d,\kappa_n| + |d,\kappa_n\rangle\!\rangle\langle\!\langle k,\kappa_n|]dk \, , \end{aligned} \tag{10}$$

where $|\cdot\rangle\!\rangle$ denotes the vector in the composite space $\mathcal{F}$.

In Equation (10), the first two terms represent the strong coupling between the TLS and the driving field in the Floquet composite basis. We note that the first two terms can be diagonalized in terms of the *Wannier-Stark basis* [50] given by

$$|\phi_d^{(n)}\rangle\!\rangle = \sum_{m=-\infty}^{\infty} e^{-i(n-m)\theta} J_{n-m}(a)|d,\kappa_m\rangle\!\rangle \, , \tag{11}$$

where $J_n(x)$ is the n-th order Bessel function of the first kind and $a \equiv A/\omega$. Please note that the Stark state is represented as a coherent sum of the bare discrete states in terms of the Floquet modes. Then, the Floquet Hamiltonian $\hat{H}_F$ can be rewritten as

$$\begin{aligned}\hat{H}_F = &\sum_{n=-\infty}^{\infty}\left\{(\Delta_0 + n\omega)|\phi_d^{(n)}\rangle\!\rangle\langle\!\langle\phi_d^{(n)}| + \int(\omega_k + n\omega)|k,\kappa_n\rangle\!\rangle\langle\!\langle k,\kappa_n|dk\right\} \\ &+ \lambda\sum_{n,m=-\infty}^{\infty}\int \mathcal{C}_k J_{m-n}(a)\left(e^{-i(m-n)\theta}|k,\kappa_n\rangle\!\rangle\langle\!\langle\phi_d^{(m)}| + e^{i(m-n)\theta}|\phi_d^{(m)}\rangle\!\rangle\langle\!\langle k,\kappa_n|\right)dk \, , \end{aligned} \tag{12}$$

where the first and second terms represent the diagonal Floquet energies for the Stark states and the continuous states, respectively. The last term of Equation (12) shows that that the TLS couples with the radiation field with different Floquet modes and the nonlinear interaction depends on the initial phase of the driving field. Please note that this coupling represented by the Bessel function is nonlinear in terms of the driving field amplitude a.

3. Complex Eigenvalue Problem of the Floquet Hamiltonian

The original time-dependent problem now becomes a time-independent eigenvalue problem, where we may employ the established method to solve the complex eigenvalue problem of the Hamiltonian. The difficulty arises, however, when we try to solve the eigenvalue problem by keeping the unstable discrete states in the spectrum in ordinary Hilbert space in the composite space because of the resonance singularity in the interaction between the TLS and the free radiation field [54].

To solve the problem of the resonance singularity, we extend the eigenvector subspace to the *extended Hilbert space*, where the norm of the eigenvector vanishes [41,42,44,46–48,55]. The complex eigenvalue problems of $\hat{H}_F$ read

$$\hat{H}_F|\Phi_\zeta^{(n)}\rangle\!\rangle = z_\zeta^{(n)}|\Phi_\zeta^{(n)}\rangle\!\rangle, \quad \langle\!\langle\tilde{\Phi}_\zeta^{(n)}|\hat{H}_F = z_\zeta^{(n)}\langle\!\langle\tilde{\Phi}_\zeta^{(n)}| \, , \tag{13}$$

where the right-eigenstate $|\Phi_\zeta^{(n)}\rangle\!\rangle$ and left-eigenstate $\langle\!\langle\tilde{\Phi}_\zeta^{(n)}|$ have the same complex eigenvalue $z_\zeta^{(n)}$.

The complex eigenbasis of $|\Phi^{(n)}_\zeta\rangle\rangle$ and $\langle\langle\tilde{\Phi}^{(n)}_\zeta|$ satisfy the bi-completeness and bi-orthonormal relation [41,46,53]:

$$\sum_{n=-\infty}^{\infty}\sum_\zeta |\Phi^{(n)}_\zeta\rangle\rangle\langle\langle\tilde{\Phi}^{(n)}_\zeta| = \hat{I}, \quad (14)$$

$$\langle\langle\tilde{\Phi}^{(n)}_\zeta|\Phi^{(n')}_{\zeta'}\rangle\rangle = \delta^{\text{Kr}}_{\zeta,\zeta'}\delta^{\text{Kr}}_{n,n'}, \quad (15)$$

where $\delta^{\text{Kr}}_{i,j}$ is the Kronecker delta.

The eigenvalue problem of the Floquet Hamiltonian was solved in terms of the Brillouin–Wigner–Feshbach projection method, as shown in Appendix B [46–48,53].

In the weak coupling case $\lambda \ll 1$, the right-resonance eigenstate is given by

$$|\Phi^{(n)}_d\rangle\rangle = \langle\langle\phi^{(n)}_d|\Phi^{(n)}_d\rangle\rangle\left\{|\phi^{(n)}_d\rangle\rangle + \lambda\sum_m\int dk\mathcal{C}_k\frac{J_{n-m}(a)e^{-i(n-m)\theta}}{[z-(\omega_k+m\omega)]^+}\Big|_{z=z^{(n)}_d}|k,\kappa_m\rangle\rangle\right\}, \quad (16)$$

where the $+$ sign in the denominator of Equation (16) indicates the analytic continuation of z from the upper half of the complex energy plane [41]. The second term of the curl bracket shows that the resonance states are given by the superposition of the discrete Stark state $|\phi^{(n)}_d\rangle\rangle$ and the free radiation field $|k,\kappa_m\rangle\rangle$ belonging to the different Floquet modes with the laser phase-dependent weighted sum of the Bessel function. The left-resonance eigenstates are also obtained by first taking the Hermite conjugate, and then the same analytic continuation with the $+$ index instead of the opposite analytic continuation [41,46,53]. The complex eigenvalue of the resonance state is obtained by solving the nonlinear dispersion equation

$$z^{(n)}_d = \chi^+_{n,n}(z^{(n)}_d) = \Delta_0 + n\omega + \lambda^2\sum_{m=-\infty}^{+\infty}\sigma^+(z^{(n)}_d - m\omega)J^2_{n-m}(a), \quad (17)$$

where the dynamical self-energy $\chi^+_{n,n}(z^{(n)}_d)$ is defined by Equation (A17) and the scalar self-energy function is given by Equation (A18). Of special importance is the fact that the self-energy in the right-hand-side depends on the eigenvalues, which originate in the nonlinearity of the eigenvalue problem of the effective Hamiltonian, as shown in Equation (A19). It should be emphasized that only if we take into account this nonlinearity will the eigenvalues of the non-Hermitian effective Hamiltonian coincide with the Hermitian total Hamiltonian [41]. We have solved this dispersion equation iteratively to obtain the complex eigenvalues of the Floquet Hamiltonian, and we have thereby considered the nonlinearity of the eigenvalue problem of the effective Hamiltonian, as shown in Equation (A14). The resonance state decays exponentially, with the decay rate given by the imaginary part of $z^{(n)}_d$.

The dressed continuous right-eigenstates are also obtained in Appendix B.2 as

$$|\Phi^{(n)}_k\rangle\rangle = |k,\kappa_n\rangle\rangle + \lambda\mathcal{C}_k\sum_m\frac{J_{m-n}(a)e^{i(m-n)\theta}}{\omega_k + n\omega + i0^+ - \chi^+_{m,m,D}(\omega_k+n\omega)}$$

$$\times\left\{|\phi^{(m)}_d\rangle\rangle + \lambda\sum_{m'}\int\frac{\mathcal{C}_{k'}J_{m-m'}(a)e^{-i(m-m')\theta}|k',\kappa_{m'}\rangle\rangle dk'}{\omega_k - \omega_{k'} + (n-m')\omega + i0^+}\right\}, \quad (18)$$

where $\chi^+_{n,n,D}(\varepsilon)$ is the dynamical self-energy, defined by Equation (A17), with the delayed analytic continuation from the upper half plane [41,46]. The continuous left-eigenstate has been similarly obtained without taking the delayed analytic continuation. The dressed continuous right-eigenstate $|\Phi^{(n)}_k\rangle\rangle$ and the left-eigenstate $\langle\langle\tilde{\Phi}^{(n)}_k|$ have the same real eigenvalues of $z^{(n)}_k = \omega_k + n\omega$ that is equivalent to unperturbed energy. The right- and left-eigenstates of the resonance states and the dressed continuous states satisfy the bi-completeness relation in the composite space $\mathcal{F}$ as

$$\hat{I}_\mathcal{F} = \sum_n \left\{ |\Phi_d^{(n)}\rangle\rangle\langle\langle\tilde{\Phi}_d^{(n)}| + \int dk |\Phi_k^{(n)}\rangle\rangle\langle\langle\tilde{\Phi}_k^{(n)}| \right\} . \tag{19}$$

This decomposition of the identity makes it possible to represent any state vector of the total system in terms of the complex spectral expansion.

Using Equation (14), the state vector at time t in the $\mathcal{R}$ space is given by the Floquet eigenstates as [50]

$$|\Psi(t)\rangle = \sum_\zeta \sum_n e^{-iz_\zeta^{(n)} t} |\Phi_\zeta^{(n)}(t)\rangle \langle\tilde{\Phi}_\zeta^{(p)}(0)|\Psi(0)\rangle \tag{20}$$

$$= \sum_\zeta \sum_n e^{-i(z_\zeta^{(0)} + n\omega)t} e^{in\omega t} |\Phi_\zeta^{(0)}(t)\rangle \langle\tilde{\Phi}_\zeta^{(n)}(0)|\Psi(0)\rangle \tag{21}$$

$$= \sum_\zeta e^{-iz_\zeta^{(0)} t} |\Phi_\zeta^{(0)}(t)\rangle \langle\tilde{\Phi}_\zeta^{(0)}(0)|\Psi(0)\rangle , \tag{22}$$

where we have used the Floquet mode-translational symmetry Equation (9). Using Equation (19), the state vector is given by

$$|\Psi(t)\rangle = e^{-iz_d^{(0)} t} |\Phi_d^{(0)}(t)\rangle \langle\tilde{\Phi}_d^{(0)}(0)|\Psi(0)\rangle + \int dk e^{-i\omega_k t} |\Phi_k^{(0)}(t)\rangle \langle\tilde{\Phi}_k^{(0)}(0)||\Psi(0)\rangle . \tag{23}$$

It should be noted that the wave function of the emitted single photon is described as the superposition of the single photon states with different frequencies, as shown in the second term of Equation (23).

4. HHG Spectrum

In this work, the HHG spectrum is studied in the case where the TLS is excited from $|g\rangle$ to $|e\rangle$ at $t = 0$ by a single-photon pulse. In this case, the spontaneous HHG single-photon emission spectrum, defined as the probability of detecting an emitted photon with frequency ω_k during the observation interval t, is obtained by

$$S(\omega_k, t) = \langle a_k^\dagger a_k \rangle_t = |\langle k|\Psi(t)\rangle|^2 , \tag{24}$$

with $|\Psi(0)\rangle = |d\rangle$ [31,56,57]. Substituting the right- and left-eigenstates of $\hat{H}_F$ in Equation (23), the analytical expression for the spectral amplitude is obtained as

$$\langle k|\Psi(t)\rangle = -\lambda \mathcal{C}_k \sum_{m=-\infty}^{\infty} e^{-iz_d^{(m)} t} \frac{J_m(a) e^{-im\theta}}{[\omega_k - z]_{z=z_d^{(m)}}^+} + \lambda \mathcal{C}_k e^{-i\omega_k t} \sum_{m=-\infty}^{\infty} \frac{J_m(a) e^{-im\theta}}{\omega_k + i0^+ - \chi_{m,m}^+(\omega_k)}$$

$$+ \frac{i}{2\pi} \lambda \mathcal{C}_k \sum_{n,l} \int_\Gamma d\omega' \rho(\omega') \frac{\mathcal{C}_{\omega'}^2 J_l^2(a) e^{-i(\omega' - m\omega)t}}{\sum_m J_{l-m}^2(a) \mathcal{C}_{\omega' - m\omega}^2 \rho(\omega' - m\omega)}$$

$$\times \frac{1}{\omega' - \chi_{l,l}^\Gamma(\omega')} \frac{J_{l-n}(a) e^{-i(l-n)\theta}}{\omega' + i0^+ - (\omega_k + n\omega)} \tag{25}$$

$$\equiv s_R(k, t) + s_C(k, t) + s_{BR}(k, t) , \tag{26}$$

where $\rho(\omega)$ is the density of states of the free radiation field, and $\mathcal{C}_k^2 dk = \mathcal{C}^2(\omega)\rho(\omega)d\omega$ with $\omega = c|k|$.

Equation (25) is the principal result of this paper: the contributions of the resonance state and the dressed continuous states are analytically decomposed in the first (s_R) and second (s_C) terms, respectively. While the first term decays exponentially with time, the second term does not decay over time, giving a stationary HHG spectrum. The third term (s_{BR}) represents the *branch point effect* [41,44], where the contour of the integral denoted by Γ is taken in the different Riemann sheets at the branch point. This term represents the non-Markovian effect, only contributing to the very short time known as *Zeno time*, or the very long time known as the long-time tale [44]. The contribution of the third term is very small in the present case, with a large amplitude of the driving laser field.

In Figure 2, we show the calculated results of the long-time HHG spectrum $S_\infty(\omega_k) \equiv \lim_{t \to \infty} S(\omega_k, t)$ for the following parameters: $\Delta_0 = 20, a = 10, \lambda = 0.06$, and $\theta = 0$ in (a) and $\theta = -\pi/2$ in (b). We take these parameters to approximately represent the experiments of WSe$_2$, such as the driving laser field frequency $2\pi/\omega \simeq 20$ THz, the band gap between the valence and the conduction bands $\Delta_0 \simeq 400$ THz [27]. The driving laser amplitude $a \equiv A/\omega$ is determined so as to agree with the cutoff energy of the experiments $n_{\text{cutoff}}/\hbar\omega \simeq 10$.

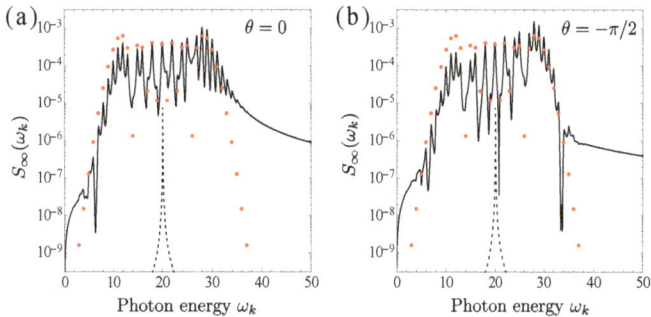

Figure 2. Stationary HHG spectrum $S_\infty(\omega_k)$ for $\Delta_0 = 20, a = 10, \lambda = 0.06$. (a) $\theta = 0$ and (b) $\theta = -\pi/2$. The fundamental spontaneous emission spectrum at $\omega_k = 20$ is shown by the dashed lines. The red marks indicate the absolute value of the Bessel function $|J_l(a)|^2$.

The intensity of the *m*th-order high-harmonics of the long-time HHG spectrum is mostly determined by the absolute values of the Bessel function $|J_m(a)|^2$, as shown by the red marks in the figures. The characteristic features of the HHG spectrum, such as the plateau and cutoff, are explained by the behavior of the Bessel functions. Because the ratio of the successive order of the Bessel function is evaluated as $|J_m(a)/J_{m-1}(a)|^2 \simeq (a/2m)^2$ for $m \gtrsim a$, the intensity of the high-harmonics sharply drops at $m \sim a$. Consequently, the cutoff energy is determined by the amplitude of the laser field a, and not by the intensity a^2, underlining the typical feature of the HHG spectrum from solids [17–19,21].

The cross terms of the different Floquet modes in $S_\infty(\omega_k)$ represent the quantum interferences of the photon emissions from them. Because of this interference effect, Fano-type dip structures appear in the plateau region, as shown in Figure 2. Because the coefficients in the summation in Equation (25) include the initial phase of the laser, the spectral profile of the stationary HHG spectrum is also affected, as shown in Figure 2a,b. Hence, it is possible to quantum mechanically control the HHG photon emission by changing the initial phase of the driving laser field.

Within the decay time of an excited state, the resonance state components crucially contribute to the temporal profile of the HHG spectrum. As seen from Equation (25) the first and second terms have opposite signs; hence, the spectral amplitude cancels out at $t = 0$, except for the small branch point effect. As the resonance component decays exponentially with time, the spectral cancellation weakens, approaching the stationary HHG spectrum. In Figure 3, the temporal change of the HHG spectrum is shown, where the components of the resonance and dressed-continuous states are separately depicted. Although spectral cancellation of the resonance and dressed field states has been studied in configurational space for a simple spontaneous emission system [44], the present result demonstrates spectral cancellation in the frequency domain under a strong driving field.

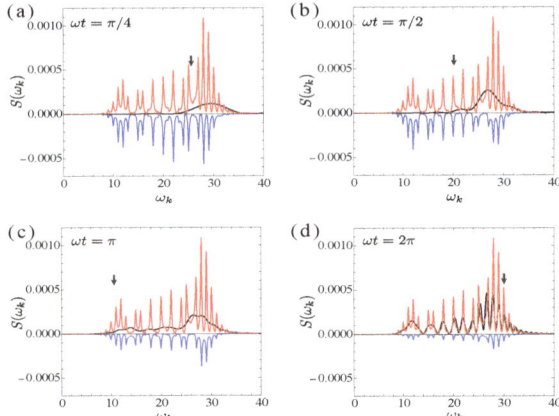

Figure 3. The temporal spectral profile of HHG $S(\omega_k, t)$ (black line) for $\omega t = \pi/4$ (**a**), $\pi/2$ (**b**), π (**c**), and 2π (**d**), where the same parameters of Figure 2a are used. The resonance state $|s_R(k,t)|^2$ and the dressed continuous state $(-1)|s_C(k,t)|^2$ components are depicted by blue and red lines, respectively. The excited state energies $E_e(t) = \Delta_0 + A\cos(\omega t + \theta)$ are indicated by the arrows.

The resonance state components of the HHG not only reduce its intensity in time, but also change its spectral shape as a result of the interference of the Floquet resonance modes, as shown by the blue curves in Figure 3, while the dressed-continuous state components retain their spectral shape. Because of the interference of the Floquet resonance states, the peak position of the HHG spectrum adiabatically follows the temporal excited state energy $E_e(t) = \Delta_0 + A\cos(\omega t + \theta)$, as shown in Figure 4. In time, the adiabatic behavior of the transient HHG asymptotically approaches the stationary HHG spectrum.

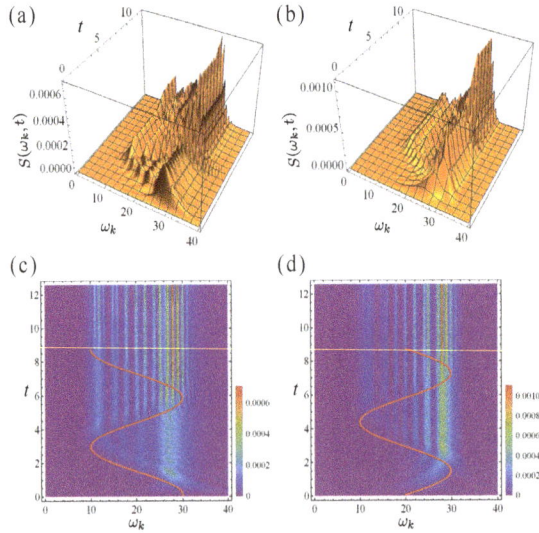

Figure 4. The transient HHG spectrum for $\theta = 0$ (**a**,**c**), and $\theta = -\pi/2$ (**b**,**d**). The parameters are the same as in Figure 2. The red curves in the contour maps (**c**,**d**) indicate $E_e(t)$.

5. Concluding Remarks

We have studied the HHG from a TLS driven by a monochromatic phase-locked laser field in terms of complex spectral analysis for the total system, including the free radiation field, where we have treated the spontaneous HHG photon emission as a coherent quantum process. We have obtained the complex eigenstates of the Floquet Hamiltonian in the extended Hilbert space with the use of the Wannier–Stark basis, going far beyond the ordinary perturbation method. The decomposition of the identity in the extended Hilbert space is represented by the exponentially decaying Floquet resonance states, with complex eigenvalues, and the stable dressed radiation field, with real eigenvalues. These eigenstates are written as a superposition of the different Floquet mode states. The time evolution of the quantum state is then described by the eigenstate expansion of the total system; thereby, the quantum coherence is retained.

We have obtained the analytical expression of the time-frequency resolved spectral amplitude for a HHG single-photon observation. The amplitude is decomposed into the Floquet resonance states and the dressed radiation field, where the former and the latter give the transient and the long-time HHG spectra, respectively. The calculated long-time spectrum shows a typical HHG spectral feature with the plateau and the cutoff, where the spectral cutoff is not proportional to the driving field intensity but the amplitude, as seen in the HHG from solids [17–21]. It is interesting to see that the simple TLS system captures the characteristics of the HHG from solids that possess various electronic excitations. It is likely that the two-level state excitation corresponds to the optically allowed excitation at the Γ point from a valence band to a conduction band [58,59].

Recent experiments have observed multiple plateau structures in the HHG spectrum from solids as a consequence of the quantum interference of the different electronic excitations in solids [18,19]. In this work, we have revealed the other type of quantum interference in the HHG process: the Fano interference between the different Floquet modes, i.e., different high-order harmonics. This quantum interference is caused when a single emitted photon with different frequencies interferes via a common free radiation field, similar to the quantum interference involving different energy states of a single quanta of light [60,61]. This type of interference might be smeared out under a phenomenological assumption.

Within the decay time, the Floquet resonance states contribute to the transient behavior of the HHG spectrum. The transient HHG spectrum changes as if a photon emission occurs from the driven excited state, and the emitted photon energy adiabatically follows the temporal change of the excitation energy. We have shown that this temporal behavior of HHG is understood as a result of the quantum interference between the Floquet resonance states and the dressed field states. Our calculation also shows that the spectrum asymptotically approaches the long-time HHG spectrum, as the resonance state contribution decays exponentially over time.

In the present method, the decay process of the excited state is consistently described with the HHG process because the whole process is treated as a coherent quantum process. We find that the decay rate increases with the amplitude of the driving field, as shown in Figure A1. This is because more Floquet resonance is involved in the decaying process as the amplitude of the driving field increases. In our calculation of the HHG spectrum, we have used the decay rate $2|\mathrm{Im} z_d^{(0)}(a=10)| \simeq 0.2885$, which corresponds to the lifetime of the excited state of 22 fs. (Please note that the radiative lifetime is considered to be much longer than this value [62]). This is much shorter than the pulse width of the driving laser used in the experiments ($\simeq$100 fs) [27]. As long as the pulse width of the driving field is longer than the lifetime of the excited state, the HHG spectrum does not depend on its pulse width.

The conventional theories of the HHG are attempting to solve the time-dependent coupled Schrödinger equation of the electron and the radiation field as an initial value problem [31,37,57]. The problem with these theories is the validity of the Markovian approximation in deriving the kinetic equation of the electron, as its applicability remains uncertain for the far-from-equilibrium situation caused by the driving field [63]. Conversely, the present method attempts to solve the *stationary* eigenvalue problem in the Floquet space, independent of the initial condition [46,53], where the irreversible time-symmetry breaking is not derived as a result of the Markov approximation for the equation but as a rigorous result of the dynamics caused by the resonance singularity [41,42,44,55]. The present method is an extension of the complex eigenvalue problem of the total Hamiltonian to the Floquet space. Because we have dealt with the HHG photon emission as a coherent quantum process including the radiation field, we may study the time evolution of the quantum coherence of a single photon with different modes $\langle \hat{a}_q^\dagger \hat{a}_k \rangle_t$, in terms of which we analyze the creation of the quantum coherence through the nonlinear interaction of the electron and the driving field.

In this work, we have assumed the delta-pulse for the excitation pump pulse, which equally excites all the Floquet modes by white excitation. In a real situation, the excitation pulse has a finite pulse width that is as long as 10~100 fs. It will be interesting to study the effect of a finite pulse width of the excitation, whereby the frequency correlation between the excitation light and the HHG photon can be clarified. Another interesting subject is the competition between the Raman scattering process and the luminescence process in coherent resonant scattering spectroscopy under an intense driving laser field [64–67]. A study of the effect of the excitation pulse width on the HHG is now underway.

Author Contributions: H.Y. did theoretical analyses and numerical calculations, and wrote the first draft. S.T. supervised this program and finalized the manuscript.

Funding: This research was funded by JSPS KAKENHI grants number JP16H04003, No. JP16K05481, and No. JP17K05585.

Acknowledgments: We are very grateful T. Petrosky, K. Noba, K. Kanki, S. Garmon, Y. Kayanuma, and M. Domina for fruitful discussions.

Conflicts of Interest: The authors declare no conflict of interest. The founding sponsors had no role in the design of the study; in the collection, analyses, or interpretation of data; in the writing of the manuscript, and in the decision to publish the results.

Abbreviations

The following abbreviations are used in this manuscript:

HHG   High-Harmonic Generation
TLS   Two-Level System
CEP   Carrier-Envelope-Phase

Appendix A. Hamiltonian of the Driven TLS

In this section, we shall derive the driven TLS Hamiltonian Equation (4) starting from an off-diagonal coupling of the TLS with the radiation field.

We consider spontaneous photon emission from a driven two-level system (TLS) consisting of the ground state $|1\rangle$ and an excited state $|2\rangle$ with the excitation energy $\Delta_0 \equiv E_2 - E_1$. The TLS is driven energetically by a monochromatic phase-locked laser field, as shown in Figure 1. The TLS is excited from the ground state to the excited state at $t = 0$ by a delta-function pulse, followed by spontaneous emission under the energy driving. The Hamiltonian is given by Equation (1):

$$\hat{H}(t) = E_1 |1\rangle\langle 1| + (E_1 + \Delta_0)|2\rangle\langle 2| + \mathcal{E}(t) \left(|1\rangle\langle 2| + |2\rangle\langle 1| \right)$$
$$+ \int \omega_k \hat{a}_k^\dagger \hat{a}_k dk + \lambda \int \mathcal{C}_k \left(|1\rangle\langle 2| + |2\rangle\langle 1| \right) (\hat{a}_k + \hat{a}_k^\dagger) dk \tag{A1}$$

$$\equiv \hat{H}_{\text{TLS}}(t) + \hat{H}_R + \hat{H}_{\text{TLS,R}} \tag{A2}$$

We solve the adiabatic eigenvalue problem of the driven TLS:

$$\hat{H}_{\text{TLS}}(t)|\phi_j(t)\rangle = \lambda_j(t)|\phi_j(t)\rangle \, , \, (j = \pm) \, . \tag{A3}$$

The adiabatic eigenvalues are given by

$$\lambda_\pm(t) = \frac{\Delta_0}{2} \pm \frac{1}{2}\sqrt{\Delta_0^2 + 4\mathcal{E}^2(t)}, \tag{A4}$$

and the corresponding eigenstates are

$$\begin{aligned}|\phi_+(t)\rangle &= \cos\varphi(t)|1\rangle + \sin\varphi(t)|2\rangle, \\ |\phi_-(t)\rangle &= -\sin\varphi(t)|1\rangle + \cos\varphi(t)|2\rangle,\end{aligned} \tag{A5}$$

where

$$\tan\varphi(t) = -\frac{\mathcal{E}(t)}{\Delta_0} \tag{A6}$$

We consider the situation where the energy gap between the ground state and the excited state is much larger than the amplitude of the driving field:

$$\Delta_0 \gg |\mathcal{E}_0|, \tag{A7}$$

and the amplitude is much larger than the energy quanta of the driving field,

$$\mathcal{E}_0 \gg \omega. \tag{A8}$$

Under these conditions, we can rewrite $\hat{H}(t)$ in terms of the adiabatic eigenstates as

$$\begin{aligned}\hat{H}(t) = &-\frac{2\mathcal{E}^2(t)}{\Delta_0}|\phi_-(t)\rangle\langle\phi_-(t)| + \left(\Delta_0 + \frac{2\mathcal{E}^2(t)}{\Delta_0}\right)|\phi_+(t)\rangle\langle\phi_+(t)| \\ &+ \int \omega_k \hat{a}_k^\dagger \hat{a}_k dk + \lambda \int C_k \left(|\phi_+(t)\rangle\langle\phi_-(t)| + |\phi_-(t)\rangle\langle\phi_+(t)|\right)\left(\hat{a}_k + \hat{a}_k^\dagger\right) dk.\end{aligned} \tag{A9}$$

Shifting the energy origin to $-2\mathcal{E}^2(t)/\Delta_0$, we have

$$\begin{aligned}\hat{H}(t) = &\left(\Delta_0 + \frac{4\mathcal{E}_0^2}{\Delta_0} + \frac{2\mathcal{E}_0^2}{\Delta_0}\cos(\omega t + \theta)\right)|\phi_+(t)\rangle\langle\phi_+(t)| \\ &+ \int \omega_k \hat{a}_k^\dagger \hat{a}_k dk + \lambda \int C_k \left(|\phi_+(t)\rangle\langle\phi_-(t)| + |\phi_-(t)\rangle\langle\phi_+(t)|\right)\left(\hat{a}_k + \hat{a}_k^\dagger\right) dk.\end{aligned} \tag{A10}$$

With the use of the rotating wave approximation for the TLS and the free radiation field and defining

$$|e\rangle \equiv |\phi_+(t)\rangle, \quad |g\rangle \equiv |\phi_-(t)\rangle, \quad \frac{2\mathcal{E}_0^2}{\Delta_0} \equiv A, \tag{A11}$$

we have

$$\hat{H}_{\text{ad}}(t) = E_0|g\rangle\langle g| + (E_e + A\cos(\omega t + \theta))|e\rangle\langle e| + \int \omega_k \hat{a}_k^\dagger \hat{a}_k dk + \lambda \int C_k \left(|e\rangle\langle g|\hat{a}_k + |g\rangle\langle e|\hat{a}_k^\dagger\right) dk. \tag{A12}$$

With the definition of Equation (A11), the above conditions for Equations (A7) and (A8) reduce to Equation (3).

Appendix B. Complex Eigenvalue Problem of the Floquet Hamiltonian

Appendix B.1. Discrete Floquet Resonance State

To solve the complex eigenvalue problem for the atom in the $\mathcal{F}$-space, we use the Feshbach—Brilloiun–Wigner projection method with the projection operators

$$\hat{P}_d \equiv \sum_{n=-\infty}^{\infty} |\phi_d^{(n)}\rangle\!\rangle\langle\!\langle\phi_d^{(n)}|,$$

$$\hat{Q}_d \equiv 1 - \hat{P}_d = \sum_{n=-\infty}^{\infty} \int_{-\infty}^{\infty} |k,\kappa_n\rangle\!\rangle\langle\!\langle k,\kappa_n| dk, \tag{A13}$$

where $\hat{P}_d$ is the projection operators on the Stark basis set $\{|\phi_d^{(n)}\rangle\!\rangle\}$. Acting these projection operators on Equation (13), we have a closed form of the eigenvalue problem of the effective Hamiltonian in the P-subspace as

$$\hat{\mathcal{H}}_{\text{eff},d}(z_\alpha^{(n)})\hat{P}_d|\Phi_\alpha^{(n)}\rangle\!\rangle = z_\alpha^{(n)}\hat{P}_d|\Phi_\alpha^{(n)}\rangle\!\rangle, \tag{A14}$$

$$\langle\!\langle\tilde{\Phi}_\alpha^{(n)}|\hat{P}_d\hat{\mathcal{H}}_{\text{eff},d}(z_\alpha^{(n)}) = z_\alpha^{(n)}\langle\!\langle\tilde{\Phi}_\alpha^{(n)}|\hat{P}_d, \tag{A15}$$

where the effective Hamiltonian $\hat{\mathcal{H}}_{\text{eff},d}(z)$ is given by

$$\begin{aligned}\hat{\mathcal{H}}_{\text{eff},d}(z) &= \hat{P}_d\hat{H}_\text{F}\hat{P}_d + \hat{P}_d\hat{H}_\text{F}\hat{Q}_d\frac{1}{z-\hat{Q}_d\hat{H}_\text{F}\hat{Q}_d}\hat{Q}_d\hat{H}_\text{F}\hat{P}_d, \\ &= \sum_{n,n'=-\infty}^{\infty} \chi_{n,n'}^+(z)|\phi_d^{(n)}\rangle\!\rangle\langle\!\langle\phi_d^{(n')}|.\end{aligned} \tag{A16}$$

The dynamical self-energy $\chi_{n,n'}(z)$ is given by

$$\chi_{n,n'}^+(z) \equiv (\Delta_0 + n\omega)\delta_{n,n'} + \lambda^2 \sum_m J_{n-m}(a)J_{n'-m}(a)\sigma^+(z-m\omega)e^{i(n-n')\theta}, \tag{A17}$$

with the self-energy $\sigma^+(z)$ represented as the Cauchy integral,

$$\sigma^+(z) = \int_{-\infty}^{\infty} \frac{\mathcal{C}_k^2}{z-\omega_k} dk. \tag{A18}$$

Because of the resonance singularity in the self-energy, the effective Hamiltonian becomes non-Hermitian with the complex eigenvalues. We would emphasize that the eigenvalue problem is nonlinear because the effective Hamiltonian depends on its own eigenvalue in the eigenvalue problem Equation (A19). When this nonlinearity is taken into account, the eigenvalues of the effective Hamiltonian are the same as those of the total Hamiltonian [41,47].

In our previous work, we have solved the eigenvalue problem of the Floquet Hamiltonian using the continued fraction expansion [46]; here, the strong coupling with the driving field has been fully incorporated in the Stark basis. With the Floquet translational symmetry Equation (9), it is enough to consider the eigenstates for the principal mode $n = 0$. In the week coupling case $\lambda \ll 1$, we can neglect the off-diagonal component of $\chi_{n,n'}(z)$ so that the eigenvalue problem of $\hat{\mathcal{H}}_{\text{eff},d}(z)$ has been solved as

$$\hat{\mathcal{H}}_{\text{eff},d}(z)|\phi_d^{(0)}\rangle\!\rangle = z_d^{(0)}|\phi_d^{(0)}\rangle\!\rangle, \tag{A19}$$

where the complex eigenvalue of the resonance state is obtained by iteratively solving the nonlinear dispersion equation

$$z_d^{(0)} = \chi_{0,0}^+(z_d^{(0)}) = \Delta_0 + \lambda^2 \sum_{m=-\infty}^{+\infty} \sigma^+(z_d^{(0)} - m\omega) J_{-m}^2(a) \,. \tag{A20}$$

The imaginary part of $z_d^{(0)}$ is given by the decay rate of the excited state.

It should be noted that the decay rate is given by the weighted sum of the self-energy with the Bessel function. In Figure A1, we show the calculated results of the decay rate as a function of $a = A/\omega$. It can be shown that if the bandwidth of the free radiation field is on the same order as ω, such as for a photonic crystal, the decay rate may completely vanish as the coherent destruction tunneling from $|e\rangle$ to $|g\rangle$ [46].

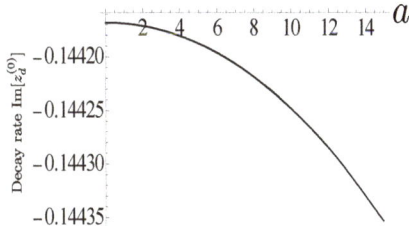

Figure A1. $\text{Im}[z_d^{(0)}]$ as a function of the driving field amplitude for $\lambda = 0.06$, $\Delta_0 = 20$, $\omega = 1$.

Now that we have solved the eigenvalue problem of $\hat{\mathcal{H}}_{\text{eff},d}(z)$ in the P-subspace, the eigenstate of the *total* Hamiltonian with the same eigenvalue is obtained by adding the Q-component.

$$\begin{aligned} |\Phi_d^{(0)}\rangle\!\rangle &= |\phi_d^{(0)}\rangle\!\rangle + \frac{1}{z_d^{(0)} - \hat{Q}_d \hat{H}_F \hat{Q}_d} \hat{Q}_d \hat{H}_F \hat{P}_d |\phi_d^{(0)}\rangle\!\rangle \\ &= \langle\!\langle \phi_d^{(0)} | \Phi_d^{(0)} \rangle\!\rangle \left\{ |\phi_d^{(0)}\rangle\!\rangle + \lambda \sum_{m=-\infty}^{\infty} \int dk C_k \frac{J_{-m}(a) e^{im\theta}}{[z - (\omega_k + m\omega)]_{z=z_d^{(0)}}^+} |k, \kappa_m\rangle\!\rangle \right\} \,, \end{aligned} \tag{A21}$$

where we take the analytic continuation toward these complex poles from the upper complex plane in the Cauchy integral [41]. The left resonance state is similarly obtained as

$$\begin{aligned} \langle\!\langle \tilde{\Phi}_d^{(0)} | &= \langle\!\langle \phi_d^{(0)} | + \langle\!\langle \phi_d^{(0)} | \hat{P}_d \hat{H}_F \hat{Q}_d \frac{1}{z_d^{(0)} - \hat{Q}_d \hat{H}_F \hat{Q}_d} \\ &= \langle\!\langle \tilde{\Phi}_d^{(0)} | \phi_d^{(0)} \rangle\!\rangle \left\{ \langle\!\langle \phi_d^{(0)} | + \lambda \sum_{m=-\infty}^{\infty} \int dk C_k \frac{J_{-m}(a) e^{-im\theta}}{[z - (\omega_k + m\omega)]_{z=z_d^{(0)}}^+} \langle\!\langle k, \kappa_m | \right\} \,, \end{aligned} \tag{A22}$$

where it should be noted that we need to take the same direction in the analytic continuation of the Cauchy integral as with the right-resonance state to obtain the same complex eigenvalue.

The normalization constants of $\langle\!\langle \phi_d^{(0)} | \Phi_d^{(0)} \rangle\!\rangle$ and $\langle\!\langle \tilde{\Phi}_d^{(0)} | \phi_d^{(0)} \rangle\!\rangle$ are determined so as to satisfy the bi-normalization condition of

$$\langle\!\langle \tilde{\Phi}_d | \Phi_d \rangle\!\rangle = 1 \,. \tag{A23}$$

With use of Equations (A17), (A21), and (A22), the bi-normalization condition reads

$$1 = \langle\!\langle \phi_d^{(0)} | \Phi_d^{(0)} \rangle\!\rangle \langle\!\langle \widetilde{\Phi}_d^{(0)} | \phi_d^{(0)} \rangle\!\rangle \left\{ 1 + \lambda^2 \sum_m \int dk \mathcal{C}_k^2 \frac{J_{-m}^2(a)}{([z-(\omega_k+m\omega)]^+)^2} \right\} \quad \text{(A24)}$$

$$= \langle\!\langle \phi_d^{(0)} | \Phi_d^{(0)} \rangle\!\rangle \langle\!\langle \widetilde{\Phi}_d^{(0)} | \phi_d^{(0)} \rangle\!\rangle \left(1 - \frac{d}{dz}\chi_{0,0}(z)\bigg|_{z=z_d^{(0)}} \right). \quad \text{(A25)}$$

Appendix B.2. Dressed Radiation Field

We take the same procedure for the radiation field of the Floquet Hamiltonian. The projection for the continuum state is taken as

$$\hat{P}_k \equiv \sum_n |k,\kappa_n\rangle\!\rangle\langle\!\langle k,\kappa_n|,$$

$$\hat{Q}_k \equiv \sum_n \left(|\phi_d^{(n)}\rangle\!\rangle\langle\!\langle \phi_d^{(n)}| + \sum_{k'(\neq k)} |k',\kappa_n\rangle\!\rangle\langle\!\langle k',\kappa_n| \right). \quad \text{(A26)}$$

The effective Hamiltonian can be obtained as with the resonant state,

$$\hat{\mathcal{H}}_{\text{eff},k} = \sum_n (\epsilon_k + n\omega)|k,\kappa_n\rangle\!\rangle\langle\!\langle k,\kappa_n|. \quad \text{(A27)}$$

Therefore, we get the eigenvalue problem of the effective Hamiltonian (A27)

$$\hat{\mathcal{H}}_{\text{eff},k}|k,\kappa_n\rangle\!\rangle = (\epsilon_k + n\omega)|k,\kappa_n\rangle\!\rangle, \quad \text{(A28a)}$$

$$\langle\!\langle k,\kappa_n|\hat{\mathcal{H}}_{\text{eff},k} = (\epsilon_k + n\omega)\langle\!\langle k,\kappa_n|. \quad \text{(A28b)}$$

We have obtained the expression for the dressed radiation field as

$$|\Phi_k^{(n)}\rangle\!\rangle = |k,\kappa_n\rangle\!\rangle + \frac{1}{\omega_k + n\omega - \hat{Q}_k\hat{H}_F\hat{Q}_k}\hat{Q}_k\hat{H}_F\hat{P}_k|k,\kappa_n\rangle\!\rangle,$$

$$= |k,\kappa_n\rangle\!\rangle + \lambda\mathcal{C}_k \sum_l J_{l-n}(a)e^{i(l-n)\theta}\sum_m G_{m,l}^{k(dd)}(\omega_k+n\omega)|\phi_d^{(m)}\rangle\!\rangle \quad \text{(A29)}$$

$$+ \lambda\mathcal{C}_k \sum_l J_{l-n}(a)e^{i(l-n)\theta}\sum_m \sum_{k'(\neq k)} \frac{\lambda V_{k'}}{(\omega_k+n\omega)-(\omega_{k'}+m\omega)+i0^+}\sum_{m'} J_{m'-m}(a) G_{m',l}^{k(dd)}(\omega_k+n\omega)|k',\kappa_m\rangle\!\rangle,$$

where Green's function $G_{m,l}^{k(dd)}$ is determined by

$$G_{m,l}^{k(dd)}(z) \equiv \langle\!\langle \phi_d^{(m)}|\frac{1}{z-\hat{Q}_k\hat{H}_F\hat{Q}_k}|\phi_d^{(l)}\rangle\!\rangle = \frac{1}{z-(\Delta_0+l\omega)}\left\{\delta_{m,l} + \sum_{m'} \chi_{m,m'}(z) G_{m',l}^{k(dd)}(z)\right\}. \quad \text{(A30)}$$

In the weak coupling case, we may neglect the off-diagonal terms, approximating

$$G_{m,l}^{k(dd)}(z) \simeq \frac{\delta_{m,l}^{\text{Kr}}}{z - \chi_{l,l}(z)}. \quad \text{(A31)}$$

Therefore, we finally obtain

$$|\Phi_k^{(n)}\rangle\rangle = |k,\kappa_n\rangle\rangle + \sum_l \frac{\lambda \mathcal{C}_k J_{l-n}(a) e^{i(l-n)\theta}}{\omega_k + n\omega + i0^+ - \chi_{l,l,DL}^+(\omega_k + n\omega)}$$

$$\times \left[|\phi_d^{(l)}\rangle\rangle + \sum_{k'(\neq k)} \sum_m \frac{\lambda \mathcal{C}_{k'} J_{l-m}(a) e^{-i(l-m)\theta}}{\omega_k + n\omega + i0^+ - (\omega_{k'} + m\omega)}|k',\kappa_m\rangle\rangle\right], \quad \text{(A32)}$$

$$\langle\langle\widetilde{\Phi}_k^{(n)}| = \langle\langle k,\kappa_n| + \sum_l \frac{\lambda \mathcal{C}_k J_{l-n}(a) e^{-i(l-n)\theta}}{\omega_k + n\omega + i0^- - \chi_{l,l}^-(\omega_k + n\omega)}$$

$$\times \left[\langle\langle\phi_d^{(l)}| + \sum_{k'(\neq k)} \sum_m \frac{\lambda \mathcal{C}_{k'} J_{l-m}(a) e^{i(l-m)\theta}}{\omega_k + n\omega + i0^- - (\omega_{k'} + m\omega)} \langle\langle k',\kappa_m|\right], \quad \text{(A33)}$$

which give Equation (18).

References

1. Corkum, P.B.; Krausz, F. Attosecond science. *Nat. Phys.* **2007**, *3*, 381–387. [CrossRef]
2. Krausz, F.; Ivanov, M. Attosecond physics. *Rev. Mod. Phys.* **2009**, *81*, 163–234. [CrossRef]
3. Corkum, P.; Burnett, N.; Brunel, F. Above-threshold ionization in the long-wavelength limit. *Phys. Rev. Lett.* **1989**, *62*, 1259–1262. [CrossRef] [PubMed]
4. Mohideen, U.; Sher, M.; Tom, H.; Aumiller, G.; Wood, O.; Freeman, R.; Boker, J.; Bucksbaum, P. High intensity above-threshold ionization of He. *Phys. Rev. Lett.* **1993**, *71*, 509–512. [CrossRef] [PubMed]
5. Corkum, P.B. Plasma perspective on strong field multiphoton ionization. *Phys. Rev. Lett.* **1993**, *71*, 1994–1997. [CrossRef] [PubMed]
6. Schafer, K.J.; Kulander, K.C. High Harmonic Generation from Ultrafast Pump Lasers. *Phys. Rev. Lett.* **1997**, *78*, 638–641. [CrossRef]
7. Krause, J.; Schafer, K.; Kulander, K. High-order harmonic generation from atoms and ions in the high intensity regime. *Phys. Rev. Lett.* **1992**, *68*, 3535–3538. [CrossRef] [PubMed]
8. Brabec, T.; Krausz, F. Intense few-cycle laser fields: Frontiers of nonlinear optics. *Rev. Mod. Phys.* **2000**, *72*, 545–591. [CrossRef]
9. Popmintchev, T.; Chen, M.C.; Arpin, P.; Murnane, M.M.; Kapteyn, H.C. The attosecond nonlinear optics of bright coherent X-ray generation. *Nat. Photonics* **2010**, *4*, 822–832. [CrossRef]
10. Popmintchev, T.; Chen, M.C.; Popmintchev, D.; Arpin, P.; Brown, S.; Ališauskas, S.; Andriukaitis, G.; Balčiunas, T.; Mücke, O.D.; Pugzlys, A.; et al. Bright Coherent Ultrahigh Harmonics in the keV X-ray Regime from Mid-Infrared Femtosecond Lasers. *Science* **2012**, *336*, 1287–1291. [CrossRef] [PubMed]
11. Kim, K.T.; Zhang, C.; Ruchon, T.; Hergott, J.F.; Auguste, T.; Villeneuve, D.M.; Corkum, P.B.; Quéré, F. Photonic streaking of attosecond pulse trains. *Nat. Photonics* **2013**, *7*, 651–656. [CrossRef]
12. Gruson, V.; Barreau, L.; Jiménez-Galan, Á.; Risoud, F.; Caillat, J.; Maquet, A.; Carré, B.; Lepetit, F.; Hergott, J.F.; Ruchon, T.; et al. Attosecond dynamics through a Fano resonance: Monitoring the birth of a photoelectron. *Science* **2016**, *354*, 734–738. [CrossRef] [PubMed]
13. Kaldun, A.; Blättermann, A.; Stooß, V.; Donsa, S.; Wei, H.; Pazourek, R.; Nagele, S.; Ott, C.; Lin, C.D.; Burgdörfer, J.; et al. Observing the ultrafast buildup of a Fano resonance in the time domain. *Science* **2016**, *354*, 738–741. [CrossRef] [PubMed]
14. Cirelli, C.; Marante, C.; Heuser, S.; Petersson, C.; Galán, Á.; Argenti, L.; Zhong, S.; Busto, D.; Isinger, M.; Nandi, S.; et al.. Anisotropic photoemission time delays close to a Fano resonance. *Nat. Commun.* **2018**, *9*. [CrossRef] [PubMed]
15. Nisoli, M.; Decleva, P.; Calegari, F.; Palacios, A.; Martín, F. Attosecond Electron Dynamics in Molecules. *Chem. Rev.* **2017**, *117*, 10760–10825. [CrossRef] [PubMed]
16. Young, L.; Ueda, K.; Gühr, M.; Bucksbaum, P.; Simon, M.; Mukamel, S.; Rohringer, N.; Prince, K.; Masciovecchio, C.; Meyer, M.; et al. Roadmap of ultrafast x-ray atomic and molecular physics. *J. Phys. B* **2018**, *51*. [CrossRef]
17. Ghimire, S.; Dichiara, A.; Sistrunk, E.; Agostini, P.; Dimauro, L.; Reis, D. Observation of high-order harmonic generation in a bulk crystal. *Nat. Phys.* **2011**, *7*, 138–141. [CrossRef]

18. Schubert, O.; Hohenleutner, M.; Langer, F.; Urbanek, B.; Lange, C.; Huttner, U.; Golde, D.; Meier, T.; Kira, M.; Koch, S.W.; et al. Sub-cycle control of terahertz high-harmonic generation by dynamical Bloch oscillations. *Nat. Photonics* **2014**, *8*, 119–123. [CrossRef]
19. Hohenleutner, M.; Langer, F.; Schubert, O.; Knorr, M.; Huttner, U.; Koch, S.; Kira, M.; Huber, R. Real-time observation of interfering crystal electrons in high-harmonic generation. *Nature* **2015**, *523*, 572–575. [CrossRef] [PubMed]
20. Vampa, G.; Hammond, T.; Thiré, N.; Schmidt, B.; Légaré, F.; McDonald, C.; Brabec, T.; Corkum, P. Linking high harmonics from gases and solids. *Nature* **2015**, *522*, 462–464. [CrossRef] [PubMed]
21. Ndabashimiye, G.; Ghimire, S.; Wu, M.; Browne, D.A.; Schafer, K.J.; Gaarde, M.B.; Reis, D.A. Solid-state harmonics beyond the atomic limit. *Nature* **2016**, *534*, 520–523. [CrossRef] [PubMed]
22. Bauer, D.; Hansen, K.K. High-Harmonic Generation in Solids with and without Topological Edge States. *Phys. Rev. Lett.* **2018**, *120*, 177401. [CrossRef] [PubMed]
23. McDonald, C.R.; Amin, K.S.; Aalmalki, S.; Brabec, T. Enhancing High Harmonic Output in Solids through Quantum Confinement. *Phys. Rev. Lett.* **2017**, *119*, 183902. [CrossRef] [PubMed]
24. Luu, T.T.; Garg, M.; Kruchinin, S.Y.; Moulet, A.; Hassan, M.T.; Goulielmakis, E. Extreme ultraviolet high-harmonic spectroscopy of solids. *Nature* **2015**, *521*, 498–502. [CrossRef] [PubMed]
25. You, Y.S.; Yin, Y.; Wu, Y.; Chew, A.; Ren, X.; Zhuang, F.; Gholam-Mirzaei, S.; Chini, M.; Chang, Z.; Ghimire, S. High-harmonic generation in amorphous solids. *Nat. Commun.* **2017**, *8*, 724. [CrossRef] [PubMed]
26. Zaks, B.; Liu, R.B.; Sherwin, M.S. Experimental observation of electron–hole recollisions. *Nature* **2012**, *483*, 580–583. [CrossRef] [PubMed]
27. Langer, F.; Hohenleutner, M.; Schmid, C.P.; Poellmann, C.; Nagler, P.; Korn, T.; Schüller, C.; Sherwin, M.S.; Huttner, U.; Steiner, J.T.; et al. Lightwave-driven quasiparticle collisions on a subcycle timescale. *Nature* **2016**, *533*, 225–229. [CrossRef] [PubMed]
28. Uchida, K.; Otobe, T.; Mochizuki, T.; Kim, C.; Yoshita, M.; Akiyama, H.; Pfeiffer, L.N.; West, K.W.; Tanaka, K.; Hirori, H. Subcycle Optical Response Caused by a Terahertz Dressed State with Phase-Locked Wave Functions. *Phys. Rev. Lett.* **2016**, *117*, 277402. [CrossRef] [PubMed]
29. Uchida, K.; Otobe, T.; Mochizuki, T.; Kim, C.; Yoshita, M.; Tanaka, K.; Akiyama, H.; Pfeiffer, L.N.; West, K.W.; Hirori, H. Coherent detection of THz-induced sideband emission from excitons in the nonperturbative regime. *Phys. Rev. B* **2018**, *97*, 165122. [CrossRef]
30. Luu, T.; Wörner, H. High-order harmonic generation in solids: A unifying approach. *Phys. Rev. B* **2016**, *94*. [CrossRef]
31. Milonni, P.W. *The Quantum Vacuum: An Introduction to Quantum Electrodynamics*; Academic Press: New York, NY, USA, 1994.
32. Mukamel, S. *Principles of Nonlinear Optical Spectroscopy*; Oxford Series in Optical and Imaging Sciences; Oxford University Press: Oxford, UK, 1995.
33. Kulander, K.C.; Schafer, K.J. Time-dependent calculations of electron and photon emission from an atom in an intense laser field. In *Atoms and Molecules in Intense Fields*; Springer: Berlin/Heidelberg, Germany, 1997; pp. 149–172.
34. Tancogne-Dejean, N.; Mücke, O.D.; Kärtner, F.X.; Rubio, A. Impact of the Electronic Band Structure in High-Harmonic Generation Spectra of Solids. *Phys. Rev. Lett.* **2017**, *118*, 087403. [CrossRef] [PubMed]
35. Vampa, G.; McDonald, C.R.; Orlando, G.; Corkum, P.B.; Brabec, T. Semiclassical analysis of high harmonic generation in bulk crystals. *Phys. Rev. B* **2015**, *91*, 064302. [CrossRef]
36. Golde, D.; Meier, T.; Koch, S.W. High harmonics generated in semiconductor nanostructures by the coupled dynamics of optical inter- and intraband excitations. *Phys. Rev. B* **2008**, *77*, 075330. [CrossRef]
37. Weisskopf, V.; Wigner, E. Berechnung der natürlichen Linienbreite auf Grund der Diracschen Lichttheorie. *Zeitschrift für Physik* **1930**, *63*, 54–73. [CrossRef]
38. Heitler, W. *The Quantum Theory of Radiation*; International Series of Monographs on Physics; Oxford University Press: Oxford, UK, 1947.
39. Compagno, G.; Passante, R.; Persico, F. *Atom-Field Interactions and Dressed Atoms*; Cambridge Studies in Modern Optics; Cambridge University Press: Cambridge, UK, 1995.
40. Cohen-Tannoudji, C.; Dupont-Roc, J.; Grynberg, G. Atom—Photon Interactions: Basic Process and Applications. In *Atom—Photon Interactions*; Wiley: Weinheim, Germany, 2008.

41. Petrosky, T.; Prigogine, I.; Tasaki, S. Quantum theory of non-integrable systems. *Phys. A Stat. Mech. Appl.* **1991**, *173*, 175–242. [CrossRef]
42. Petrosky, T.; Prigogine, I. The Liouville space extension of quantum mechanics. *Adv. Chem. Phys.* **1997**, *99*, 1–120.
43. Ordonez, G.; Petrosky, T.; Prigogine, I. Quantum transitions and dressed unstable states. *Phys. Rev. A* **2001**, *63*, 052106. [CrossRef]
44. Petrosky, T.; Ordonez, G.; Prigogine, I. Space-time formulation of quantum transitions. *Phys. Rev. A* **2001**, *64*, 062101/1–062101/21. [CrossRef]
45. Tanaka, S.; Garmon, S.; Petrosky, T. Nonanalytic enhancement of the charge transfer from adatom to one-dimensional semiconductor superlattice and optical absorption spectrum. *Phys. Rev. B* **2006**, *73*, 115340. [CrossRef]
46. Yamada, N.; Noba, K.I.; Tanaka, S.; Petrosky, T. Dynamical suppression and enhancement of instability for an unstable state by a periodic external field. *Phys. Rev. B* **2012**, *86*. [CrossRef]
47. Tanaka, S.; Garmon, S.; Kanki, K.; Petrosky, T. Higher-order time-symmetry-breaking phase transition due to meeting of an exceptional point and a Fano resonance. *Phys. Rev. A* **2016**, *94*, 022105. [CrossRef]
48. Fukuta, T.; Garmon, S.; Kanki, K.; Noba, K.i.; Tanaka, S. Fano absorption spectrum with the complex spectral analysis. *Phys. Rev. A* **2017**, *96*, 052511. [CrossRef]
49. Cohen-Tannoudji, C.; Dupont-Roc, J.; Grynberg, G. *Atom-Photon Interactions: Basic Processes and Applications*; Wiley: Hoboken, NJ, USA, 1998.
50. Grifoni, M.; Hänggi, P. Driven quantum tunneling. *Phys. Rep.* **1998**, *304*, 229–354. [CrossRef]
51. Shirley, J. Solution of the schrödinger equation with a hamiltonian periodic in time. *Phys. Rev.* **1965**, *138*, B979–B987. [CrossRef]
52. Sambe, H. Steady states and quasienergies of a quantum-mechanical system in an oscillating field. *Phys. Rev. A* **1973**, *7*, 2203–2213. [CrossRef]
53. Yamane, H.; Tanaka, S.; Domina, M.; Passante, R.; Petrosky, T. Analysis of high-harmonic generation in terms of complex Floquet spectral analysis. In Proceedings of the 2017 Progress in Electromagnetics Research Symposium, Singapore, 19–22 November 2017; pp. 1437–1444.
54. Friedrichs, K. On the perturbation of continuous spectra. *Commun. Pure Appl. Math.* **1948**, *1*, 361–406. [CrossRef]
55. Petrosky, T.; Ordonez, G.; Prigogine, I. Quantum transitions and nonlocality. *Phys. Rev. A* **2000**, *62*, 042106. [CrossRef]
56. Glauber, R.J. Optical Coherence and Photon Statistics. In *Quantum Theory of Optical Coherence*; de Witt, C., Blandin, A., Cohen-Tannoudji, C., Eds.; WileyWiley: Hoboken, NJ, USA, 2007; Chapter 2, pp. 23–182.
57. Carmichael, H. *Statistical Methods in Quantum Optics 1: Master Equations and Fokker-Planck Equations*; Theoretical and Mathematical Physics; Springer: Berlin/Heidelberg, Germany, 2013.
58. Korbman, M.; Kruchinin, S.Y.; Yakovlev, V.S. Quantum beats in the polarization response of a dielectric to intense few-cycle laser pulses. *New J. Phys.* **2013**, *15*, 013006. [CrossRef]
59. Wu, M.; Browne, D.A.; Schafer, K.J.; Gaarde, M.B. Multilevel perspective on high-order harmonic generation in solids. *Phys. Rev. A* **2016**, *94*, 063403. [CrossRef]
60. Clemmen, S.; Farsi, A.; Ramelow, S.; Gaeta, A.L. Ramsey Interference with Single Photons. *Phys. Rev. Lett.* **2016**, *117*, 223601. [CrossRef] [PubMed]
61. Whiting, D.J.; Šibalić, N.; Keaveney, J.; Adams, C.S.; Hughes, I.G. Single-Photon Interference due to Motion in an Atomic Collective Excitation. *Phys. Rev. Lett.* **2017**, *118*, 253601. [CrossRef] [PubMed]
62. Tripathi, L.N.; Iff, O.; Betzold, S.; Dusanowski, Ł.; Emmerling, M.; Moon, K.; Lee, Y.J.; Kwon, S.H.; Höfling, S.; Schneider, C. Spontaneous Emission Enhancement in Strain-Induced WSe2 Monolayer-Based Quantum Light Sources on Metallic Surfaces. *ACS Photonics* **2018**, *5*, 1919–1926. [CrossRef]
63. Browne, D.E.; Keitel, C.H. Resonance fluorescence in intense laser fields. *J. Mod. Opt.* **2000**, *47*, 1307–1337. [CrossRef]
64. Hizhyakov, V.; Tehver, I. Theory of Resonant Secondary Radiation due to Impurity Centres in Crystals. *Phys. Status Solidi B* **1967**, *21*, 755–768. [CrossRef]
65. Toyozawa, Y. Resonance and Relaxation in Light Scattering. *J. Phys. Soc. Jpn.* **1976**, *41*, 400–411. [CrossRef]

66. Kotani, A. On the Relationship between Resonant Light Scattering and Luminescence—A Singular Aspect in Localized Electron-Phonon System with Linear Interaction. *J. Phys. Soc. Jpn.* **1978**, *44*, 965–972. [CrossRef]
67. Kayamura, Y. Resonant Secondary Radiation in Strongly Coupled Localized Electron-Phonon System. *J. Phys. Soc. Jpn.* **1988**, *57*, 292–301. [CrossRef]

© 2018 by the authors. Licensee MDPI, Basel, Switzerland. This article is an open access article distributed under the terms and conditions of the Creative Commons Attribution (CC BY) license (http://creativecommons.org/licenses/by/4.0/).

Article

Three-Body Dispersion Potentials Involving Electric Octupole Coupling

Stefan Yoshi Buhmann [1,2] and A. Salam [1,2,3],*

1. Physikalisches Institut, Albert-Ludwigs-Universität Freiburg, Herman-Herder-Straße 3, 79104 Freiburg, Germany; stefan.buhmann@physik.uni-freiburg.de
2. Freiburg Institute for Advanced Studies (FRIAS), Albert-Ludwigs-Universität Freiburg, Albertstraße 19, 79104 Freiburg, Germany
3. Department of Chemistry, Wake Forest University, Winston-Salem, NC 27109, USA
* Correspondence: salama@wfu.edu; Tel.: +1-336-758-3713

Received: 14 July 2018; Accepted: 13 August 2018; Published: 16 August 2018

Abstract: Non-pairwise additive three-body dispersion potentials dependent upon one or more electric octupole moments are evaluated using the theory of molecular quantum electrodynamics. To simplify the perturbation theory calculations, an effective two-photon interaction Hamiltonian operator is employed. This leads to only third-order theory being required to evaluate energy shifts instead of the usual sixth-order formula, and the summation over six time-ordered sequences of virtual photon creation and annihilation events. Specific energy shifts computed include DD-DD-DO, DD-DO-DO, DO-DO-DO, and DD-DO-OO terms, where D and O are electric dipole and octupole moments, respectively. The formulae obtained are applicable to an arbitrary arrangement of the three particles, and we present explicit results for the equilateral triangle and collinear configurations, which complements the recently published DD-DD-OO potential. In this last case it was found that the contribution from the octupole weight-1 term could be viewed as a higher-order correction to the triple-dipole dispersion potential DD-DD-DD. In a similar fashion the octupole moment is decomposed into its irreducible components of weights-1 and -3, enabling insight to be gained into the potentials obtained in this study. Dispersion interaction energies proportional to mixed dipole-octupole polarisabilities, for example, are found to depend only on the weight-1 octupole moment for isotropic species and are retarded. Additional approximations are necessary in the evaluation of wave vector integrals for these cases in order to yield energy shifts that are valid in the near-zone.

Keywords: molecular quantum electrodynamics; dispersion potentials; octupole coupling

1. Introduction

It is well-known that the contribution to the total interaction energy arising from the non-pairwise additive three-body van der Waals dispersion potential is very small [1]. In a few cases, however, it can be significant enough to warrant consideration and eventual inclusion [2,3], as sometimes also occurs in Casimir–Polder [4] and Casimir–Lifshitz [5] interactions. One of the best-known examples is provided by the crystal energy of rare gas atoms, in which the closely packed structure found in the solid phase is stabilised by an additional 5–10% of the total energy when the triple-dipole dispersion energy term is accounted for. Another area, of wider applicability, in which many-body effects are known to be important, is the design and computation of inter-particle potential energy functions. This is motivated by the desire for ever-greater accuracy, and incorporating a host of chemical and physical phenomena in order to improve the transferability of the surface generated. Recent efforts have been spurred on by advances in ultracold spectroscopy and dynamics, especially three-atom and atom-molecule collisional processes involving alkali and alkaline Earth elements [6]. In a similar vein, interaction potentials among three Group 8 elements have been studied [7]. In this last work, the electric dipole approximation was relaxed and dispersion energies in which the perturbation operator included electric quadrupole and octupole coupling were computed. These couplings were taken to be static, and therefore applicable in the near-zone, that is, for separation distances between pairs of species that are a lot smaller than characteristic reduced transition wavelengths in atomic and molecular systems. Because the signal propagating between individual centres is instantaneous in this approximation, the coupling is unphysical and unable to treat dispersion interactions at larger separation distances, where the finite speed of light must be properly accounted for since inter-atomic/molecular forces are fundamentally a manifestation of electromagnetic effects. This means that the correct form of perturbation operator coupling centres should represent the intrinsic electrodynamic nature of the interaction between particles.

A physical theory that furnishes a description in terms of photons and includes the electromagnetic field from the outset is quantum electrodynamics (QED) [8]. Its non-relativistic formulation applicable to slowly moving bound electrons in atoms and molecules, and termed *molecular* QED, has been rigorously developed and applied with success to linear and nonlinear spectroscopic processes and inter-particle interactions [9–12]. A *macroscopic* version [13] has been used to calculate dispersion forces between objects such as plates, surfaces, slabs, spheres and bodies with other geometries so as to better understand Casimir effects, as well as Casimir–Polder and van der Waals forces that respectively involve one or two microscopic particles interacting with a body. A key difference between QED and various semi-classical theories of radiation-matter interaction is that both the electromagnetic field and the system of particles is subject to quantum mechanical laws in the former, with light taken to be a classical external perturbation in the latter treatment.

Very recently, molecular QED theory has been applied to calculate higher-electric multipole moment contributions to the dispersion energy shift between three particles [14–16]. These have included potentials between two electric dipole polarisable species, and a third that is either electric quadrupole or electric octupole polarisable, as well as the interaction energy of an electric dipole polarisable molecule with two electric quadrupole polarisable molecules. The potentials obtained hold for all separation distances outside the region of wave function overlap and extending out to infinity, for oriented and isotropic systems. Approximating the speed of light to be infinite resulted in the reproduction of the potentials computed using static multipolar couplings, applicable in the near-zone [6,7]. Retardation corrected forms, applicable at very long-range, were obtained on taking the far-zone asymptote, in which virtual photons with low frequency contribute most significantly in mediating the interaction. In addition to formulae being given for arbitrary triangular arrangements of the three bodies, energy shifts for particular configurations were evaluated. These included equilateral triangle geometry, and when all three particles lie on the same line. Taken together, these works extended the leading contribution to the non-pairwise additive dispersion energy, namely the retarded triple dipole dispersion potential, first calculated by Aub and Zienau in 1960 [17], and rederived by others [18–23], and extended to systems containing excited atoms [24,25]. It is worth pointing out that these genuine non-pairwise additive three-body contributions to the dispersion interaction energy are distinct from the sum of the three pair dispersion energy shifts that also contribute to the total interaction energy in the pairwise additive approximation. The three-body term is expected to grow in importance as the density of the ensemble increases. Of historical interest is that the non-retarded result for atoms in the ground state, obtained via third-order perturbation theory and static dipolar coupling operators, was first computed by Axilrod and Teller [26], and by Muto [27]. Their energy shift exhibited inverse cubic separation distance dependence on each inter-particle displacement, and hence inverse ninth power law behaviour in the case of an equilateral triangle set up. Results were also given for right-angled triangle and collinear geometries. Interestingly, the sign of three-body dispersion potentials is geometry dependent. Much later this non-retarded three-body shift was related to the polarisation [28].

A particularly interesting feature arises in dispersion potentials when the electric octupole coupling term is included in the perturbation operator [14,15,29,30], and is revealed on decomposing the octupole moment into irreducible components of weights-1 and -3. It was found that the weight-1 term is only present when the interaction is retarded, unlike the weight-3 component, which appears in both static and retarded couplings. Furthermore, because the weight-1 octupole moment has three independent components, and transformation properties similar to that of a vector, the weight-1 dependent part of the dipole-dipole-octupole energy shift was interpreted as a higher-order correction to the triple dipole dispersion potential. This aspect was actually first noticed on computation of the pair dispersion potential between an electric dipole polarisable molecule and an electric octupole polarisable one [29], and in a recent study of dispersion interaction energies involving a DD-DO, and a DO-DO pair [30]. Similar features were also found in the rate of resonant transfer of excitation energy between an electric dipole donor moiety and an electric octupole acceptor species [31]. While the electric octupole moment is a factor of the fine structure constant squared smaller than the dipole moment, and gives rise to weak spectroscopic signals, selection rules will ultimately determine whether transitions vanish or not.

In the case of three bodies that are in fixed orientation with respect to each other, the dispersion energy when two of them, A and B, are electric dipole polarisable, and the third, C, is electric octupole polarisable, is given by [14,15]

$$\Delta E^{DD-DD-OO} = -\frac{\hbar c}{64\pi^4 \varepsilon_0^3} \int_0^\infty du\, \alpha_{ij}^{DD}(A; iu) \alpha_{kl}^{DD}(B; iu) \alpha_{m_1 m_2 m_3 n_1 n_2 n_3}^{OO}(C; iu) \\ \times L_{km_1 m_2 m_3}(ua) L_{ln_1 n_2 n_3}(ub) F_{jl}(uc). \quad (1)$$

In expression (1), the pure electric dipole polarisability tensor of species A evaluated at the imaginary frequency $\omega = icu$ is defined by

$$\alpha_{ij}^{DD}(A; iu) = \sum_x \left\{ \frac{\mu_i^{0x}(A)\mu_j^{x0}(A)}{E_{x0} - i\hbar cu} + \frac{\mu_j^{0x}(A)\mu_i^{x0}(A)}{E_{x0} + i\hbar cu} \right\}, \quad (2)$$

where $\mu_i^{0x}(A) = \langle 0| \mu_i(A) |x\rangle$ is the i-th Cartesian component of the transition electric dipole moment, $\vec{\mu}(A)$, taken between ground $|0\rangle$ and virtual excited state $|x\rangle$ of particle A, with difference in energy between these states denoted by $E_{x0} = E_x - E_0$. A similar definition holds for the electric dipole polarisability tensor of particle B, whose complete set of intermediate electronic levels is denoted by $|y\rangle$. The Roman sub-scripts denote Cartesian tensor components in the space-fixed frame of reference. Einstein summation convention is assumed for repeating indices. Analogously to formula (2), the sixth-rank pure electric octupole polarisability tensor of molecule C is defined as

$$\alpha_{ijklmn}^{OO}(C; iu) = \sum_z \left\{ \frac{O_{ijk}^{0z}(C) O_{lmn}^{z0}(C)}{E_{z0} - i\hbar cu} + \frac{O_{lmn}^{0z}(C) O_{ijk}^{z0}(C)}{E_{z0} + i\hbar cu} \right\}, \quad (3)$$

expressed in terms of transition electric octupole moments, whose reducible component operator form is defined as

$$O_{ijk}(C) = -\frac{e}{3!}(\vec{q} - \vec{R}_C)_i (\vec{q} - \vec{R}_C)_j (\vec{q} - \vec{R}_C)_k, \quad (4)$$

where $-e$ is the electronic charge, $\vec{q}$ is the generalised electron coordinate, and $\vec{R}_C$ is the point in centre C about which the multipolar expansion is made. Virtual electronic states of C are designated by $|z\rangle$. Also appearing in the result (1) are two geometric tensors, $F_{ij}(uR)$ and $L_{ijkl}(uR)$, which will feature later on in this work, and whose definitions are now conveniently introduced as

$$F_{ij}(uR) \equiv (-\nabla^2 \delta_{ij} + \nabla_i \nabla_j) \frac{e^{-uR}}{R} = -[(\delta_{ij} - \hat{R}_i \hat{R}_j) u^2 R^2 + (\delta_{ij} - 3\hat{R}_i \hat{R}_j)(uR+1)] \frac{e^{-uR}}{R^3}, \quad (5)$$

and

$$
\begin{aligned}
L_{ijkl}(uR) &\equiv (-\nabla^2 \delta_{ij} + \nabla_i \nabla_j) \nabla_k \nabla_l \frac{e^{-uR}}{R} \\
&= \{[(\delta_{ij}\delta_{kl} + \delta_{ik}\delta_{jl} + \delta_{il}\delta_{jk}) \\
&\quad - 5(\delta_{ij}\hat{R}_k\hat{R}_l + \delta_{ik}\hat{R}_j\hat{R}_l + \delta_{il}\hat{R}_j\hat{R}_k + \delta_{jk}\hat{R}_i\hat{R}_l + \delta_{jl}\hat{R}_i\hat{R}_k + \delta_{kl}\hat{R}_i\hat{R}_j) \\
&\quad + 35\hat{R}_i\hat{R}_j\hat{R}_k\hat{R}_l](3 + 3uR + u^2R^2) + [\delta_{ij}(\delta_{kl} - 3\hat{R}_k\hat{R}_l) \\
&\quad - (\delta_{ij}\hat{R}_k\hat{R}_l + \delta_{ik}\hat{R}_j\hat{R}_l + \delta_{il}\hat{R}_j\hat{R}_k + \delta_{jk}\hat{R}_i\hat{R}_l + \delta_{jl}\hat{R}_i\hat{R}_k + \delta_{kl}\hat{R}_i\hat{R}_j) \\
&\quad + 10\hat{R}_i\hat{R}_j\hat{R}_k\hat{R}_l](u^2R^2 + u^3R^3) - (\delta_{ij} - \hat{R}_i\hat{R}_j)\hat{R}_k\hat{R}_l u^4 R^4 \} \frac{e^{-uR}}{R^5}.
\end{aligned} \tag{6}
$$

The distances a, b and c appearing implicitly in the energy shift formula (1) are the side lengths of the scalene triangle formed by the three objects A, B and C. They are defined as $a = |\vec{R}_B - \vec{R}_C|$, $b = |\vec{R}_C - \vec{R}_A|$ and $c = |\vec{R}_A - \vec{R}_B|$. Multiplying the geometric tensors in Equation (1) using formula (5) and (6) gives the potential for an arbitrary triangular configuration, from which specific geometrical arrangements then follow on inserting the appropriate distance and angular variables. These have been obtained for equilateral triangle and collinear geometries [14].

The three-body dispersion potentials considered in the literature thus far have all been between species that are characterised by electric polarisability tensors that contain multipole moments of one particular type, for example pure electric dipole or pure electric quadrupole moments. In this paper we aim to study dispersion forces among three particles in which one or more entities is described by mixed electric dipole-octupole polarisability. This quantity is non-vanishing for all molecules but is zero for atoms that undergo transitions via these two multipole moments from the ground state to the same virtual electronic level. For instance, an interaction of identical order of magnitude to Equation (1) would occur between an electric dipole polarisable molecule, and two species with mixed electric dipole-octupole polarisability. A systematic series of calculations are carried out in this work, progressing from one, to two, to three molecules possessing mixed dipole-octupole polarisable characteristics, with the other entities or entity in the first two cases being purely electric dipole polarisable. We also compute the dispersion potential between an electric dipole polarisable molecule, an electric octupole polarisable one, and a species with mixed dipole-octupole polarisability, since this is of the same order as that arising between three species with mixed dipole-octupole polarisability. This complements previous studies [14,15] wherein the effect of including electric quadrupole (Q) coupling was accounted for by evaluating the DD-DD-QQ and the DD-QQ-QQ three-body dispersion energy shifts. The first of these is comparable to the DD-DD-DO potential and the second is of the same order of magnitude as the DD-DO-DO energy shift, both of which are to be calculated in what follows, and the previously computed DD-DD-OO interaction energy given by Equation (1). While all molecules possess a non-zero pure electric quadrupole polarisability, symmetry dictates that only non-centrosymmetric species will support a non-vanishing mixed electric dipole-quadrupole polarisability tensor. Key questions to be answered include whether higher-order corrections to the triple-dipole potential arise from energy shifts involving mixed multipole moment polarisabilities, and the role played by octupole weight-1 and -3 components in such interactions. While the magnetic dipole moment, which is a similar order of magnitude to the electric quadrupole moment, and which features in the paramagnetic susceptibility tensor, $\chi^{mm}(\omega)$, i.e., the magnetic dipole analogue of Equation (2), and the magnetic quadrupole moment, which is comparable in magnitude to O, magnetic couplings have been excluded from the present work since they are difficult to measure and to compute. Whether, and to what extent, magnetic transitions need to be accounted for in a given case depends not only on the general order of magnitude, but also more importantly on the specific atomic species involved and their electronic wave functions. Attention is therefore confined to electric dipole and octupole contributions to facilitate ready comparison with previous work.

The paper is structured as follows. A very brief summary of molecular QED theory is given in Section 2, along with the form of the interaction Hamiltonian when electric octupole coupling is accounted for, and the calculation of the non-pairwise additive three-body dispersion potential. The next four sections contain specific results for dispersion energy shifts for each of the four cases mentioned above. Potentials applicable to equilateral triangle and collinear arrangements are also presented in the respective section devoted to each specific interaction. A summary is given in Section 7. Useful integrals required to obtain asymptotically limiting forms of energy shifts dependent upon one, two or three atomic polarisabilities applicable at short-range are given in the Appendices.

2. Molecular QED Calculation of the 3-Body Dispersion Potential

Consider three ground state atoms or molecules A, B and C, positioned at $\vec{R}_A$, $\vec{R}_B$ and $\vec{R}_C$, respectively. The total molecular QED Hamiltonian operator, for which the electromagnetic field forms an intrinsic part of the complete system, is given by [9–11]

$$H = \sum_\xi H_{\text{mol}}(\xi) + H_{\text{rad}} + \sum_\xi H_{\text{int}}(\xi), \; \xi = A, B, C. \tag{7}$$

$H_{\text{mol}}(\xi)$ is the familiar molecular Hamiltonian of quantum chemistry. The second term of Equation (7) signifies the radiation field Hamiltonian. The energy of the electromagnetic field is represented by a sum of independent simple harmonic oscillators, whose quantisation is rudimentary. Photons are the resulting particles that describe the elementary excitations of the radiation field. In the occupation number representation that follows from effecting second quantisation techniques, bosonic annihilation and creation operators, $a^{(\lambda)}(\vec{k})$ and $a^{\dagger(\lambda)}(\vec{k})$, are introduced and are used to express H_{rad} as

$$H_{\text{rad}} = \sum_{\vec{k},\lambda} \{a^{\dagger(\lambda)}(\vec{k})a^{(\lambda)}(\vec{k}) + \frac{1}{2}\}\hbar\omega, \tag{8}$$

where the sum is taken over radiation field modes denoted by $\vec{k}, \lambda$ corresponding to the direction of propagation and index of polarisation, respectively. Quantisation of the radiation field is carried out in a cube of volume V, thereby restricting the possible modes. ω is the circular frequency, defined according to $\omega = ck$; k is the modulus of the wave vector. One possible choice of eigenstates for the radiation field is number states, $|n(\vec{k}, \lambda)\rangle$, with the number operator n defined as $a^{\dagger(\lambda)}(\vec{k})a^{(\lambda)}(\vec{k})$ such that $n|n(\vec{k},\lambda)\rangle = a^{\dagger(\lambda)}(\vec{k})a^{(\lambda)}(\vec{k})|n(\vec{k},\lambda)\rangle$. Thus the creation and annihilation operators respectively increase or decrease by one the number of photons of a particular mode in the electromagnetic field. As expected, the eigenvalues of the radiation field are identical to those of the harmonic oscillator, namely $E_{\text{rad}} = (n + \frac{1}{2})\hbar\omega$, $n = 0, 1, 2, ...$, with n restricted to positive integer values and zero. This last value of n corresponds to the vacuum state of the field, that is, all modes have vanishing occupation number. This is an important feature of the theory, giving rise to observable phenomena [32], one of the best known being the dispersion force.

The final term of Equation (7) designates the interaction Hamiltonian, representing the coupling between radiation and matter. In the multipolar version of molecular QED theory, atoms and molecules engage with Maxwell field operators via their electric, magnetic and diamagnetic multipole moment distributions. Restricting to the first few moments of the electric polarisation field, in light of the applications to follow, the interaction Hamiltonian is written as

$$H_{\text{int}}^{\text{elec}}(\xi) = -\varepsilon_0^{-1}\mu_i(\xi)d_i^\perp(\vec{R}_\xi) - \varepsilon_0^{-1}Q_{ij}(\xi)\nabla_j d_i^\perp(\vec{R}_\xi) - \varepsilon_0^{-1}O_{ijk}(\xi)\nabla_j\nabla_k d_i^\perp(\vec{R}_\xi) + ..., \tag{9}$$

in which $\mu_i(\xi)$, $Q_{ij}(\xi)$, and $O_{ijk}(\xi)$ are the electric dipole, quadrupole and octupole moment operators of particle ξ. These moments couple directly, or through the application of one or more gradient operators, to the transverse electric displacement field operator, $\vec{d}^\perp(\vec{r})$, whose Fourier series mode expansion is of the form

$$\vec{d}^{\perp}(\vec{r}) = i \sum_{\vec{k},\lambda} \left(\frac{\hbar c k \varepsilon_0}{2V}\right)^{1/2} [\vec{e}^{(\lambda)}(\vec{k}) a^{(\lambda)}(\vec{k}) e^{i\vec{k}\cdot\vec{r}} - \overline{\vec{e}}^{(\lambda)}(\vec{k}) a^{\dagger(\lambda)}(\vec{k}) e^{-i\vec{k}\cdot\vec{r}}]. \tag{10}$$

Rather than the electric field, $\vec{e}(\vec{r})$, matter couples to $\vec{d}^{\perp}(\vec{r})$ in the multipolar framework because the field momentum canonically conjugate to the coordinate variable is proportional to the transverse electric displacement field in this coupling scheme. The field operator Equation (10) is linear in the photon creation and annihilation operators, and the normalisation pre-factor ensures that the operator correctly reproduces the energy of the electromagnetic field. In Equation (10), $\vec{e}^{(\lambda)}(\vec{k})$ is the complex unit electric polarisation vector for mode $(\vec{k}, \lambda)$ radiation, and the overbar denotes the complex conjugate quantity.

Solutions to the Schrödinger equation with Equation (7) as the Hamiltonian operator are frequently derived via perturbation theory, with the sum of the molecular and radiation field Hamiltonians constituting the unperturbed Hamiltonian. Solutions to each sub-system are taken to be known, and because the unperturbed Hamiltonian is itself separable, the base states employed to study the influence of the perturbation operator on the coupled system are product molecule-radiation field states $|E_p^\xi, E_q^{\xi'}, ...; n(\vec{k}, \lambda), n'(\vec{k}', \lambda'), ...\rangle$, in which $|E_p^\xi\rangle$ and $|E_q^{\xi'}\rangle$ are energy eigenstates for species ξ and ξ' when in electronic states labelled by quantum numbers p and q, respectively, and n and n' denote the number of photons of mode $(\vec{k}, \lambda)$ and $(\vec{k}', \lambda')$ present in the electromagnetic field. The effect of the perturbation is to cause transitions between states or a shift in energy. Standard time-dependent perturbation theory yields a series expansion in powers of H_{int} for the probability amplitude for a process to occur between specified initial and final total system states.

For the particular problem at hand, namely the dispersion interaction between three molecules, the initial and final states are identical to one another and represent each of the three species in the ground electronic state, with no photons of any mode being present in the electromagnetic field. Hence the ket $|0\rangle = |E_0^A, E_0^B, E_0^C\rangle$ may be employed unambiguously. As for dispersion interactions between pairs of particles [9–11], the three-body term contributing to the interaction energy is mediated by the exchange of two virtual photons between each coupled centre. Hence the use of Equation (9) requires that the sixth-order term in the perturbation theory expansion of the energy shift in series of powers of H_{int} be employed in the computation. A consequence is that the number of contributory terms that arise from Feynman-like diagrams that have to be evaluated and then summed over is excessively large, amounting to 360 time-ordered sequences in the case where each species is electric dipole polarisable. To circumvent such aspects, the Craig–Power Hamiltonian [33], which is quadratic in the displacement field, has been used to compute the leading and first few higher-order corrections to the retarded three-body dispersion energy [14,15,22]. In the electric dipole approximation the coupling Hamiltonian for species A is of the form

$$H_{\text{int}}^{\text{DD}}(A) = -\frac{1}{2\varepsilon_0^2} \sum_{\text{modes}} \alpha_{ij}^{\text{DD}}(A;k) d_i^{\perp}(\vec{R}_A) d_j^{\perp}(\vec{R}_A), \tag{11}$$

where the dynamic electric dipole polarisability in Equation (11) is evaluated at the real frequency $\omega = ck$. The coupling Hamiltonian Equation (11) was first used to calculate the Casimir–Polder potential [10,33]. It may be obtained by carrying out a unitary transformation with the generator chosen such that the $-\varepsilon_0^{-1} \vec{\mu}(\xi) \cdot \vec{d}^{\perp}(\vec{R}_\xi)$ term is cancelled except on the energy shell. Explicit demonstrations have been given in the Appendix of the paper by E. A. Power and T. Thirunamachandran, *Chem. Phys.* **171**, 1 (1993) and in Ref [22]. Higher-order multipole terms may be derived in a similar manner. Because Equation (11) represents an effective two-photon coupling vertex, second rather than fourth-order perturbation theory could be employed together with two instead of twelve time-ordered diagrams to yield the pair dispersion energy shift. Even greater advantages accrue on using the interaction Hamiltonian (11) to compute the potential between three atoms or molecules [22]. Only six topologically distinct diagrams are required to be summed over at third-order of perturbation theory using the formula

$$\Delta E = \sum_{I,II} \frac{\langle 0| H_{int}^{DD}(\xi)|II\rangle \langle II| H_{int}^{DD}(\xi)|I\rangle \langle I| H_{int}^{DD}(\xi)|0\rangle}{E_{II0}E_{I0}}, \qquad (12)$$

where the sums are taken over complete sets of intermediate states that result on excitation due to virtual transitions and return the total system to the ground state. Denominators signify differences between intermediate and ground energy levels. Overall, three different virtual photons are exchanged between the interacting particles. The collapsed two-photon coupling vertex at each centre accounts for absorption of two different virtual photons, or emission of two different virtual photons, or emission of one type of virtual photon and absorption of another mode or vice versa [34]. One of the six possible time-ordered sequences containing effective two-photon interaction vertices is shown in Figure 1. As characteristic of field theories, particles with zero or integer spin—the bosons, mediate interactions between material particles possessing half-integer spin—the fermions. In the case of QED, the electromagnetic force is mediated between electrons by the exchange of virtual photons [35,36]. By definition these types of photons are unobservable. They appear from and return to the electromagnetic vacuum with energy and lifetime dictated by Heisenberg's time-energy uncertainty relation.

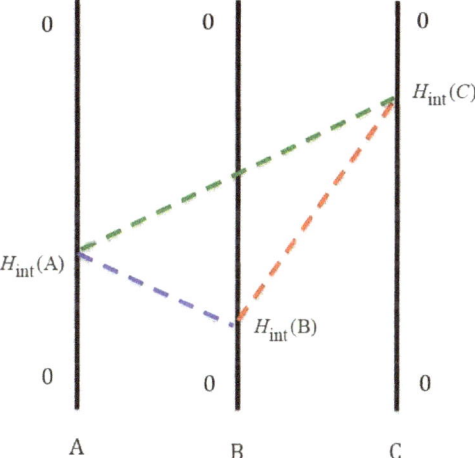

Figure 1. One of the six possible time-ordered sequences containing effective two-photon interaction vertices.

For the particular three-body dispersion interactions involving electric octupole coupling to be studied in the remaining, the appropriate effective two-photon interaction Hamiltonians to be used in the diagrammatic perturbation theory calculation are as follows. For a molecule ξ that is mixed electric dipole-octupole polarisable, coupling to the electromagnetic field occurs via the two-photon interaction operator

$$H_{int}^{DO}(\xi) = -\frac{1}{2\epsilon_0^2} \sum_{modes} \alpha_{ijkl}^{DO}(\xi;k) d_i^\perp(\vec{R}_\xi) \nabla_k \nabla_l d_j^\perp(\vec{R}_\xi), \qquad (13)$$

where the mixed dipole-octupole polarisability tensor at real wave number is given by

$$\alpha_{ijkl}^{DO}(\xi;k) = \sum_t \left\{ \frac{\mu_i^{0t}(\xi)O_{jkl}^{t0}(\xi)}{E_{t0} - \hbar ck} + \frac{O_{jkl}^{0t}(\xi)\mu_i^{t0}(\xi)}{E_{t0} + \hbar ck} \right\}. \tag{14}$$

For the same molecule that is pure electric octupole polarisable, the Craig–Power form of the interaction Hamiltonian is

$$H_{int}^{OO}(\xi) = -\frac{1}{2\varepsilon_0^2} \sum_{modes} \alpha_{ijklmn}^{OO}(\xi;k) \nabla_j \nabla_k d_i^\perp(\vec{R}_\xi) \nabla_m \nabla_n d_l^\perp(\vec{R}_\xi), \tag{15}$$

with $\alpha_{ijklmn}^{OO}(\xi;k)$ defined for molecule C by Equation (3).

At this stage it is convenient to introduce the decomposition of the octupole moment into its irreducible components of weights-1 and -3. Whence

$$O_{ijk} = O_{ijk}^{(1)} + O_{ijk}^{(3)}, \tag{16}$$

where

$$O_{ijk}^{(1)} = -\frac{e}{30}\vec{q}^2(\vec{q}_i \delta_{jk} + \vec{q}_j \delta_{ik} + \vec{q}_k \delta_{ij}), \tag{17}$$

and

$$O_{ijk}^{(3)} = -\frac{e}{6}[\vec{q}_i \vec{q}_j \vec{q}_k - \frac{1}{5}\vec{q}^2(\vec{q}_i \delta_{jk} + \vec{q}_j \delta_{ik} + \vec{q}_k \delta_{ij})], \tag{18}$$

for the multipole moment defined with respect to the origin. $O_{ijk}^{(1)}$ has three independent components and the transformation properties of a vector, while $O_{ijk}^{(3)}$ has seven independent components. When any two of the Cartesian tensor components are equal, Equation (18) vanishes. Similarly, from the form of the electric octupole coupling to the electric displacement field in either form of interaction Hamiltonian, be it Equation (9) or Equation (13) or Equation (15), the contribution is zero when the suffix of the field is equal to that of any of the gradient operators immediately preceding it. This is because for a neutral entity, the electric displacement field is exclusively transverse in nature outside of the source. It is for the same reason that the trace of the electric quadrupole moment does not contribute to the coupling in the interaction term $-\varepsilon_0^{-1} Q_{ij}(\xi)\nabla_j d_i^\perp(\vec{R}_\xi)$.

It is clear from the partitioning Equation (16) that any mixed polarisability tensor containing an electric octupole moment, or pure electric octupole polarisability Equation (3), may also be separated into contributions that are explicitly dependent upon scalar weight-1 and -3 octupole moment terms. For instance, inserting Equation (16) into the mixed electric dipole-octupole polarisability Equation (14) produces

$$\alpha_{ijkl}^{DO}(\xi;k) = \alpha_{ijkl}^{DO^1}(\xi;k) + \alpha_{ijkl}^{DO^3}(\xi;k)$$
$$= \sum_t \left\{ \frac{\mu_i^{0t}(\xi)O_{jkl}^{(1)t0}(\xi)}{E_{t0} - \hbar ck} + \frac{O_{jkl}^{(1)0t}(\xi)\mu_i^{t0}(\xi)}{E_{t0} + \hbar ck} \right\} + \sum_t \left\{ \frac{\mu_i^{0t}(\xi)O_{jkl}^{(3)t0}(\xi)}{E_{t0} - \hbar ck} + \frac{O_{jkl}^{(3)0t}(\xi)\mu_i^{t0}(\xi)}{E_{t0} + \hbar ck} \right\}. \tag{19}$$

Analogously, the pure electric octupole polarisability tensor $\alpha_{ijklmn}^{OO}(\xi;k)$, Equation (3), has octupole weight 1-1, 1-3, 3-1 and 3-3 dependent contributions,

$$\alpha_{ijklmn}^{OO}(\xi;k) = \alpha_{ijklmn}^{O^1O^1}(\xi;k) + \alpha_{ijklmn}^{O^1O^3}(\xi;k) + \alpha_{ijklmn}^{O^3O^1}(\xi;k) + \alpha_{ijklmn}^{O^3O^3}(\xi;k). \tag{20}$$

3. DD-DD-DO Energy Shift

The first dispersion potential to be evaluated is the leading correction involving the octupole interaction term, namely that between two electric dipole polarisable molecules A and B, and a third, C, that is mixed electric dipole-octupole polarisable and characterised by Equation (14) or Equation (19). Hence the interaction Hamiltonian for the system of three coupled ground state particles is

$$H_{\text{int}} = H_{\text{int}}^{\text{DD}}(A) + H_{\text{int}}^{\text{DD}}(B) + H_{\text{int}}^{\text{DO}}(C), \tag{21}$$

with the pure electric dipole Craig–Power coupling Hamiltonian given by Equation (11). Emulating the calculational procedure recently employed in evaluating three-body energy shifts [14,15] yields

$$\Delta E^{\text{DD-DD-DO}} = -\frac{\hbar c}{64\pi^4 \varepsilon_0^3}(-\nabla^2 \delta_{km} + \nabla_k \nabla_m)^a (-\nabla^2 \delta_{ip} + \nabla_i \nabla_p)^b (\nabla_q \nabla_r)^b$$
$$\times (-\nabla^2 \delta_{jl} + \nabla_j \nabla_l)^c \frac{1}{abc} \int_0^\infty du\, \alpha_{ij}^{\text{DD}}(A;iu)\alpha_{kl}^{\text{DD}}(B;iu)\alpha_{mpqr}^{\text{DO}}(C;iu)e^{-u(a+b+c)}, \tag{22}$$

for an arbitrary triangular configuration of the three particles, with side lengths a, b and c defined earlier. Each polarisability is evaluated at the imaginary frequency. Utilising the definitions of $F_{ij}(uR)$ and $L_{ijkl}(uR)$ introduced in Equations (5) and (6), allows the energy shift Equation (22) to be written more succinctly as

$$\Delta E^{\text{DD-DD-DO}} = -\frac{\hbar c}{64\pi^4 \varepsilon_0^3} \int_0^\infty du\, \alpha_{ij}^{\text{DD}}(A;iu)\alpha_{kl}^{\text{DD}}(B;iu)\alpha_{mpqr}^{\text{DO}}(C;iu) F_{km}(ua) L_{ipqr}(ub) F_{jl}(uc). \tag{23}$$

Both expressions for the energy shift hold for molecules in fixed relative orientation to one another.

To obtain the potential applicable to isotropic molecules, a rotational average of Equation (23) is required. This may be done as separate averages over each particle. For electric dipole polarisable species A and B, the randomly averaged tensor, enclosed in angular brackets, is given by

$$<\alpha_{ij}^{\text{DD}}(\xi;iu)> = \frac{1}{3}\delta_{ij}\delta_{\rho\sigma}\alpha_{\rho\sigma}^{\text{DD}}(\xi;iu) = \delta_{ij}\alpha^{\text{DD}}(\xi;iu),\ \xi = A, B, \tag{24}$$

where the Greek subscripts denote Cartesian tensor components in the molecule-fixed frame of reference, and a factor of 1/3 has been absorbed into the definition of the isotropic polarisability. From expression (19) it is seen that the mixed dipole-octupole polarisability is a sum of weight-1 and -3 octupole moments, and overall is a Cartesian tensor of rank four. An orientational average of such an object is obtained via [37]

$$<T_{ijkl}> = I_{ijkl;\lambda\mu\nu\pi}^{(4)} T_{\lambda\mu\nu\pi}, \tag{25}$$

where T_{ijkl} is a fourth-rank tensor, and $I_{ijkl;\lambda\mu\nu\pi}^{(4)}$ is given by

$$I_{ijkl;\lambda\mu\nu\pi}^{(4)} = \tfrac{1}{30}[\delta_{ij}\delta_{kl}(4\delta_{\lambda\mu}\delta_{\nu\pi} - \delta_{\lambda\nu}\delta_{\mu\pi} - \delta_{\lambda\pi}\delta_{\mu\nu})$$
$$+\delta_{ik}\delta_{jl}(-\delta_{\lambda\mu}\delta_{\nu\pi} + 4\delta_{\lambda\nu}\delta_{\mu\pi} - \delta_{\lambda\pi}\delta_{\mu\nu}) \tag{26}$$
$$+\delta_{il}\delta_{jk}(-\delta_{\lambda\mu}\delta_{\nu\pi} - \delta_{\lambda\nu}\delta_{\mu\pi} + 4\delta_{\lambda\pi}\delta_{\mu\nu})].$$

From the form of the mixed dipole-octupole coupling Equation (13), it is seen that $H_{\text{int}}^{\text{DO}}(C)$ vanishes when $j = k$ and when $j = l$. Hence the second and third terms within square brackets of Equation (26) do not contribute. Similarly, the mixed dipole-octupole coupling is zero when $\mu = \nu$ and when $\mu = \pi$, so that the second and third terms within each of the three terms written in parentheses do not contribute to the orientational average. Therefore the orientationally averaged mixed electric dipole-octupole polarisability of molecule C appearing in formula (23) is

$$<\alpha_{mpqr}^{\text{DO}}(C;iu)> = \frac{2}{15}\delta_{mp}\delta_{qr}\delta_{\lambda\mu}\delta_{\nu\pi}\alpha_{\lambda\mu\nu\pi}^{\text{DO}}(C;iu) = \frac{2}{15}\delta_{mp}\delta_{qr}\alpha_{\lambda\lambda\mu\mu}^{\text{DO}^1}(C;iu). \tag{27}$$

From relation (16) it is seen that the octupole moment appearing in Equation (27) is composed only of the weight-1 term, with the weight-3 contribution vanishing identically. The superscript "1" serves to label the contributing weight, as in expression (19). On employing relation (24) twice and Equation (27) in Equation (23) yields the energy shift for isotropic molecules

$$\Delta E^{DD-DD-DO^1} = -\frac{\hbar c}{64\pi^4\varepsilon_0^3} \int_0^\infty du\, \alpha^{DD}(A;iu)\alpha^{DD}(B;iu)\alpha^{DO^1}_{\lambda\lambda\mu\mu}(C;iu) F_{jk}(ua)L_{ijll}(ub)F_{ik}(uc), \quad (28)$$

where a factor of 2/15 has been absorbed into the definition of the isotropic mixed dipole-octupole polarisability. Multiplying the geometric tensors using Equations (5) and (6) produces for the dispersion potential.

$$\begin{aligned}
\Delta E^{DD-DD-DO^1} &= \frac{\hbar c}{64\pi^4\varepsilon_0^3}\frac{1}{a^3b^5c^3}\int_0^\infty du\, e^{-u(a+b+c)}\alpha^{DD}(A;iu)\alpha^{DD}(B;iu)\alpha^{DO^1}_{\lambda\lambda\mu\mu}(C;iu) \\
&\times \Big\{ [(\hat a\cdot\hat b)^2 + (\hat b\cdot\hat c)^2 + (\hat c\cdot\hat a)^2 - (\hat a\cdot\hat b)(\hat b\cdot\hat c)(\hat c\cdot\hat a)]u^8 a^2 b^4 c^2 \\
&\quad + [-2 + (\hat a\cdot\hat b)^2 + 3[(\hat b\cdot\hat c)^2 + (\hat c\cdot\hat a)^2] - 3(\hat a\cdot\hat b)(\hat b\cdot\hat c)(\hat c\cdot\hat a)](u^7 a^2 b^4 c + u^6 a^2 b^4) \\
&\quad + [-2 + 3[(\hat a\cdot\hat b)^2 + (\hat c\cdot\hat a)^2] + (\hat b\cdot\hat c)^2 - 3(\hat a\cdot\hat b)(\hat b\cdot\hat c)(\hat c\cdot\hat a)](u^7 a^2 b^3 c^2 + u^6 a^2 b^2 c^2) \\
&\quad + [-2 + 3[(\hat a\cdot\hat b)^2 + (\hat c\cdot\hat a)^2] + (\hat b\cdot\hat c)^2 - 3(\hat a\cdot\hat b)(\hat b\cdot\hat c)(\hat c\cdot\hat a)](u^7 a b^4 c^2 + u^6 b^4 c^2) \\
&\quad + [-4 + 3[(\hat a\cdot\hat b)^2 + 3(\hat b\cdot\hat c)^2 + (\hat c\cdot\hat a)^2] - 9(\hat a\cdot\hat b)(\hat b\cdot\hat c)(\hat c\cdot\hat a)](u^6 a^2 b^3 c + u^5 a^2 b^2(b+c) + u^4 a^2 b^2) \\
&\quad + [-4 + 3[(\hat a\cdot\hat b)^2 + (\hat b\cdot\hat c)^2 + 3(\hat c\cdot\hat a)^2] - 9(\hat a\cdot\hat b)(\hat b\cdot\hat c)(\hat c\cdot\hat a)](u^6 a b^4 c + u^5 b^4(a+c) + u^4 b^4) \\
&\quad + [-4 + 3[3(\hat a\cdot\hat b)^2 + (\hat b\cdot\hat c)^2 + (\hat c\cdot\hat a)^2] - 9(\hat a\cdot\hat b)(\hat b\cdot\hat c)(\hat c\cdot\hat a)](u^6 a b^3 c^2 + u^5 b^2 c^2(a+b) + u^4 b^2 c^2) \\
&\quad + [-6 + 9[(\hat a\cdot\hat b)^2 + (\hat b\cdot\hat c)^2 + (\hat c\cdot\hat a)^2] - 27(\hat a\cdot\hat b)(\hat b\cdot\hat c)(\hat c\cdot\hat a)] \\
&\quad \times (u^5 a b^3 c + u^4 b^2(ab+bc+ac) + u^3 b^2(a+b+c)) \Big\},
\end{aligned} \quad (29)$$

which applies to a scalene triangle geometry. The circumflex denotes a unit vector. It is interesting to note that each of the eight terms contained within square brackets inside the braces, namely involving direction cosines, appears distinctly in the corresponding expression for the triple dipole dispersion potential (Equation (57) from Ref. [15]). Because $O^{(1)}_{ijk}(\xi)$ has transformation properties equivalent to that of an electric dipole, result (29) may be interpreted as a higher-order correction to $\Delta E^{DD-DD-DD}$.

Expressions for specific configurations follow straightforwardly from Equation (29). For an equilateral triangle, $a = b = c = R$, and $\hat a \cdot \hat b = \hat b \cdot \hat c = \hat c \cdot \hat a = -\cos 60° = -\frac{1}{2}$. Whence

$$\begin{aligned}
\Delta E^{DD-DD-DO^1}_{Eq} &= \frac{\hbar c}{512\pi^4\varepsilon_0^3 R^{11}} \int_0^\infty du\, e^{-3uR}\alpha^{DD}(A;iu)\alpha^{DD}(B;iu)\alpha^{DO^1}_{\lambda\lambda\mu\mu}(C;iu) \\
&\quad \times [7(uR)^8 + 3(uR)^7 + 24(uR)^6 + 75(uR)^5 + 120(uR)^4 + 99(uR)^3],
\end{aligned} \quad (30)$$

whose coefficients preceding each power of uR are identical to corresponding terms appearing in the triple dipole energy shift when the triangle is equilateral (see Equation (48) of Ref. [14]).

Another noteworthy feature of Equation (29) (and consequently result (30)), is that there is no term independent of u, leading to no direct near-zone asymptote. This is due to its dependence solely upon the octupole weight-1 term, and the absence of a contribution from the weight-3 term. Nevertheless, by making the following approximations we may arrive at a short-range limiting form for the interaction energy. Retaining the $(uR)^3$ term, with $e^{-3uR} \sim 1$ for $uR << 1$, and using result (A15) from Appendix C, we see that Equation (30) results in an R^{-8} near-zone limiting dependence on separation distance for an equilateral triangle arrangement:

$$\Delta E^{DD-DD-DO^1}_{Eq}(NZ) \approx -\frac{11}{480\hbar^3 c^3 \pi^4 \varepsilon_0^3 R^8} \sum_{x,y,z} |\vec{\mu}^{0x}(A)|^2 |\vec{\mu}^{0y}(B)|^2 \mu^{0z}_\lambda(C) O^{(1)z0}_{\lambda\mu\mu}(C) E_{x0} E_{y0} E_{z0}$$

$$\times \left\{ \frac{E_{x0}^2(E_{y0}^2-E_{z0}^2)\ln(3E_{x0}R/\hbar c) + E_{y0}^2(E_{z0}^2-E_{x0}^2)\ln(3E_{y0}R/\hbar c) + E_{z0}^2(E_{x0}^2-E_{y0}^2)\ln(3E_{z0}R/\hbar c)}{(E_{y0}^2-E_{z0}^2)(E_{z0}^2-E_{x0}^2)(E_{x0}^2-E_{y0}^2)} \right\}. \quad (31)$$

This compares with an inverse ninth power dependence on R in a true near-zone limit for the triple-dipole potential.

The far-zone limiting form of the interaction energy for an equilateral triangle configuration follows directly from Equation (30) on taking the polarisabilities to be static, corresponding to the zero frequency limit $\omega = icu \to 0$, since in the far-zone $uR \gg 1$ with $R \to \infty$, and evaluating the ensuing u-integral using the standard integral

$$\int_0^\infty x^n e^{-\eta x}\, dx = n!\eta^{-n-1},\ \mathrm{Re}\eta > 0. \tag{32}$$

This leads to the far-zone asymptote

$$\Delta E_{\mathrm{Eq}}^{\mathrm{DD-DD-DO^1}}(\mathrm{FZ}) = \frac{122638\hbar c}{512 \times 3^7 \pi^4 \varepsilon_0^3 R^{12}} \alpha^{\mathrm{DD}}(A;0)\alpha^{\mathrm{DD}}(B;0)\alpha_{\lambda\lambda\mu\mu}^{\mathrm{DO^1}}(C;0), \tag{33}$$

and which exhibits inverse twelfth power separation distance dependence. Note that for this particular three-particle configuration, the force is repulsive.

For a collinear arrangement, in which the mixed dipole-octupole polarisable species C lies mid-way between A and B, $2a = 2b = c = R$, and $\theta_A = \theta_B = 0°$, and $\theta_C = 180°$, so that $\hat{a} \cdot \hat{b} = 1$, and $\hat{b} \cdot \hat{c} = \hat{c} \cdot \hat{a} = -1$, the dispersion energy shift from Equation (29) is

$$\begin{aligned}\Delta E_{\mathrm{Coll}}^{\mathrm{DD-DD-DO^1}} =\ & \frac{\hbar c}{8\pi^4 \varepsilon_0^3 R^{11}} \int_0^\infty du\, e^{-2uR} \alpha^{\mathrm{DD}}(A;iu)\alpha^{\mathrm{DD}}(B;iu)\alpha_{\lambda\lambda\mu\mu}^{\mathrm{DO^1}}(C;iu) \\ & \times [(uR)^8 + 5(uR)^7 + 17(uR)^6 + 16(uR)^5 - 36(uR)^4 - 96(uR)^3].\end{aligned} \tag{34}$$

Again the coefficients appearing in the polynomial function are identical to that found for the collinear geometry of three dipole polarisable molecules (Equation (52) of Ref. [14]), as well as in the weight-1 dependent contribution to the DD-DD-OO dispersion energy shift (see Equation (51) of Ref. [14]). As for the equilateral triangle case, inverse eighth power law behaviour is found in the near-zone on using result (A15). It is given by

$$\begin{aligned}\Delta E_{\mathrm{Coll}}^{\mathrm{DD-DD-DO^1}}(\mathrm{NZ}) \approx\ & \frac{64}{45\hbar^3 c^3 \pi^4 \varepsilon_0^3 R^8} \sum_{x,y,z} |\vec{\mu}^{0x}(A)|^2 |\vec{\mu}^{0y}(B)|^2 \mu_\lambda^{0z}(C) O_{\lambda\mu\mu}^{(1)z0}(C) E_{x0} E_{y0} E_{z0} \\ & \times \left\{ \frac{E_{x0}^2(E_{y0}^2 - E_{z0}^2)\ln(2E_{x0}R/\hbar c) + E_{y0}^2(E_{z0}^2 - E_{x0}^2)\ln(2E_{y0}R/\hbar c) + E_{z0}^2(E_{x0}^2 - E_{y0}^2)\ln(2E_{z0}R/\hbar c)}{(E_{y0}^2 - E_{z0}^2)(E_{z0}^2 - E_{x0}^2)(E_{x0}^2 - E_{y0}^2)} \right\}.\end{aligned} \tag{35}$$

In the radiation zone Equation (34) reduces to

$$\Delta E_{\mathrm{Coll}}^{\mathrm{DD-DD-DO^1}}(\mathrm{FZ}) = \frac{3837\hbar c}{128\pi^4 \varepsilon_0^3 R^{12}} \alpha^{\mathrm{DD}}(A;0)\alpha^{\mathrm{DD}}(B;0)\alpha_{\lambda\lambda\mu\mu}^{\mathrm{DO^1}}(C;0), \tag{36}$$

which varies as R^{-12}. The potentials for the collinear arrangement are positive in sign.

4. DD-DO-DO Dispersion Potential

The next dispersion energy involving octupole moments to be examined is that between an electric dipole polarisable molecule, A, and two mixed electric dipole-octupole polarisable species, B and C. This interaction is of the same order of magnitude as that between two electric dipole polarisable particles, and a third that is pure electric octupole polarisable, and which has previously been published [14,15]. The calculation is similar to that outlined in the last section and to other dispersion interactions between three bodies.

Relative to Equation (21) the interaction Hamiltonian is

$$H_{\mathrm{int}} = H_{\mathrm{int}}^{\mathrm{DD}}(A) + H_{\mathrm{int}}^{\mathrm{DO}}(B) + H_{\mathrm{int}}^{\mathrm{DO}}(C). \tag{37}$$

Summing over the six time-ordered diagrams at third-order of perturbation theory produces the following result for molecules in fixed relative orientation to one another, and which depends on octupole weight-1 and -3 terms:

$$\Delta E^{\text{DD-DO-DO}} = -\frac{\hbar c}{64\pi^4 \varepsilon_0^3} \int_0^\infty du\, \alpha_{ij}^{\text{DD}}(\text{A}; iu) \alpha_{klmn}^{\text{DO}}(\text{B}; iu) \alpha_{pqrs}^{\text{DO}}(\text{C}; iu) F_{kp}(ua) L_{iqrs}(ub) L_{jlmn}(uc). \quad (38)$$

Orientational averaging using Equations (24) and (27) produces the energy shift expression for isotropic molecules

$$\Delta E^{\text{DD-DO}^1\text{-DO}^1} = -\frac{\hbar c}{64\pi^4 \varepsilon_0^3} \int_0^\infty du\, \alpha^{\text{DD}}(\text{A}; iu) \alpha_{\lambda\lambda\mu\mu}^{\text{DO}^1}(\text{B}; iu) \alpha_{\nu\nu\pi\pi}^{\text{DO}^1}(\text{C}; iu) F_{ij}(ua) L_{jkll}(ub) L_{ikmm}(uc), \quad (39)$$

on absorbing factors of 2/15 into each of the isotropic mixed dipole-octupole polarisabilities, and on making use of the relation $L_{kjll}(uR) = L_{jkll}(uR)$. After averaging, the weight-3 octupole moment makes no further contribution to the interaction energy, with only the weight-1 term in play. The product of geometrical tensors produces a result identical in form to that occurring in the octupole weight-1 dependent term of the DD-DD-OO potential given in Equation (54) of Ref. [15] for a scalene triangle, and to the geometrical part of the triple dipole result, given by Equation (57) of Ref. [15]. Thus $\Delta E^{\text{DD-DO}^1\text{-DO}^1}$ is another higher-order correction to the $\Delta E^{\text{DD-DD-DD}}$ energy shift. Explicitly,

$$\Delta E^{\text{DD-DO}^1\text{-DO}^1} = \frac{\hbar c}{64\pi^4 \varepsilon_0^3} \frac{1}{a^3 b^5 c^5} \int_0^\infty du\, e^{-u(a+b+c)} \alpha^{\text{DD}}(\text{A}; iu) \alpha_{\lambda\lambda\mu\mu}^{\text{DO}^1}(\text{B}; iu) \alpha_{\nu\nu\pi\pi}^{\text{DO}^1}(\text{C}; iu)$$
$$\times \{ [(\hat{a}\cdot\hat{b})^2 + (\hat{b}\cdot\hat{c})^2 + (\hat{c}\cdot\hat{a})^2 - (\hat{a}\cdot\hat{b})(\hat{b}\cdot\hat{c})(\hat{c}\cdot\hat{a})]u^{10}a^2 b^4 c^4$$
$$+ [-2 + (\hat{a}\cdot\hat{b})^2 + 3[(\hat{b}\cdot\hat{c})^2 + (\hat{c}\cdot\hat{a})^2] - 3(\hat{a}\cdot\hat{b})(\hat{b}\cdot\hat{c})(\hat{c}\cdot\hat{a})](u^9 a^2 b^4 c^3 + u^8 a^2 b^4 c^2)$$
$$+ [-2 + 3[(\hat{a}\cdot\hat{b})^2 + (\hat{b}\cdot\hat{c})^2] + (\hat{c}\cdot\hat{a})^2 - 3(\hat{a}\cdot\hat{b})(\hat{b}\cdot\hat{c})(\hat{c}\cdot\hat{a})](u^9 a^2 b^3 c^4 + u^8 a^2 b^2 c^4)$$
$$+ [-2 + 3[(\hat{a}\cdot\hat{b})^2 + (\hat{c}\cdot\hat{a})^2] + (\hat{b}\cdot\hat{c})^2 - 3(\hat{a}\cdot\hat{b})(\hat{b}\cdot\hat{c})(\hat{c}\cdot\hat{a})](u^9 ab^4 c^4 + u^8 b^4 c^4) \quad (40)$$
$$+ [-4 + 3[(\hat{a}\cdot\hat{b})^2 + 3(\hat{b}\cdot\hat{c})^2 + (\hat{c}\cdot\hat{a})^2] - 9(\hat{a}\cdot\hat{b})(\hat{b}\cdot\hat{c})(\hat{c}\cdot\hat{a})](u^8 a^2 b^3 c^3 + u^7 a^2 b^2 c^2(b+c) + u^6 a^2 b^2 c^2)$$
$$+ [-4 + 3[(\hat{a}\cdot\hat{b})^2 + (\hat{b}\cdot\hat{c})^2 + 3(\hat{c}\cdot\hat{a})^2] - 9(\hat{a}\cdot\hat{b})(\hat{b}\cdot\hat{c})(\hat{c}\cdot\hat{a})](u^8 ab^4 c^3 + u^7 b^4 c^2 (a+c) + u^6 b^4 c^2)$$
$$+ [-4 + 3[3(\hat{a}\cdot\hat{b})^2 + (\hat{b}\cdot\hat{c})^2 + (\hat{c}\cdot\hat{a})^2] - 9(\hat{a}\cdot\hat{b})(\hat{b}\cdot\hat{c})(\hat{c}\cdot\hat{a})](u^8 ab^3 c^4 + u^7 b^2 c^4 (a+b) + u^6 b^2 c^4)$$
$$+ [-6 + 9[(\hat{a}\cdot\hat{b})^2 + (\hat{b}\cdot\hat{c})^2 + (\hat{c}\cdot\hat{a})^2] - 27(\hat{a}\cdot\hat{b})(\hat{b}\cdot\hat{c})(\hat{c}\cdot\hat{a})](u^7 ab^3 c^3 + u^6 b^2 c^2 (ab + bc + ac)$$
$$+ u^5 b^2 c^2 (a+b+c) + u^4 b^2 c^2) \}.$$

From this last equation, which applies to a scalene triangle arrangement of the three atoms, the dispersion potential for an equilateral triangle geometry is readily found to be

$$\Delta E_{\text{Eq}}^{\text{DD-DO}^1\text{-DO}^1} = \frac{\hbar c}{512\pi^4 \varepsilon_0^3 R^{13}} \int_0^\infty du\, e^{-3uR} \alpha^{\text{DD}}(\text{A}; iu) \alpha_{\lambda\lambda\mu\mu}^{\text{DO}^1}(\text{B}; iu) \alpha_{\nu\nu\pi\pi}^{\text{DO}^1}(\text{C}; iu)$$
$$\times [7(uR)^{10} + 3(uR)^9 + 24(uR)^8 + 75(uR)^7 + 120(uR)^6 + 99(uR)^5 + 33(uR)^4]. \quad (41)$$

Because there is no u-independent term, a true near-zone limit does not exist. One may be found by retaining the $(uR)^4$ term and using the integral result (A16). This gives a potential with an R^{-9} short-range dependence,

$$\Delta E_{\text{Eq}}^{\text{DD-DO}^1\text{-DO}^1}(\text{NZ}) \approx \frac{11}{7200\hbar^4 c^4 \pi^3 \varepsilon_0^3 R^9} \sum_{x,y,z} |\vec{\mu}^{0x}(\text{A})|^2 \mu_\lambda^{0y}(\text{B}) O_{\lambda\mu\mu}^{(1)y0}(\text{B}) \mu_\nu^{0z}(\text{C}) O_{\nu\pi\pi}^{(1)z0}(\text{C})$$
$$\times E_{x0} E_{y0} E_{z0} \frac{[E_{x0} E_{y0} + E_{x0} E_{z0} + E_{y0} E_{z0}]}{(E_{x0} + E_{y0})(E_{x0} + E_{z0})(E_{y0} + E_{z0})}. \quad (42)$$

At very large separation distances between nuclei, Equation (41) reduces to the far-zone asymptote

$$\Delta E_{\text{Eq}}^{\text{DD}-\text{DO}^1-\text{DO}^1}(\text{FZ}) = \frac{716608\hbar c}{512 \times 3^7 \pi^4 \varepsilon_0^3 R^{14}} \alpha^{\text{DD}}(A;0) \alpha_{\lambda\lambda\mu\mu}^{\text{DO}^1}(B;0) \alpha_{\nu\nu\pi\pi}^{\text{DO}^1}(C;0), \qquad (43)$$

displaying R^{-14} behaviour, and for which the polarisabilities are static.

For three molecules in a straight line, C lying in the centre,

$$\begin{aligned}\Delta E_{\text{Coll}}^{\text{DD}-\text{DO}^1-\text{DO}^1} &= \frac{\hbar c}{8\pi^4\varepsilon_0^3 R^{13}} \int_0^\infty du\, e^{-2uR} \alpha^{\text{DD}}(A;iu)\alpha_{\lambda\lambda\mu\mu}^{\text{DO}^1}(B;iu)\alpha_{\nu\nu\pi\pi}^{\text{DO}^1}(C;iu) \\ &\quad \times [(uR)^{10} + 5(uR)^9 + 17(uR)^8 + 16(uR)^7 - 36(uR)^6 - 96(uR)^5 - 48(uR)^4],\end{aligned} \qquad (44)$$

whose near-zone limiting form exhibits inverse ninth power dependence on using Equation (A16),

$$\begin{aligned}\Delta E_{\text{Coll}}^{\text{DD}-\text{DO}^1-\text{DO}^1}(\text{NZ}) &\approx -\frac{32}{225\hbar^4 c^4 \pi^3 \varepsilon_0^3 R^9} \sum_{x,y,z} |\vec{\mu}^{0x}(A)|^2 \mu_\lambda^{0y}(B) O_{\lambda\mu\mu}^{(1)y0}(B) \mu_\nu^{0z}(C) O_{\nu\pi\pi}^{(1)z0}(C) \\ &\quad \times E_{x0}E_{y0}E_{z0} \frac{[E_{x0}E_{y0}+E_{x0}E_{z0}+E_{y0}E_{z0}]}{(E_{x0}+E_{y0})(E_{x0}+E_{z0})(E_{y0}+E_{z0})}.\end{aligned} \qquad (45)$$

In the far-zone Equation (44) gives rise to an asymptotic energy shift

$$\Delta E_{\text{Coll}}^{\text{DD}-\text{DO}^1-\text{DO}^1}(\text{FZ}) = \frac{48114\hbar c}{128\pi^4\varepsilon_0^3 R^{14}} \alpha^{\text{DD}}(A;0)\alpha_{\lambda\lambda\mu\mu}^{\text{DO}^1}(B;0)\alpha_{\nu\nu\pi\pi}^{\text{DO}^1}(C;0), \qquad (46)$$

with identical power law dependence to that found for an equilateral triangle geometry, Equation (43). It is interesting to note that the polynomial terms within square brackets of the u-integrals (41) and (44) are identical to the octupole weight-1 dependent terms occurring in the DD-DD-OO dispersion potential, given by the first integral terms of Equations (47) and (51) of Ref. [14], respectively, and are therefore necessarily higher-order corrections to the triple dipole potential, as seen by comparing Equations (41) and (44) with Equations (48) and (52) of Ref. [15].

5. DO-DO-DO Interaction Energy

The next three-body dispersion energy shift involving electric octupole coupling to be studied is that between three mixed electric dipole-octupole polarisable species. The interaction Hamiltonian is the same for each centre, namely

$$H_{\text{int}} = \sum_{\xi=A,B,C} H_{\text{int}}^{\text{DO}}(\xi), \qquad (47)$$

with $H_{\text{int}}^{\text{DO}}(\xi)$ given explicitly by Equation (13). Standard calculational procedure leads to the following formula applicable for molecules with locked-in relative configurations,

$$\Delta E^{\text{DO}-\text{DO}-\text{DO}} = -\frac{\hbar c}{64\pi^4\varepsilon_0^3} \int_0^\infty du\, \alpha_{ijkl}^{\text{DO}}(A;iu)\alpha_{mnpq}^{\text{DO}}(B;iu)\alpha_{rstu}^{\text{DO}}(C;iu) L_{nrpq}(ua) L_{istu}(ub) L_{jmkl}(uc), \qquad (48)$$

and is dependent upon octupole weight-1 and -3 contributions. After random orientational averaging there is no dependence on $O_{ijk}^{(3)}(\xi)$, $\xi = A, B, C$, and the energy shift simplifies to

$$\Delta E^{\text{DO}^1-\text{DO}^1-\text{DO}^1} = -\frac{\hbar c}{64\pi^4\varepsilon_0^3} \int_0^\infty du\, \alpha_{\lambda\lambda\mu\mu}^{\text{DO}^1}(A;iu)\alpha_{\nu\nu\pi\pi}^{\text{DO}^1}(B;iu)\alpha_{\rho\rho\sigma\sigma}^{\text{DO}^1}(C;iu) L_{jkll}(ua) L_{ikmm}(ub) L_{ijnn}(uc), \qquad (49)$$

exhibiting a dependence solely on octupole weight-1 moment as found in each of the previous cases involving mixed dipole-octupole polarisability. A factor of 2/15 has been taken into each of the isotropic mixed polarisability tensors of Equation (49). On multiplying the product of L_{ijkl} tensors using Equation (6), the isotropic potential for a scalene triangle is

$$\Delta E^{DO^1-DO^1-DO^1} = \frac{\hbar c}{8\pi^4\varepsilon_0^3} \frac{1}{a^5 b^5 c^5} \int_0^\infty du e^{-u(a+b+c)} \alpha^{DO^1}_{\lambda\lambda\mu\mu}(A;iu)\alpha^{DO^1}_{\nu\nu\pi\pi}(B;iu)\alpha^{DO^1}_{\rho\rho\sigma\sigma}(C;iu)$$

$$\times \{[(\hat{a}\cdot\hat{b})^2 + (\hat{b}\cdot\hat{c})^2 + (\hat{c}\cdot\hat{a})^2 - (\hat{a}\cdot\hat{b})(\hat{b}\cdot\hat{c})(\hat{c}\cdot\hat{a})]u^{12}a^4b^4c^4$$
$$+[-2 + (\hat{a}\cdot\hat{b})^2 + 3[(\hat{b}\cdot\hat{c})^2 + (\hat{c}\cdot\hat{a})^2] - 3(\hat{a}\cdot\hat{b})(\hat{b}\cdot\hat{c})(\hat{c}\cdot\hat{a})](u^{11}a^4b^4c^3 + u^{10}a^4b^4c^2)$$
$$+[-2 + 3[(\hat{a}\cdot\hat{b})^2 + (\hat{b}\cdot\hat{c})^2] + (\hat{c}\cdot\hat{a})^2 - 3(\hat{a}\cdot\hat{b})(\hat{b}\cdot\hat{c})(\hat{c}\cdot\hat{a})](u^{11}a^4b^3c^4 + u^{10}a^4b^2c^4)$$
$$+[-2 + 3[(\hat{a}\cdot\hat{b})^2 + (\hat{c}\cdot\hat{a})^2] + (\hat{b}\cdot\hat{c})^2 - 3(\hat{a}\cdot\hat{b})(\hat{b}\cdot\hat{c})(\hat{c}\cdot\hat{a})](u^{11}a^3b^4c^4 + u^{10}a^2b^4c^4) \quad (50)$$
$$+[-4 + 3[(\hat{a}\cdot\hat{b})^2 + 3(\hat{b}\cdot\hat{c})^2 + (\hat{c}\cdot\hat{a})^2] - 9(\hat{a}\cdot\hat{b})(\hat{b}\cdot\hat{c})(\hat{c}\cdot\hat{a})](u^{10}a^4b^3c^3 + u^9a^4b^2c^2(b+c) + u^8a^4b^2c^2)$$
$$+[-4 + 3[(\hat{a}\cdot\hat{b})^2 + (\hat{b}\cdot\hat{c})^2 + 3(\hat{c}\cdot\hat{a})^2] - 9(\hat{a}\cdot\hat{b})(\hat{b}\cdot\hat{c})(\hat{c}\cdot\hat{a})](u^{10}a^3b^4c^3 + u^9a^2b^4c^2(a+c) + u^8a^2b^4c^2)$$
$$+[-4 + 3[3(\hat{a}\cdot\hat{b})^2 + (\hat{b}\cdot\hat{c})^2 + (\hat{c}\cdot\hat{a})^2] - 9(\hat{a}\cdot\hat{b})(\hat{b}\cdot\hat{c})(\hat{c}\cdot\hat{a})](u^{10}a^3b^3c^4 + u^9a^2b^2c^4(a+b) + u^8a^2b^2c^4)$$
$$+[-6 + 9[(\hat{a}\cdot\hat{b})^2 + (\hat{b}\cdot\hat{c})^2 + (\hat{c}\cdot\hat{a})^2] - 27(\hat{a}\cdot\hat{b})(\hat{b}\cdot\hat{c})(\hat{c}\cdot\hat{a})](u^9a^3b^3c^3 + u^8a^2b^2c^2(ab+bc+ac)$$
$$+ u^7a^2b^2c^2(a+b+c) + u^6a^2b^2c^2)\}.$$

The eight individual direction cosine terms within braces are identical to those featuring in the triple dipole dispersion potential between three particles in arbitrary geometrical arrangement.

For three atoms or molecules positioned in an equilateral triangle configuration, Equation (50) yields the potential

$$\Delta E^{DO^1-DO^1-DO^1}_{Eq} = \frac{\hbar c}{512\pi^4\varepsilon_0^3 R^{15}} \int_0^\infty du e^{-3uR} \alpha^{DO^1}_{\lambda\lambda\mu\mu}(A;iu)\alpha^{DO^1}_{\nu\nu\pi\pi}(B;iu)\alpha^{DO^1}_{\rho\rho\sigma\sigma}(C;iu)$$
$$\times [7(uR)^{12} + 3(uR)^{11} + 24(uR)^{10} + 75(uR)^9 + 120(uR)^8 + 99(uR)^7 + 33(uR)^6]. \quad (51)$$

The coefficients match those computed in Equation (41). Even though there is no u-independent term, a short-range asymptote may nonetheless be extracted from Equation (51). This is done by substituting Equation (19) for the octupole weight-1 contribution to the mixed dipole-octupole polarisability evaluated at imaginary frequency and making the approximation $E_{x0}, E_{y0}, E_{z0} \ll \hbar cu$, so that the product of energy denominators simplifies to $(\hbar cu)^6$. The u-integral in Equation (51) therefore becomes

$$\int_0^\infty du \frac{e^{-3uR}}{u^6}[7(uR)^{12} + 3(uR)^{11} + 24(uR)^{10} + 75(uR)^9 + 120(uR)^8 + 99(uR)^7 + 33(uR)^6], \quad (52)$$

which may be evaluated using the integral result (32). This produces $10112R^5/243$, yielding a near-zone asymptote

$$\Delta E^{DO^1-DO^1-DO^1}_{Eq}(NZ) \approx \frac{10112}{243} \frac{1}{(15)^3(\hbar c)^5} \frac{1}{\pi^4\varepsilon_0^3 R^{10}} \sum_{x,y,z} E_{x0}E_{y0}E_{z0}$$
$$\times \mu^{0x}_\lambda(A) O^{(1)x0}_{\lambda\mu\mu}(A) \mu^{0y}_\nu(B) O^{(1)y0}_{\nu\nu\pi\pi}(B) \mu^{0z}_\rho(C) O^{(1)z0}_{\rho\rho\sigma\sigma}(C), \quad (53)$$

which exhibits R^{-10} behaviour.

At very long-range, the limiting form of the energy shift is

$$\Delta E^{DO^1-DO^1-DO^1}_{Eq}(FZ) = \frac{23709440\hbar c}{512 \times 3^8 \pi^4\varepsilon_0^3 R^{16}} \alpha^{DO^1}_{\lambda\lambda\mu\mu}(A;0)\alpha^{DO^1}_{\nu\nu\pi\pi}(B;0)\alpha^{DO^1}_{\rho\rho\sigma\sigma}(C;0), \quad (54)$$

with inverse separation distance exponent of sixteen.

For a collinear arrangement, the energy shift is

$$\Delta E^{DO^1-DO^1-DO^1}_{Coll} = \frac{\hbar c}{8\pi^4\varepsilon_0^3 R^{15}} \int_0^\infty du e^{-2uR} \alpha^{DO^1}_{\lambda\lambda\mu\mu}(A;iu)\alpha^{DO^1}_{\nu\nu\pi\pi}(B;iu)\alpha^{DO^1}_{\rho\rho\sigma\sigma}(C;iu)$$
$$\times [(uR)^{12} + 5(uR)^{11} + 17(uR)^{10} + 16(uR)^9 - 36(uR)^8 - 96(uR)^7 - 48(uR)^6], \quad (55)$$

with identical coefficients to that given in Equation (44). A limiting form of the energy shift which is dominant at short-range may be obtained in an identical manner to that carried out for the equilateral triangle case. Approximating $(k_{x0}^2 + u^2)(k_{y0}^2 + u^2)(k_{z0}^2 + u^2)$ in the energy denominators of the polarisabilities by u^6, the u-integral in Equation (55) is evaluated using Equation (32) to give

$$\int_0^\infty du \frac{e^{-2uR}}{u^6}[(uR)^{12} + 5(uR)^{11} + 17(uR)^{10} + 16(uR)^9 - 36(uR)^8 - 96(uR)^7 - 48(uR)^6] = -\frac{93}{4}R^5, \quad (56)$$

and a near-zone asymptote

$$\Delta E_{\text{Coll}}^{\text{DO}^1-\text{DO}^1-\text{DO}^1}(\text{NZ}) \approx -\frac{186}{(15)^3(\hbar c)^5} \frac{8}{\pi^4 \varepsilon_0^3 R^{10}} \sum_{x,y,z} E_{x0} E_{y0} E_{z0} \\ \times \mu_\lambda^{0x}(A) O_{\lambda\mu\mu}^{(1)x0}(A) \mu_\nu^{0y}(B) O_{\nu\pi\pi}^{(1)y0}(B) \mu_\rho^{0z}(C) O_{\rho\sigma\sigma}^{(1)z0}(C). \quad (57)$$

From Equation (55) the far-zone limit of the potential is

$$\Delta E_{\text{Coll}}^{\text{DO}^1-\text{DO}^1-\text{DO}^1}(\text{FZ}) = \frac{2207925\hbar c}{128\pi^4 \varepsilon_0^3 R^{16}} \alpha_{\lambda\lambda\mu\mu}^{\text{DO}^1}(A;0) \alpha_{\nu\nu\pi\pi}^{\text{DO}^1}(B;0) \alpha_{\rho\rho\sigma\sigma}^{\text{DO}^1}(C;0), \quad (58)$$

which displays an R^{-16} dependence.

6. DD-DO-OO Dispersion Potential

The final dispersion interaction energy to be computed involving the electric octupole moment is that between an electric dipole polarisable molecule, A, an electric dipole-octupole polarisable species, B, and a purely electric octupole polarisable particle, C. This potential is of the same order of magnitude as the energy shift considered in the previous section between three mixed electric dipole-octupole polarisable objects. In the present case the interaction Hamiltonian is

$$H_{\text{int}} = H_{\text{int}}^{\text{DD}}(A) + H_{\text{int}}^{\text{DO}}(B) + H_{\text{int}}^{\text{OO}}(C), \quad (59)$$

with the last contribution given by Equation (15). From third-order perturbation theory and summing the contributions from six time-ordered graphs, the potential for molecules in fixed mutual orientation is

$$\Delta E^{\text{DD}-\text{DO}-\text{OO}} = -\frac{\hbar c}{64\pi^4 \varepsilon_0^3} \int_0^\infty du \, \alpha_{ij}^{\text{DD}}(A;iu) \alpha_{klmn}^{\text{DO}}(B;iu) \alpha_{pqrstu}^{\text{OO}}(C;iu) L_{kpqr}(ua) L_{istu}(ub) L_{jlmn}(uc), \quad (60)$$

where $\alpha_{pqrstu}^{\text{OO}}(C;iu)$ is the pure electric octupole polarisability of C, Equation (3).

To obtain the interaction energy for randomly oriented molecules requires an average of $\alpha_{pqrstu}^{\text{OO}}(C;iu)$, a sixth-rank Cartesian tensor. Utilising the form of the octupole moment and the nature of its coupling to the transverse electric displacement field, the averaged quantity is [37]

$$<\alpha_{pqrstu}^{\text{OO}}(C;iu)> = \tfrac{14}{210}[\delta_{ps}\delta_{qr}\delta_{tu}\alpha_{\lambda\mu\mu\lambda\nu\nu}^{\text{OO}}(C;iu) \\ +2(\delta_{ps}\delta_{qt}\delta_{ru} + \delta_{ps}\delta_{qu}\delta_{rt} + \delta_{pt}\delta_{qs}\delta_{ru} + \delta_{pt}\delta_{qu}\delta_{rs} + \delta_{pu}\delta_{qs}\delta_{rt} + \delta_{pu}\delta_{qt}\delta_{rs})\alpha_{\lambda\mu\nu\lambda\mu\nu}^{\text{OO}}(C;iu)]. \quad (61)$$

Contracting tensor indices after multiplying factors from the average over each molecule, and using the relation $\alpha_{\lambda\mu\mu\lambda\nu\nu}^{\text{OO}}(C;iu) = \alpha_{\lambda\mu\mu\lambda\nu\nu}^{O^1O^1}(C;iu)$ and $\alpha_{\lambda\mu\nu\lambda\mu\nu}^{\text{OO}}(C;iu) = \tfrac{3}{5}\alpha_{\lambda\mu\mu\lambda\nu\nu}^{O^1O^1}(C;iu) + \alpha_{\lambda\mu\nu\lambda\mu\nu}^{O^3O^3}(C;iu)$, which follow from Equations (16)–(18), an explicit expression for the energy shift (60) in terms of octupole weights is

$$\Delta E^{\text{DD-DO-OO}} = -\frac{14\hbar c}{210 \times 32\pi^4 \varepsilon_0^3} \int_0^\infty du\, \alpha^{\text{DD}}(A; iu)\alpha^{\text{DO}^1}_{\rho\rho\sigma\sigma}(B; iu)$$
$$\times [\alpha^{O^1 O^1}_{\lambda\mu\mu\lambda\nu\nu}(C; iu) L_{jkll}(ua) L_{ikmm}(ub) L_{ijnn}(uc) + 4[\tfrac{3}{5}\alpha^{O^1 O^1}_{\lambda\mu\mu\lambda\nu\nu}(C; iu) + \alpha^{O^3 O^3}_{\lambda\mu\nu\lambda\mu\nu}(C; iu)] \qquad (62)$$
$$\times \left\{ L_{jklm}(ua)[L_{iklm}(ub) + L_{ilkm}(ub) + L_{imkl}(ub)] L_{ijnn}(uc) \right\}].$$

It is interesting to note that apart from pre-factors, Equation (62) is identical to the DD-DD-OO dispersion potential given by Equation (53) of Ref. [15], or if expressed in terms of reducible components of the octupole moment, is equivalent to Equation (46) of Ref. [15]. This recognition is arrived at on realising that $L_{ijnn}(uR) = u^2 F_{ij}(uR)$. Thus energy shift formulae for particular geometrical arrangements may be written down immediately from the results presented in Section VI of Ref. [14].

For the pure electric octupole polarisability having implicit dependence upon octupole weight-1 and -3 dependent terms, where 1/3 is factored into $\alpha^{\text{DD}}(A; iu)$, a factor of 2/15 is absorbed into $\alpha^{\text{DO}^1}_{\rho\rho\sigma\sigma}(B; iu)$, but the factor 14/210 is retained explicitly, the dispersion energy for an equilateral triangle configuration is

$$\Delta E^{\text{DD-DO}^1\text{-OO}}_{\text{Eq}} = \frac{\hbar c}{128 \times 60\pi^4 \varepsilon_0^3 R^{15}} \int_0^\infty du\, u^2 e^{-3uR} \alpha^{\text{DD}}(A; iu)\alpha^{\text{DO}^1}_{\rho\rho\sigma\sigma}(B; iu)\alpha^{\text{OO}}_{\lambda\mu\mu\lambda\nu\nu}(C; iu)$$
$$\times [7(uR)^{10} + 3(uR)^9 + 24(uR)^8 + 75(uR)^7 + 120(uR)^6 + 99(uR)^5 + 33(uR)^4]$$
$$+ \frac{\hbar c}{128 \times 60\pi^4 \varepsilon_0^3 R^{15}} \int_0^\infty du\, u^2 e^{-3uR} \alpha^{\text{DD}}(A; iu)\alpha^{\text{DO}^1}_{\rho\rho\sigma\sigma}(B; iu)\alpha^{\text{OO}}_{\lambda\mu\nu\lambda\mu\nu}(C; iu) \qquad (63)$$
$$\times [13(uR)^{10} + 119(uR)^9 + 785(uR)^8 + 2784(uR)^7 + 5307(uR)^6 + 3789(uR)^5$$
$$- 1446(uR)^4 + 9441(uR)^3 + 46332(uR)^2 + 58725uR + 19575],$$

with coefficients identical to that found in the DD-DD-OO interaction energy. The additional factor u^2/R^2 in each integral term of Equation (63) ensures the potential is entirely retarded, containing no uR-independent terms, as expected since the mixed dipole-octupole polarisability of B is independent of the octupole weight-3 term. A form applicable at very short range may be obtained on retaining the u-independent term in the second integral of Equation (63) and using the integral result (A14). This is found to be

$$\Delta E^{\text{DD-DO}^1\text{-OO}}_{\text{Eq}}(\text{NZ}) \approx \frac{29}{64\hbar^2 c^2 \pi^3 \varepsilon_0^3 R^{15}} \sum_{x,y,z} |\vec{\mu}^{0x}(A)|^2 \mu^{0y}_\rho(B) O^{(1)y0}_{\rho\sigma\sigma}(B) O^{0z}_{\lambda\mu\nu}(C) O^{z0}_{\lambda\mu\nu}(C)$$
$$\times \frac{E_{x0} E_{y0} E_{z0}}{(E_{x0}+E_{y0})(E_{x0}+E_{z0})(E_{y0}+E_{z0})}, \qquad (64)$$

displaying inverse fifteenth power dependence.

With similar definitions for the isotropic polarisabilities, the dispersion potential for collinear geometry is

$$\Delta E^{\text{DD-DO}^1\text{-OO}}_{\text{Coll}} = \frac{\hbar c}{120\pi^4 \varepsilon_0^3 R^{15}} \int_0^\infty du\, u^2 e^{-2uR} \alpha^{\text{DD}}(A; iu)\alpha^{\text{DO}^1}_{\rho\rho\sigma\sigma}(B; iu)\alpha^{\text{OO}}_{\lambda\mu\mu\lambda\nu\nu}(C; iu)$$
$$\times [(uR)^{10} + 5(uR)^9 + 17(uR)^8 + 16(uR)^7 - 36(uR)^6 - 96(uR)^5 - 48(uR)^4]$$
$$+ \frac{\hbar c}{30\pi^4 \varepsilon_0^3 R^{15}} \int_0^\infty du\, u^2 e^{-2uR} \alpha^{\text{DD}}(A; iu)\alpha^{\text{DO}^1}_{\rho\rho\sigma\sigma}(B; iu)\alpha^{\text{OO}}_{\lambda\mu\nu\lambda\mu\nu}(C; iu) \qquad (65)$$
$$\times [(uR)^{10} + 21(uR)^9 + 273(uR)^8 + 2498(uR)^7 + 14790(uR)^6 + 352880(uR)^5$$
$$+ 127576(uR)^4 + 50688(uR)^3 - 57344(uR)^2 - 1382400uR - 691200],$$

which also contains no uR-independent term. Using the result (A14) a short-range limiting form of Equation (65) may be obtained as

$$\Delta E_{\text{Coll}}^{\text{DD-DO}^1-\text{OO}}(\text{NZ}) \approx -\frac{4096}{\hbar^2 c^2 \pi^3 \varepsilon_0^3 R^{15}} \sum_{x,y,z} |\vec{\mu}^{0x}(A)|^2 \mu_\rho^{0y}(B) O_{\rho\sigma\sigma}^{(1)y0}(B) O_{\lambda\mu\nu}^{0z}(C) O_{\lambda\mu\nu}^{z0}(C) \times \frac{E_{x0} E_{y0} E_{z0}}{(E_{x0}+E_{y0})(E_{x0}+E_{z0})(E_{y0}+E_{z0})},$$ (66)

which has identical power law behaviour as result Equation (64).

7. Summary

A systematic study has been peformed of dispersion interactions between three molecules when the effects of electric octupole coupling have been accounted for, supplementing a previously published result involving two electric dipole polarisable species, and a third that is pure electric octupole polarisable. This has been carried out using the theory of molecular QED, in which the electromagnetic field is quantised and interactions between non-relativistic microscopic particles take place via the exchange of one or more virtual photons. As in the case of pair dispersion potentials, the transfer of two virtual photons between each interacting pair mediates coupling between three molecules in the ground electronic state, with the radiation field in the vacuum state. To simplify the computations, for instance by considerably reducing the number of time-ordered diagrams that have to be summed over, an extension to higher multipoles of the Craig–Power Hamiltonian operator was adopted instead of the usual interaction Hamiltonian that is linear in the Maxwell field operator. This alternate perturbation operator, which has the form of an effective two-photon coupling operator, enables third-order perturbation theory to be used in the evaluation of the three-body dispersion potential.

Specific energy shifts calculated include that between two electric dipole polarisable molecules and one that is mixed electric dipole-octupole polarisable; one electric dipole polarisable molecule and two mixed dipole-octupole polarisable molecules; and three mixed electric dipole-octupole polarisable molecules. Also computed was the potential between an electric dipole polarisable molecule, an octupole polarisable species, and a mixed dipole-octupole polarisable molecule, which is of the same order as the DO-DO-DO interaction. Important insight into the results obtained was gained by decomposing the octupole moment into its irreducible components of weights-1 and -3. The weight-1 dependent contributions to each of the potentials contained sums of direction cosine terms that preceded polynomial terms in various powers of u, a, b, and c that were identical to that found in the leading non-pairwise additive triple-dipole contribution to the energy shift, with the DD-DO^1-DO^1 and previously obtained DD-DD-O^{1}O^1 contributions being viewed as higher-order correction terms to the DD-DD-DD potential.

Interestingly, for isotropic energy shifts involving mixed dipole-octupole polarisable species, the interaction energies are wholly retarded, containing no static terms. Furthermore, the octupole weight-3 term of this tensor vanishes on random orientational averaging, leaving a dependence solely on the weight-1 contribution. Nevertheless, evaluation of the u-integral for small displacements of the three particles may be used to obtain an energy shift valid in the near-zone. Explicit expressions for dispersion potentials were also given for equilateral triangle and collinear arrangements of the three molecules for each of the multipole moment combinations considered. The hierarchy of emerging power laws in the near-zone can be understood from the fact that the nonretarded DD-DD-DD potential is proportional to R^{-9}, where each replacement of a dipole with an octupole leads to a factor of the order $(a/R)^2 \ll 1$, where a represents the extent of the electronic wave function. On top of this, the absence of true static terms leads to factors $(kR)^m \ll 1$, where m is zero or a positive integer and k is the wave number of the radiation exchanged between the molecules. Note that DO-DO and DO-DO-DO interactions are special cases where the exact balance between $(a/R)^4$ and $(a/R)^6$, respectively with $(kR)^4$ and $(kR)^6$ leads to an additional factor R^{-1} arising from a Casimir–Polder type integral. The emerging power laws for pair and three-body interactions are shown in Table 1.

Table 1. Short-range dependences of dispersion potentials: Near-zone limiting behaviour of various equally displaced two- and three-body dispersion energy shifts involving electric dipole (D), quadrupole (Q) and octupole (O) couplings.

Multipole Coupling	Near-Zone Power Law	Ref.
DD-DD	R^{-6}	[9–11]
DD-QQ	$R^{-6} \times (a/R)^2 \times (kR)^0 \sim R^{-8}$	[29]
DD-DO	$R^{-6} \times (a/R)^2 \times (kR)^2 \sim R^{-6}$	[30]
DO-DO	$R^{-6} \times (a/R)^4 \times (kR)^4 \times R^{-1} \sim R^{-7}$	[30]
DD-OO	$R^{-6} \times (a/R)^4 \times (kR)^3 \sim R^{-7}$	[29]
DD-DD-DD	R^{-9}	[26]
DD-DD-QQ	$R^{-9} \times (a/R)^2 \times (kR)^0 \sim R^{-11}$	[14]
DD-DD-DO	$R^{-9} \times (a/R)^2 \times (kR)^3 \sim R^{-8}$	
DD-QQ-QQ	$R^{-9} \times (a/R)^4 \times (kR)^0 \sim R^{-13}$	[14]
DD-DO-DO	$R^{-9} \times (a/R)^4 \times (kR)^4 \sim R^{-9}$	
DO-DO-DO	$R^{-9} \times (a/R)^6 \times (kR)^6 \times R^{-1} \sim R^{-10}$	
DD-DO-OO	$R^{-9} \times (a/R)^6 \times (kR)^0 \sim R^{-15}$	

It is also worth highlighting that the integrals over imaginary wave vector evaluated in the Appendices may be used to calculate the sub-dominant contribution to the near-zone potential between an electric dipole polarisable molecule and an electric octupole polarisable one. This two-body potential has been calculated previously [29], and is

$$\Delta E^{DD-OO} = -\frac{\hbar c}{1200\pi^3\epsilon_0^2 R^6}\int_0^\infty du\, u^4 e^{-2uR}\alpha^{DD}(A;iu)\alpha^{O^1O^1}_{\lambda\mu\mu\lambda\nu\nu}(B;iu)[(uR)^4 + 2(uR)^3 + 5(uR)^2 + 6(uR) + 3]$$
$$-\frac{\hbar c}{420\pi^3\epsilon_0^2 R^{10}}\int_0^\infty du\, e^{-2uR}\alpha^{DD}(A;iu)\alpha^{O^3O^3}_{\lambda\mu\nu\lambda\mu\nu}(B;iu)[(uR)^8 + 12(uR)^7 + 90(uR)^6 + 486(uR)^5 \qquad (67)$$
$$+1863(uR)^4 + 4950(uR)^3 + 8775(uR)^2 + 9450(uR) + 4725].$$

Strictly speaking there is no contribution to the conventional near-zone limit arising from the first term of Equation (67), that dependent upon octupole weight-1, because there is no uR-independent term. A short-range asymptote may be arrived at by approximating $\hbar c u \gg E_{x0}, E_{y0}$ in the polarisabilities and using Equation (32) to evaluate the resulting u-integral, giving

$$\Delta E^{DD-O^1O^1}(NZ) \approx -\frac{23}{3600\hbar c\pi^3\epsilon_0^2 R^7}\sum_{x,y}k_{x0}k_{y0}|\vec{\mu}^{0x}(A)|^2 O^{(1)0y}_{\lambda\mu\mu}(B)O^{(1)y0}_{\lambda\nu\nu}(B), \qquad (68)$$

exhibiting a Casimir-like inverse seventh power dependence. The typical near-zone limit, arising from the second term of Equation (67), that dependent upon octupole weight-3, in contrast displays R^{-10} behaviour. Again, the unexpected behaviour stems from an additional small factor $(kR)^3$, as shown in Table 1. In this context it is useful to remark that short- and long-range expansions of the Casimir–Polder dispersion potential, as well as all correction terms up to second order in the fine structure constant have been performed from consideration of the orbit-orbit contribution due to the Breit-Pauli Hamiltonian, including relativistic effects [38,39], and compared with recent molecular QED calculations [30].

Finally, it is worth pointing out that the ratio of the limiting forms of the triple dipole dispersion potential for an equilateral triangle configuration to the pair potential is $\Delta E_3/\Delta E_2 = \alpha(0)\epsilon_0^{-1}R^{-3}$, where $\alpha(0)$ is the static polarisability, indicating that for small values of this quantity and large separations the triple dipole energy shift is appreciably weaker than the corresponding two-body contribution. Interestingly, taking the electric dipole moment to be ea_0, where e is the proton charge and a_0 is the Bohr radius, and the transition energy to be of the order of one Rydberg, the ratio $\Delta E_3/\Delta E_2$ is unity at distances of around $3a_0$, with ΔE_3 increasing in importance at larger distances.

Funding: SYB acknowledges support by the Deutsche Forschungsgemeinschaft (grant BU 1803/3-1476), and the Freiburg Institute for Advanced Studies (FRIAS). AS acknowledges the award of a Mercator Fellowship funded by the Deutsche Forschungsgemeinschaft through the IRTG 2079/Cold Controlled Ensembles in Physics and Chemistry at the University of Freiburg.

Conflicts of Interest: The authors declare no conflict of interest.

Appendix A. Wave Vector Integrals Involving Single Polarisability

When calculating Casimir–Polder energy shifts between an atom and a macroscopic object such as a plate or a slab, wave vector integrals that feature a single atomic polarisability are encountered of the form

$$\int_0^\infty du \frac{u^n}{a^2 + u^2} e^{-uR/c} P(uR/c), \tag{A1}$$

with

$$P(x) = \sum_{n=0}^{n'} P_n x^n, \tag{A2}$$

and where $P_0 = 1$. Then for $aR/c \ll 1$, corresponding to the near-zone, we have for:
(i) $n = 0$,

$$\int_0^\infty du \frac{e^{-uR/c}}{a^2 + u^2} P(uR/c) \approx \int_0^\infty du \frac{1}{a^2 + u^2} = \frac{\pi}{2a}. \tag{A3}$$

(ii) $n = 1$,

$$\int_0^\infty du \frac{u e^{-uR/c}}{a^2 + u^2} P(uR/c) \approx \int_0^\infty dx \frac{x e^{-x}}{\left(\frac{aR}{c}\right)^2 + x^2} \approx -\ln\left(\frac{aR}{c}\right), \tag{A4}$$

on letting $x = uR/c$ and with $P(x) \sim 1$.
(iii) $n = 2$,

$$\int_0^\infty du \frac{u^2 e^{-uR/c}}{a^2 + u^2} P(uR/c) \approx \int_0^\infty du\, e^{-uR/c} P(uR/c), \tag{A5}$$

since $u^2 \gg a^2$, and which may be evaluated using Equation (32). Hence on collecting results,

$$\int_0^\infty du \frac{u^n e^{-uR/c}}{a^2 + u^2} P(uR/c) = \begin{cases} \frac{\pi}{2a}, & n = 0 \\ -\ln(aR/c), & n = 1 \\ \left(\frac{c}{R}\right)^{n-1} \int_0^\infty dx\, x^{n-2} e^{-x} P(x), & n \geq 2 \end{cases}, \tag{A6}$$

with $n \geq 2$ evaluated using Equation (32).

Appendix B. Two Polarisabilities

In calculations of the Casimir–Polder dispersion interaction between two atoms, wave vector integrals involving two polarisbilities are encountered of the form

$$\int_0^\infty du \frac{u^n}{(a^2 + u^2)(b^2 + u^2)} e^{-uR/c} P(uR/c). \tag{A7}$$

On letting aR/c and $bR/c \ll 1$, we have for:
(i) $n = 0$,

$$\int_0^\infty du \frac{1}{(a^2 + u^2)(b^2 + u^2)} e^{-uR/c} P(uR/c) \approx \int_0^\infty du \frac{1}{(a^2 + u^2)(b^2 + u^2)} = \frac{\pi}{2ab(a+b)}, \tag{A8}$$

and which is $\pi/4a^3$, for $b = a$.

(ii) $n = 1$,

$$\int_0^\infty du \frac{u}{(a^2+u^2)(b^2+u^2)} e^{-uR/c} P(uR/c) = \left(\frac{R}{c}\right)^2 \int_0^\infty dx\, x e^{-x} \frac{1}{\left[\left(\frac{aR}{c}\right)^2+x^2\right]\left[\left(\frac{bR}{c}\right)^2+x^2\right]} \approx \frac{1}{b^2-a^2} \ln\left(\frac{b}{a}\right), \quad (A9)$$

on substituting $x = uR/c$. For $b = a$, (A9) becomes $1/2a^2$ by L'Hôpital's Rule.

(iii) $n = 2$,

$$\int_0^\infty du \frac{u^2}{(a^2+u^2)(b^2+u^2)} e^{-uR/c} P(uR/c) = \left(\frac{R}{c}\right) \int_0^\infty dx\, x^2 e^{-x} \frac{1}{\left[\left(\frac{aR}{c}\right)^2+x^2\right]\left[\left(\frac{bR}{c}\right)^2+x^2\right]}$$
$$\approx \frac{R}{c} \int_0^\infty dx\, x^2 \frac{1}{\left[\left(\frac{aR}{c}\right)^2+x^2\right]\left[\left(\frac{bR}{c}\right)^2+x^2\right]} = \frac{\pi}{2(a+b)}, \quad (A10)$$

where the last integral is evaluated via

$$\int_0^\infty du \frac{u^2}{(a^2+u^2)(b^2+u^2)} = \frac{\pi}{2(a+b)}.$$

For $b = a$ (A10) simplifies to $\pi/4a$.

(iii) $n = 3$,

$$\int_0^\infty du \frac{u^3}{(a^2+u^2)(b^2+u^2)} e^{-uR/c} P(uR/c) = \int_0^\infty dx\, x^3 e^{-x} \frac{1}{\left[\left(\frac{aR}{c}\right)^2+x^2\right]\left[\left(\frac{bR}{c}\right)^2+x^2\right]} \quad (A11)$$
$$\approx \frac{1}{b^2-a^2}[a^2 \ln(aR/c) - b^2 \ln(bR/c)],$$

which tends to $-\ln(aR/c)$ for $b = a$.

(iv) $n \geq 4$,

$$\int_0^\infty du \frac{u^n}{(a^2+u^2)(b^2+u^2)} e^{-uR/c} P(uR/c) \approx \left(\frac{c}{R}\right)^{n-3} \int_0^\infty dx\, x^{n-4} e^{-x} P(x), \quad (A12)$$

which may be evaluated using

$$\int_0^\infty dx\, x^n e^{-\eta x} = n! \eta^{-n-1}, \quad \mathrm{Re}\,\eta > 0.$$

The approximation $u^2 \gg a^2, b^2$ has been made.

Appendix C. Three Polarisabilities

In three-body energy shifts involving an electric octupole moment, to arrive at a short-range limit we are required to evaluate integrals of the form

$$\int_0^\infty du \frac{u^n e^{-uR/c}}{(a^2+u^2)(b^2+u^2)(d^2+u^2)} P(uR/c), \quad (A13)$$

where $P(uR/c)$ is the polynomial function, and whose constant, u-independent term is kept, i.e., $P(uR/c) = 1$. u is transformed to $x = uR/c$, so that for $n = 2$ (A13) becomes

$$\int_0^\infty du \frac{u^2 e^{-uR/c} P(uR/c)}{(a^2+u^2)(b^2+u^2)(d^2+u^2)} = \left(\frac{R}{c}\right)^3 \int_0^\infty dx\, x^2 e^{-x} \frac{1}{\left[\left(\frac{aR}{c}\right)^2+x^2\right]\left[\left(\frac{bR}{c}\right)^2+x^2\right]\left[\left(\frac{dR}{c}\right)^2+x^2\right]} \approx \frac{\pi}{2} \frac{1}{(a+b)(a+d)(b+d)}, \quad (A14)$$

and which tends to $\pi/16a^3$ for $a = b = d$.

For $n = 3$

$$\left(\frac{R}{c}\right)^2 \int_0^\infty dx\, x^3 e^{-x} \frac{1}{\left[\left(\frac{aR}{c}\right)^2+1\right]\left[\left(\frac{bR}{c}\right)^2+1\right]\left[\left(\frac{dR}{c}\right)^2+1\right]} \approx -\left\{\frac{a^2(b^2-d^2)\ln(aR/c)+b^2(d^2-a^2)\ln(bR/c)+d^2(a^2-b^2)\ln(dR/c)}{(b^2-d^2)(d^2-a^2)(a^2-b^2)}\right\}, \quad \text{(A15)}$$

which equals $1/4a^2$ for $a = b = d$.

For $n = 4$, we have $\int_0^\infty du\, \frac{u^4 e^{-uR/c}}{(a^2+u^2)(b^2+u^2)(d^2+u^2)} P(uR/c)$. Taking the leading term of the polynomial expansion, and transforming to x, we have

$$\frac{R}{c}\int_0^\infty dx \frac{x^4 e^{-x}}{\left[\left(\frac{aR}{c}\right)^2 + x^2\right]\left[\left(\frac{bR}{c}\right)^2 + x^2\right]\left[\left(\frac{dR}{c}\right)^2 + x^2\right]} \approx \frac{\pi}{2}\left[\frac{ab+ad+bd}{(a+b)(a+d)(b+d)}\right], \quad \text{(A16)}$$

and which equals $3\pi/16a$ for $a = b = d$.

References

1. Maitland, G.C.; Rigby, M.; Smith, E.B.; Wakeham, W.A. *Intermolecular Forces*; Clarendon: Oxford, UK, 1981.
2. Stwalley, W.C.; Wang, H. Photoassociation of Ultracold Atoms: A New Spectroscopic Technique. *J. Mol. Spec.* **1999**, *195*, 194–228. [CrossRef] [PubMed]
3. Gerton, J.M.; Strekalov, D.; Prodan, I.; Hulet, R.G. Direct Observation of Growth and Collapse of a Bose-Einstein Condensate With Attractive Interactions. *Nature* **2000**, *408*, 692–695. [CrossRef] [PubMed]
4. Buhmann, S.Y.; Welsch, D.-G. Born Expansion of the Casimir-Polder Interaction of a Ground-State Atom with Dielectric Bodies. *Appl. Phys. B* **2006**, *82*, 189–201. [CrossRef]
5. Golestanian, R. Casimir-Lifshitz Interaction Between Dielectrics of Arbitrary Geometry: A Dielectric Contrast Perturbation Theory. *Phys. Rev. A* **2009**, *80*, 012509. [CrossRef]
6. Cvitas, M.J.; Soldan, P.; Hutson, J.M. Long-Range Intermolecular Forces in Triatomic Systems: Connecting the Atom-Diatom and Atom-Atom-Atom Representations. *Mol. Phys.* **2006**, *104*, 23–31. [CrossRef]
7. Tang, L.-Y.; Yan, Z.-C.; Shi, T.-Y.; Babb, J.F.; Mitroy, J. The Long-Range Non-Additive Three-Body Dispersion Interactions for the Rare Gases, Alkali, and Alkaline-Earth Atoms. *J. Chem. Phys.* **2012**, *136*, 104104. [CrossRef] [PubMed]
8. Schwinger, J. *Selected Papers on Quantum Electrodynamics*; Dover: New York, NY, USA, 1958.
9. Craig, D.P.; Thirunamachandran, T. *Molecular Quantum Electrodynamics*; Dover: New York, NY, USA, 1998.
10. Salam, A. *Molecular Quantum Electrodynamics*; John Wiley & Sons, Inc.: Hoboken, NJ, USA, 2010.
11. Salam, A. *Non-Relativistic QED Theory of the van der Waals Dispersion Interaction*; Springer: Cham, Switzerland, 2016.
12. Andrews, D.L.; Jones, G.A.; Salam, A.; Woolley, R.G. Perspective: Quantum Hamiltonians for Optical Interactions. *J. Chem. Phys.* **2018**, *148*, 040901. [CrossRef] [PubMed]
13. Buhmann, S.Y. *Dispersion Forces I*; Springer: Heidelberg, Germany, 2012.
14. Salam, A. Higher-Order Electric Multipole Contributions to Retarded Non-Additive Three-Body Dispersion Interaction Energies Between Atoms: Equilateral Triangle and Collinear Configurations. *J. Chem. Phys.* **2013**, *139*, 244105. [CrossRef] [PubMed]
15. Salam, A. Dispersion Potential Between Three-Bodies With Arbitrary Electric Multipole Polarizabilities: Molecular QED Theory. *J. Chem. Phys.* **2014**, *140*, 044111. [CrossRef] [PubMed]
16. Aldegunde, J.; Salam, A. Dispersion Energy Shifts Among N Bodies With Arbitrary Electric Multipole Polarizability: Molecular QED Theory. *Mol. Phys.* **2015**, *113*, 226–231. [CrossRef]
17. Aub, M.R.; Zienau, S. Studies on the Retarded Interaction Between Neutral Atoms. I. Three-Body London-van der Waals Interaction Between Neutral Atoms. *Proc. R. Soc. Lond. A* **1960**, *257*, 464–476. [CrossRef]
18. McLachlan, A.D. Three-Body Dispersion Forces. *Mol. Phys.* **1963**, *6*, 423–427. [CrossRef]
19. Power, E.A.; Thirunamachandran, T. The Non-Additive Dispersion Energies for N Molecules: A Quantum Electrodynamical Theory. *Proc. R. Soc. Lond. A* **1985**, *401*, 267–279. [CrossRef]

20. Power, E.A.; Thirunamachandran, T. Zero-Point Energy Differences and Many-Body Dispersion Forces. *Phys. Rev. A* **1994**, *50*, 3929–3939. [CrossRef] [PubMed]
21. Cirone, M.; Passante, R. Dressed Zero-Point Field Correlations and the Non-Additive Three-Body van der Waals Potential. *J. Phys. B At. Mol. Opt. Phys.* **1997**, *30*, 5579–5585. [CrossRef]
22. Passante, R.; Power, E.A.; Thirunamachandran, T. Radiation-Molecule Coupling Using Dynamic Polarizabilities: Application to Many-Body Forces. *Phys. Lett. A* **1998**, *249*, 77–82. [CrossRef]
23. Passante, R.; Persico, F. Virtual Photons and Three-Body Forces. *J. Phys. B At. Mol. Opt. Phys.* **1999**, *32*, 19–25. [CrossRef]
24. Passante, R.; Persico, F.; Rizzuto, L. Vacuum Field Correlations and Three-Body Casimir-Polder Potential with One Excited Atom. *J. Mod. Opt.* **2005**, *52*, 1957–1964. [CrossRef]
25. Passante, R.; Persico, F.; Rizzuto, L. Causality, Non-locality and Three-Body Casimir-Polder Energy Between Three Ground-State Atoms. *J. Phys. B At. Mol. Opt. Phys.* **2006**, *39*, S685–S694. [CrossRef]
26. Axilrod, B.M.; Teller, E. Interaction of the van der Waals Type Between Three Atoms. *J. Chem. Phys.* **1943**, *11*, 299–300. [CrossRef]
27. Muto, Y. The Force Between Nonpolar Molecules. *J. Phys. Math. Soc. Jpn.* **1943**, *17*, 629–631.
28. Li, X.; Hunt, K.L.C. Nonadditive, Three-Body Dipole Forces on Nuclei: New Interrelations and an Electrostatic Interpretation. *J. Chem. Phys.* **1996**, *105*, 4076–4093. [CrossRef]
29. Salam, A.; Thirunamachandran, T. A New Generalisation of the Casimir-Polder Potential to Higher Electric Multipole Polarisabilities. *J. Chem. Phys.* **1996**, *104*, 5094–5099. [CrossRef]
30. Salam, A. Corrections to the Casimir-Polder Potential Arising from Electric Octupole Coupling. *Mol. Phys.* **2018**. [CrossRef]
31. Salam, A. A General Formula for the Rate of Resonant Energy Transfer Between Two electric Multipole Moments of Arbitrary Order Using Molecular Quantum Electrodynamics. *J. Chem. Phys.* **2005**, *122*, 044112. [CrossRef] [PubMed]
32. Milonni, P.W. *The Quantum Vacuum*; Academic Press: San Diego, CA, USA, 1994.
33. Craig, D.P.; Power, E.A. The Asymptotic Casimir-Polder Potential From Second-Order Perturbation Theory and Its Generalization for Anisotropic Polarizabilities. *Int. J. Quant. Chem.* **1969**, *3*, 903–911. [CrossRef]
34. Salam, A. Virtual Photon Exchange, Intermolecular Interactions and Optical Response Functions. *Mol. Phys.* **2015**, *113*, 3645–3653. [CrossRef]
35. Andrews, D.L.; Bradshaw, D.S. The Role of Virtual Photons in Nanoscale Photonics. *Ann. Phys.* **2014**, *526*, 173–186. [CrossRef]
36. Salam, A. Quantum Electrodynamics Effects in Atoms and Molecules. *WIREs Comput. Mol. Sci.* **2015**, *5*, 178–201. [CrossRef]
37. Andrews, D.L.; Thirunamachandran, T. On Three-Dimensional Rotational Averages. *J. Chem. Phys.* **1977**, *67*, 5026–5033. [CrossRef]
38. Meath, W.J.; Hirschfelder, J.O. Long-Range (Retarded) Intermolecular Forces. *J. Chem. Phys.* **1966**, *44*, 3210–3215. [CrossRef]
39. Pachucki, K. Relativistic Corrections to the Long-Range Interaction Between Closed Shell Atoms. *Phys. Rev. A* **2005**, *72*, 062706. [CrossRef]

© 2018 by the authors. Licensee MDPI, Basel, Switzerland. This article is an open access article distributed under the terms and conditions of the Creative Commons Attribution (CC BY) license (http://creativecommons.org/licenses/by/4.0/).

Article

Resonance Dipole–Dipole Interaction between Two Accelerated Atoms in the Presence of a Reflecting Plane Boundary

Wenting Zhou [1,2,3,†], Roberto Passante [3,4,†] and Lucia Rizzuto [3,4,*,†]

1. Center for Nonlinear Science and Department of Physics, Ningbo University, Ningbo 315211, Zhejiang, China; zhouwenting@nbu.edu.cn
2. China Key Laboratory of Low Dimensional Quantum Structures and Quantum Control of Ministry of Education, Hunan Normal University, Changsha 410081, Hunan, China
3. Dipartimento di Fisica e Chimica, Università degli Studi di Palermo, Via Archirafi 36, I-90123 Palermo, Italy; roberto.passante@unipa.it
4. INFN, Laboratori Nazionali del Sud, I-95123 Catania, Italy
* Correspondence: lucia.rizzuto@unipa.it; Tel.: +39-091-238-91744
† These authors contributed equally to this work.

Received: 22 April 2018; Accepted: 20 May 2018; Published: 28 May 2018

Abstract: We study the resonant dipole–dipole interaction energy between two non-inertial identical atoms, one excited and the other in the ground state, prepared in a correlated *Bell-type* state, and interacting with the scalar field or the electromagnetic field nearby a perfectly reflecting plate. We suppose the two atoms move with the same uniform acceleration, parallel to the plane boundary, and that their separation is constant during the motion. By separating the contributions of radiation reaction field and vacuum fluctuations to the resonance energy shift of the two-atom system, we show that Unruh thermal fluctuations do not affect the resonance interaction, which is exclusively related to the radiation reaction field. However, non-thermal effects of acceleration in the radiation-reaction contribution, beyond the Unruh acceleration temperature equivalence, affect the resonance interaction energy. By considering specific geometric configurations of the two-atom system relative to the plate, we show that the presence of the mirror significantly modifies the resonance interaction energy between the two accelerated atoms. In particular, we find that new and different features appear with respect to the case of atoms in the free-space, related to the presence of the boundary and to the peculiar structure of the quantum electromagnetic field vacuum in the locally inertial frame. Our results suggest the possibility to exploit the resonance interaction between accelerated atoms as a probe for detecting the elusive effects of atomic acceleration on radiative processes.

Keywords: dipole–dipole interaction; Unruh effect; quantum field theory in curved space

1. Introduction

Quantum field theory in accelerated backgrounds has led to deep insights into the fundamental notions of *vacuum* and *particles*, forcing us to reconsider these basic concepts as observer-dependent notions. A prominent example of this feature is given by the Unruh effect [1], affirming that an observer moving with constant acceleration in the Minkowski vacuum feels a *thermal bath* at an Unruh temperature proportional to its proper acceleration, a:

$$T_U = \frac{\hbar}{2\pi k_B c} a, \tag{1}$$

where c is the speed of light, $\hbar$ the Planck constant, and k_B is the Boltzmann constant.

An analogous effect, in a curved space-time, is the Hawking radiation from a black hole: a free-falling observer outside a black hole should experience a bath of thermal radiation at the temperature $T_H = \hbar g/(2\pi k_B c)$, g being the local acceleration due to gravity at the event horizon [2].

As paradoxical as the concept of thermal radiation from vacuum may appear, the Unruh effect is a clear manifestation of the *non-unicity* of the notion of quantum vacuum (and of particles), as extensively discussed in the seminal paper by Fulling [3] and in following papers on the subject [4,5]. This conceptually subtle effect, merging classical general relativity and quantum field theory, has been the object of intense investigations in the literature, with different and sometimes conflicting conclusions on its physical interpretation [6–12]. Additionally, from Equation (1) (cgs units), we have

$$T_U \sim \left(10^{-23} a\right) \text{ K}, \qquad (2)$$

and therefore extremely high accelerations, of the order of 10^{23} cm/s^2, are necessary to obtain an Unruh thermal bath of a few kelvin, thus making the detection of this effect in the laboratory drastically difficult [6,8,13–18]. Whilst the absence of any experimental observation of the Unruh effect has led to question the reality of the effect [12], it has been argued that the Unruh effect is a fundamental requirement to ensure the consistency of quantum field theory [19]. In any case, a direct verification of the effect, and in general of acceleration-dependent effects, could allow us to solve some fundamental controversies about its physical interpretation.

Recently, the effects of an accelerated motion on the radiative properties of atoms/molecules in vacuum have been discussed in the literature [20–26]. Changes in the spontaneous emission rate [20,27–29] or in the Lamb shift of single uniformly accelerating atoms [21,22], as well as the dispersion Casimir–Polder interaction between a uniformly accelerated atom and a reflecting plate [30–34] or between two uniformly accelerated atoms [35,36], have been investigated, and their relation with the Unruh effect was discussed. The effect of non-equilibrium boundaries on radiative properties of atoms has been also considered [37,38].

Another, albeit related, problem, recently addressed in the literature, concerns the equivalence between acceleration and temperature. For example, it has been discussed that non-thermal features (related to a uniform acceleration) manifest in the dispersion (van der Waals/Casimir–Polder) and resonance interaction between non inertial atoms in the free-space [25,26,36,39]. These investigations reveal that the effects of a uniform acceleration are not always equivalent to Unruh thermal effects.

Motivated by these issues, in this paper, we investigate the effect of a non-inertial motion on the resonance interaction between two atoms, that accelerate with the same constant acceleration, parallel to a reflecting plate. The imposition of boundary conditions on the quantum field on the plate changes vacuum field fluctuations and the density of states of the quantized radiation field, and, thus, it can significantly influence radiative properties of atoms placed nearby [40–45]. Our aim is to investigate in detail physical manifestations of atomic acceleration in the radiation-mediated resonance interaction between the two atoms located in the proximity of a reflecting plate.

Resonance and dispersion Casimir–Polder interactions are long-range interactions involving neutral objects such as atoms or molecules [46,47], due to the zero-point fluctuations of the quantum electromagnetic field or to the source field [47–49]. When one or more atoms are in their excited state, a resonance interaction between the atoms can occur, as a result of the exchange of real photons between them. If the two atoms are prepared in a factorized state, the resonance interaction is a fourth-order effect in the coupling and scales as R^{-2} in the far-zone limit, $R \gg \lambda$ (λ is the wavelength associated to the main atomic transition, and R is the interatomic distance) [50]. These interactions, for atoms in a factorized state, have been recently investigated in the literature, also in connection with some controversial results concerning the presence or not of space oscillating terms [51–54]. Recent results show that the force on the excited state is oscillatory in space, while that on the ground state is monotonic [52,53]. A different physical phenomenon occurs if two identical atoms are prepared in a superradiant (or subradiant) Dicke-state. In this case, the resonance interaction energy is obtained at the second-order in the coupling, and it shows space oscillations in the far-zone limit. Such interaction is usually stronger than dispersion interactions and scales as R^{-1}, for very large separations ($R \gg \lambda$). Resonance interactions, and the related Förster energy transfer [55], have been extensively investigated in the literature [56]. The possibility to manipulate (enhance or inhibit) the dispersion and resonance interactions through a structured environment has been also recently investigated [57–61].

We consider two atoms moving with the same uniform proper acceleration in a direction parallel to a reflecting boundary and interacting with the quantum scalar and the electromagnetic field in the vacuum state. Following a procedure originally introduced by Dalibard, Dupont-Roc, and Cohen-Tannoudji [62,63], we identify the contribution of self reaction and vacuum fluctuations to the resonance energy shift of the two accelerated atoms [25,39,44,64]. This approach has been recently used to investigate radiative process of atoms at rest in the presence of a boundary [44,65] or in a cosmic string spacetime [66], and it has been recently generalized to the fourth order to evaluate the dispersion Casimir–Polder interaction between two atoms accelerating in the vacuum space [36]. We show that only the radiation reaction field (source field) contributes to the interatomic resonance interaction energy, while vacuum field fluctuations do not. Consequently, the resonance interaction does not show Unruh *thermal*-like terms (which are related to vacuum field fluctuations). However, non-thermal effects of acceleration appear in the source field contribution, which significantly affect the resonance interaction energy between the two accelerated atoms. To explore these effects, we consider two distinct geometric configurations of the two-atom-plate system: atoms aligned perpendicular or parallel to the plane boundary. We show that the presence of the mirror significantly modifies the character of the resonance interaction energy between the two accelerated atoms. By an appropriate choice of the orientation of the two dipole moments, we show that new effects of atomic acceleration (not present for atoms at rest) appear, yielding a non-vanishing resonance interaction energy even for specific configurations in which the interaction for stationary atoms is zero. This result also suggests new possibilities of observing the effects of a uniform acceleration through a modification of the resonance interatomic interaction between two identical entangled atoms. Thus, our findings could have relevance for a possible detection of the effect of an accelerated motion in radiation-mediated interactions between non-inertial atoms.

The paper is structured as follows. In Section 2, we briefly introduce the method used, and discuss the resonance interaction energy between two accelerating atoms interacting with a massless relativistic scalar field nearby a reflecting mirror. In Section 3, we extend our investigation for atoms interacting with the vacuum electromagnetic field. Final remarks and conclusions are given in Section 4.

Throughout the paper, we adopt units such that $\hbar = c = k_B = 1$.

2. Resonance Interaction between Two Uniformly Accelerating Atoms: The Scalar Field Case

We consider two identical atoms, A and B, interacting with a massless relativistic scalar field in the vacuum state and in the presence of a perfectly reflecting plate satisfying Dirichlet boundary conditions. The two atoms are modeled as point-like systems with two internal energy levels, $\mp\omega_0/2$, associated with the eigenstates $|g\rangle$ and $|e\rangle$, respectively. We suppose that the mirror is located at $z = 0$ and that the two atoms move in a direction parallel to the mirror, with the same uniform proper acceleration, perpendicular to their (constant) separation. The atom-field Hamiltonian in the multipolar coupling scheme and within the dipole approximation, in the locally inertial frame of the two atoms (comoving frame), is as follows [25,36,48,67]:

$$H = \omega_0 \sigma_3^A(\tau) + \omega_0 \sigma_3^B(\tau) + \sum_\mathbf{k} \omega_k a_\mathbf{k}^\dagger a_\mathbf{k} \frac{dt}{d\tau} - \lambda \left(\sigma_2^A(\tau)\phi(x_A(\tau)) + \sigma_2^B(\tau)\phi(x_B(\tau)) \right), \quad (3)$$

where $\sigma_3 = \frac{1}{2}(|e\rangle\langle e| - |g\rangle\langle g|)$ and $\sigma_2 = \frac{i}{2}(|g\rangle\langle e| - |e\rangle\langle g|)$ are the pseudospin atomic operators, $a_\mathbf{k}^\dagger$ and $a_\mathbf{k}$ are the creation and annihilation operators of the scalar field, λ is the coupling constant, and $x_\xi(\tau)$ ($\xi = A, B$) is the trajectory of atom ξ (τ is the proper time of the atoms); $\phi(x(\tau))$ is the scalar field operator, with Dirichlet boundary conditions at the surface of the plate. Equation (3) is expressed in the comoving frame of the two atoms, and we use the Heisenberg representation.

We assume two identical atoms prepared in one of the two correlated, symmetrical (superradiant), or antisymmetrical (subradiant) states ($|\psi_+\rangle$ or $|\psi_-\rangle$, respectively):

$$|\psi_\pm\rangle = \frac{1}{\sqrt{2}}(|g_A, e_B\rangle \pm |e_A, g_B\rangle). \tag{4}$$

To investigate the interatomic resonance dipole–dipole interaction energy, we exploit the procedure originally introduced in Refs. [62,63], allowing to identify the contributions of the source field and vacuum fluctuations to the interaction energy. As discussed in [25,36,62,63], this leads to the introduction of an effective Hamiltonian that governs the time evolution of the atomic observables, pertaining to atom A (B), given by the sum of two terms (similar expressions are obtained for atom B, by exchange of A and B):

$$(H_A^{eff})_{vf} = -\frac{i}{2}\lambda^2 \int_{\tau_0}^{\tau} d\tau' C^F(x_A(\tau), x_A(\tau'))[\sigma_2^{Af}(\tau), \sigma_2^{Af}(\tau')], \tag{5}$$

$$(H_A^{eff})_{sr} = -\frac{i}{2}\lambda^2 \int_{\tau_0}^{\tau} d\tau' \chi^F(x_A(\tau), x_A(\tau'))\{\sigma_2^{Af}(\tau), \sigma_2^{Af}(\tau')\} - \frac{i}{2}\lambda^2 \int_{\tau_0}^{\tau} d\tau' \left[\chi^F(x_A(\tau), x_B(\tau'))\right. \\ \left. \times \{\sigma_2^{Af}(\tau), \sigma_2^{Bf}(\tau')\}\right], \tag{6}$$

where the functions $C^F(x_A(\tau), x_A(\tau'))$ and $\chi^F(x_A(\tau), x_A(\tau'))$ are the field statistical function (symmetric correlation function and the linear susceptibility), respectively:

$$C^F(x(\tau), x(\tau')) = \frac{1}{2}\langle 0|\{\phi(x(\tau)), \phi(x(\tau'))\}|0\rangle, \tag{7}$$

$$\chi^F(x(\tau), x(\tau')) = \frac{1}{2}\langle 0|[\phi(x(\tau)), \phi(x(\tau'))]|0\rangle. \tag{8}$$

To obtain the contributions of source field and vacuum fluctuations to the energy shift of the system, we take the average values of the effective Hamiltonians $(H_{A(B)}^{eff})_{vf}$ and $(H_{A(B)}^{eff})_{sr}$ on the correlated state (4):

$$(\delta E_A)_{vf} = -i\lambda^2 \int_{\tau_0}^{\tau} d\tau' C^F(x_A(\tau), x_A(\tau'))\chi^A(\tau, \tau'), \tag{9}$$

and

$$(\delta E_A)_{sr} = -i\lambda^2 \int_{\tau_0}^{\tau} d\tau' \chi^F(x_A(\tau), x_A(\tau'))C^A(\tau, \tau') - i\lambda^2 \int_{\tau_0}^{\tau} d\tau' \chi^F(x_A(\tau), x_B(\tau'))C^{AB}(\tau, \tau'), \tag{10}$$

where $\tau_0 \to -\infty$ and $\tau \to \infty$ are the initial and final times (similar expressions are obtained for atom B); $\chi^{A(B)}(\tau, \tau')$ and $C^{A(B)}(\tau, \tau')$ are respectively the antisymmetric and symmetric statistical functions of atom A (B), while $\chi^{AB}(\tau, \tau')$ and $C^{AB}(\tau, \tau')$ refer to the collective two-atom system:

$$\chi^{AB}(\tau, \tau') = \frac{1}{2}\langle\psi_\pm|[\sigma_2^{Af}(\tau), \sigma_2^{Bf}(\tau')]|\psi_\pm\rangle, \tag{11}$$

$$C^{AB}(\tau, \tau') = \frac{1}{2}\langle\psi_\pm|\{\sigma_2^{Af}(\tau), \sigma_2^{Bf}(\tau')\}|\psi_\pm\rangle. \tag{12}$$

From expressions above, it is clear that the resonance interaction is entirely due to the source field contribution [25]. In fact, Equation (9) does not depend on the interatomic distance; it only gives the vacuum fluctuations contribution to the Lamb shift of each atom (A or B). Hence, this term does not contribute to the resonance force between the atoms. Similar considerations apply to the first term on the right-hand side of Equation (10). On the contrary, the second term on the right-hand side of

Equation (10), which depends on the distance between the two atoms, is the only contribution relevant at the second order to the interatomic interaction energy. Therefore, the interatomic resonant energy shift is obtained as

$$\delta E = -i \int_{\tau_0}^{\tau} d\tau' \chi^F(x_A(\tau), x_B(\tau')) C^{AB}(\tau, \tau') + (A \rightleftharpoons B). \tag{13}$$

This conclusion is indeed expected on a physical ground, as the resonance interaction is due to the exchange of a (real and virtual) scalar quantum between the two correlated atoms. It is thus related to the field emitted by the two atoms (source field). This property has important consequences when we consider the interaction between accelerated atoms. In fact, as discussed in [25,26], this interaction energy does not show signatures of the Unruh thermal effect (which is exclusively related to the vacuum field correlations in the locally inertial frame). However, we find that the atomic acceleration can determine a qualitative change of the interaction between the two atoms, even if not equivalent to a thermal effect.

We now apply the procedure discussed above to evaluate the resonance interaction energy between two atoms moving with uniform acceleration, interacting with the vacuum scalar field nearby a reflecting plate. We first evaluate the field's linear susceptibility. In the presence of a reflecting boundary, it can be expressed as the sum of two terms, a free term (χ_0^F) that coincides with that obtained in free-space, and a boundary-dependent term (χ_b^F), related to the presence of the reflecting plate [68]:

$$\chi^F(x_A(\tau), x_B(\tau')) = \chi_0^F(x_A(\tau), x_B(\tau')) + \chi_b^F(x_A(\tau), x_B(\tau')), \tag{14}$$

with

$$\chi_0^F(x_A(\tau), x_B(\tau')) = \frac{i}{8\pi |\Delta \mathbf{x}_-|} [\delta(\Delta t + |\Delta \mathbf{x}_-|) - \delta(\Delta t - |\Delta \mathbf{x}_-|)], \tag{15}$$

$$\chi_b^F(x_A(\tau), x_B(\tau')) = \frac{i}{8\pi |\Delta \mathbf{x}_+|} [\delta(\Delta t + |\Delta \mathbf{x}_+|) - \delta(\Delta t - |\Delta \mathbf{x}_+|)], \tag{16}$$

where $x_A(\tau) = (t, x, y, z)$, $x_B(\tau') = (t', x', y', z')$, $\Delta t = t - t'$, and $|\Delta \mathbf{x}_\mp| = [(x - x')^2 + (y - y')^2 + (z \mp z')^2]^{1/2}$.

The atomic statistical function $C^{AB}(\tau, \tau')$ can also be easily obtained [25]:

$$C^{AB}(\tau, \tau') = \pm \frac{1}{8} \left(e^{i\omega_0(\tau - \tau')} + e^{-i\omega_0(\tau - \tau')} \right), \tag{17}$$

where the ± sign respectively refers to the symmetric or antisymmetric states (Equation (4)).

Equation (14) has a general validity and can be applied to different situations, for example, two atoms at rest in the presence of a mirror or uniformly accelerating near a plane boundary, provided the appropriate atomic trajectories, $x_A(\tau)$ and $x_B(\tau)$, are given.

We now specialize our considerations to two specific cases. We suppose a mirror located at $z = 0$ and assume that the two atoms accelerate in the half-space $z > 0$, with the same uniform proper acceleration, parallel to the reflecting plate. The distance between the atoms is thus constant. We consider two different geometric configurations of the two-atom system relative to the plate: two atoms aligned along the z-axis, perpendicular to the boundary, and two atoms aligned in a direction parallel to the plate. This permits us to simplify our calculation and to discuss some relevant effects of the presence of the plate on the resonant interaction energy between the two accelerating atoms.

We first consider both atoms located along the z-direction, perpendicular to the mirror, and uniformly accelerating along the x-direction, perpendicular to their (constant) separation, as shown in Figure 1.

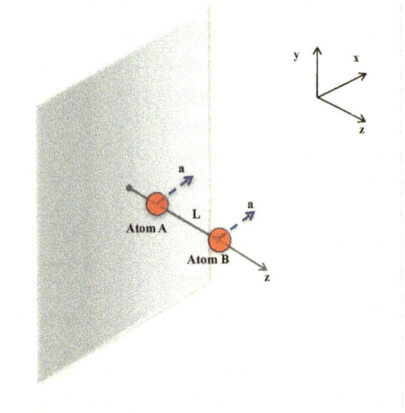

Figure 1. Pictorial description of the first geometrical configuration considered for the physical system: two atoms placed on the z-axis, perpendicular to the plate, and uniformly accelerating along the x-direction.

In the locally inertial frame of the two-atom system, the atomic trajectories, as a function of the proper time τ of both atoms, are

$$t_A(\tau) = t_B(\tau) = \frac{1}{a}\sinh(a\tau), \quad x_A(\tau) = x_B(\tau) = \frac{1}{a}\cosh(a\tau),$$
$$y_A = y_B = 0, \quad z_A = z, \, z_B = z + L. \tag{18}$$

In order to obtain the distance-dependent energy shift of the two-atom system, we first give the linear susceptibility of the scalar field on the trajectories (Equation (18)) of the two atoms. Substituting Equation (18) into the expressions of the scalar-field linear susceptibility (Equations (15) and (16)), we obtain

$$\chi_\perp^F(x_A(\tau), x_B(\tau')) = -\frac{1}{8\pi^2}\int_0^\infty d\omega (e^{i\omega\Delta\tau} - e^{-i\omega\Delta\tau})\left(\frac{\sin(\frac{2\omega}{a}\sinh^{-1}(\frac{aL}{2}))}{L\sqrt{1+\frac{1}{4}a^2L^2}} - \frac{\sin(\frac{2\omega}{a}\sinh^{-1}(\frac{a\mathcal{R}}{2}))}{\mathcal{R}\sqrt{1+\frac{1}{4}a^2\mathcal{R}^2}}\right), \tag{19}$$

where $\Delta\tau = \tau - \tau'$, L is the interatomic distance, and $\mathcal{R} = z_A + z_B = L + 2z$ is the distance between one atom and the image of the second atom relative to the mirror.

The resonance dipole–dipole interaction energy is then obtained using Equations (17) and (19) in Equation (13). We obtain

$$\delta E_\perp(z, L, a) = \mp\frac{\lambda^2}{16\pi}\left[\frac{\cos(\frac{2\omega_0}{a}\sinh^{-1}(\frac{aL}{2}))}{L\sqrt{1+\frac{1}{4}a^2L^2}} - \frac{\cos(\frac{2\omega_0}{a}\sinh^{-1}(\frac{a\mathcal{R}}{2}))}{\mathcal{R}\sqrt{1+\frac{1}{4}a^2\mathcal{R}^2}}\right], \tag{20}$$

where the $\mp$ sign refers to the symmetric or antisymmetric superposition of the atomic states, respectively.

The expression above describes the resonance dipole–dipole interaction energy in terms of the proper acceleration of the two atoms and the atom-plate distances. In the limit $a \to 0$, it reduces to that for atoms at rest. It consists of two terms: a term coinciding with the resonance interaction energy for two accelerating atoms in the free-space, discussed in [25], and a new term, depending on $\mathcal{R}$, related to the presence of the mirror. The latter term, describing the effect of the boundary on the energy shift, originates from the interaction of one atom (e.g., atom A) with the image of the other atom (B). When both atoms are very distant from the reflecting boundary, the boundary-dependent term in

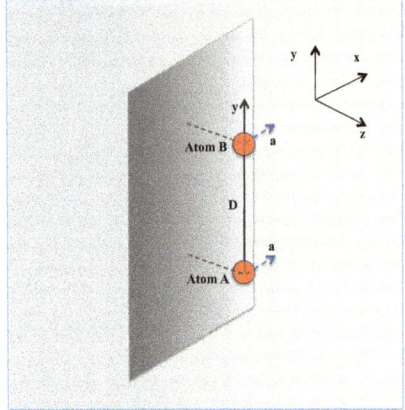

Equation (20) goes to zero, and we recover the resonance interaction between two atoms accelerating in free-space [25]. On the other hand, when the atoms are very close to the mirror, we can approximate $\mathcal{R} \sim L$, and the resonance interaction is strongly suppressed. Thus, in this limit, the interaction between the two entangled atoms can be strongly inhibited by means of the nearby plate, analogously to the case of atoms at rest discussed in [44].

Most importantly, Equation (20) shows that the effects of the atomic acceleration are not *thermal*-like. Nevertheless, the relativistic acceleration significantly affects the interaction energy, giving a different scaling of it with the interatomic distance. In fact, similarly to the results in [25,36] for atoms accelerating in the unbounded space, we can identify a characteristic length scale related to the acceleration, $z_a = 1/a$. For distances larger than z_a, the effects of relativistic acceleration can significantly change the interaction between the two non-inertial atoms; in fact, when $\mathcal{R} > L \gg z_a$, we obtain

$$\delta E_\perp(z,L,a) \sim \mp \frac{\lambda^2}{8\pi a}\left[\frac{1}{L^2}\cos(\frac{2\omega_0}{a}\ln(\frac{aL}{2})) - \frac{1}{\mathcal{R}^2}\cos(\frac{2\omega_0}{a}\ln(\frac{a\mathcal{R}}{2}))\right], \tag{21}$$

giving a different scaling law of the interaction compared to the case of inertial atoms. In the *near*-zone limit, $\mathcal{R}, L \ll z_a$, we recover the well-known result for inertial (static) atoms:

$$\delta E_\perp(z,L,a) \sim \mp \frac{\lambda^2}{16\pi}\left[\frac{1}{L}\cos(\omega_0 L) - \frac{1}{\mathcal{R}}\cos(\omega_0 \mathcal{R})\right]. \tag{22}$$

In the intermediate zone, $\mathcal{R} \gg z_a \gg L$, when the distance between the two atoms is smaller than the characteristic length z_a but their distance from the mirror is such that $\mathcal{R} \gg z_a$, we obtain

$$\delta E_\perp(z,L,a) \sim \mp \frac{\lambda^2}{8\pi}\left[\frac{1}{2L}\cos(\omega_0 L) - \frac{1}{a\mathcal{R}^2}\cos(\frac{2\omega_0}{a}\ln(\frac{a\mathcal{R}}{2}))\right]. \tag{23}$$

Thus the relativistic acceleration and the presence of the boundary affect the qualitative features of the resonance interaction, in particular, its power-law distance dependence, decreasing at large distances more rapidly than in the inertial case. Additionally, in the presence of a boundary, the non-inertial character of acceleration modifies the interatomic interaction energy, even when the separation between the two atoms is much smaller then z_a. In fact, such a result can be expected on a physical ground: the boundary-dependent term, as mentioned, can be interpreted as the interaction of one atom with the image of the other atom with respect to the plate. When the atoms are accelerating, the distance traveled by the photon emitted by one atom to reach the other one, after reflection from the mirror, increases with time; if $\mathcal{R} \gg z_a$, this effect becomes relevant and causes an overall decrease of the interaction strength between the two atoms.

We now investigate whether similar effects manifest also for a different geometric configuration of the atom-plate system. Specifically, we consider two atoms aligned in the y-direction, parallel to the mirror, as shown in Figure 2, and uniformly accelerating in the x-direction, perpendicular to their (constant) separation. In this case, the atomic trajectories are

$$\begin{aligned}t_A(\tau)=t_B(\tau)=\tfrac{1}{a}\sinh(a\tau),\quad x_A(\tau)=x_B(\tau)=\tfrac{1}{a}\cosh(a\tau),\\ y_A=0,\ y_B=D,\ z_A=z_B=z,\end{aligned} \tag{24}$$

with $D > 0$.

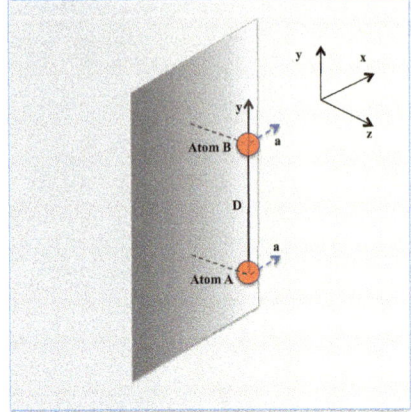

Figure 2. Pictorial description of the second geometrical configuration considered for the physical system: two atoms aligned along the y-axis, parallel to the plate, and uniformly accelerating along the x-direction.

Following the same procedure as before, we first obtain the scalar-field linear susceptibility:

$$\chi_\parallel^F(x_A(\tau), x_B(\tau')) = -\frac{1}{8\pi^2} \int_0^\infty d\omega (e^{i\omega\Delta\tau} - e^{-i\omega\Delta\tau}) \left(\frac{\sin(\frac{2\omega}{a}\sinh^{-1}(\frac{aD}{2}))}{D\sqrt{1+\frac{1}{4}a^2D^2}} - \frac{\sin(\frac{2\omega}{a}\sinh^{-1}(\frac{aR}{2}))}{R\sqrt{1+\frac{1}{4}a^2R^2}} \right), \quad (25)$$

where D is the interatomic distance, $\Delta\tau = \tau - \tau'$, and we have defined $R = R(z, D) = \sqrt{D^2 + 4z^2}$.

The substitution of Equations (25) and (17) into Equation (13) yields, after algebraic calculations, the resonance dipole–dipole interaction for accelerating atoms:

$$\delta E_\parallel(z, D, a) = \mp \frac{\lambda^2}{16\pi} \left[\frac{\cos(\frac{2\omega_0}{a}\sinh^{-1}(\frac{aD}{2}))}{D\sqrt{1+\frac{1}{4}a^2D^2}} - \frac{\cos(\frac{2\omega_0}{a}\sinh^{-1}(\frac{aR}{2}))}{R\sqrt{1+\frac{1}{4}a^2R^2}} \right]. \quad (26)$$

As before, we find that the resonance interaction energy consists of two terms. The first term on the right-hand side of Equation (26) coincides with that for atoms uniformly accelerating in free-space [25], while the second new term is related to the boundary. In the static (inertial) limit, we recover the expression of the resonance interaction for atoms at rest near the mirror for the configuration considered [44]:

$$\delta E_\parallel(z, D) = \mp \frac{\lambda^2}{16\pi} \left[\frac{\cos(\omega_0 D)}{D} - \frac{\cos(\omega_0 \sqrt{D^2 + 4z^2})}{\sqrt{D^2 + 4z^2}} \right]. \quad (27)$$

It is worth noting that the expression of $\delta E_\parallel(z, D, a)$ given by Equation (26) is formally equal to that obtained for $\delta E_\perp(z, L, a)$ in Equation (20), provided $\mathcal{R}$ is replaced by R. This is indeed expected, as the distance $R = \sqrt{D^2 + 4z^2}$ is the distance between one atom and the image of the other. In order to compare the results obtained in the two geometric configurations, in Figure 3 are plotted Equations (20) and (26) of the resonance interaction energy (in units of eV/λ^2), as a function of the atomic acceleration. In the plots, the value used for ω_0 is the ionization energy of ^{87}Rb, and the distances $L = D$ and z have been chosen in such a way that the plots cover near, intermediate, and far zones, for both perpendicular and parallel alignments of the atoms. The plots show that the resonance interaction energy depends on the acceleration and the geometric configuration of the two atoms with respect to the plate (perpendicular or parallel alignment) and that it can be enhanced or inhibited, depending on the atomic acceleration.

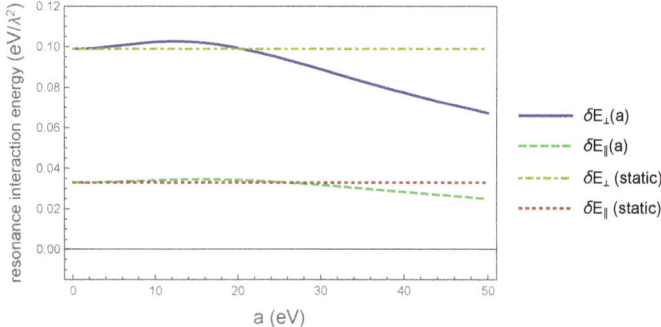

Figure 3. Resonance interaction energy between the two atoms (units: eV/λ^2, where the coupling constant λ in our units is dimensionless), as a function of the atomic acceleration, for two different geometric configurations. Blue continuous line: atoms positioned on the z-axis, which is perpendicular to the plate. Green dashed line: atoms along the y-axis, which is parallel to the plate. For comparison, the yellow dot-dashed line and the red dotted line respectively refer to the case of inertial atoms aligned in a perpendicular or parallel direction relative to the plate. The plots show that the interaction depends on the acceleration and on the geometric configuration of the two-atom system relative to the mirror. Parameters, in the units used, are chosen such that $L = D = 7.5 \times 10^{-2}$ eV^{-1}, $z = 2.0 \times 10^{-2}$ eV^{-1}, and $\omega_0 = 4.17$ eV.

3. Resonance Interaction for Two Accelerating Atoms Interacting with the Electromagnetic Field

In this section, we extend our investigations to two uniformly accelerated identical atoms interacting with the vacuum electromagnetic field, placed nearby a perfectly reflecting plate. As before, the atoms move with a uniform proper acceleration a in a direction parallel to the plane, located at $z = 0$, and their distance is constant. Our aim is to discuss whether new and further effects of acceleration may manifest in their interaction, as a consequence of the vector nature of the electromagnetic field.

We adopt the Hamiltonian in the Coulomb gauge and in the multipolar coupling scheme, within dipole approximation. In the comoving reference frame of both atoms, this is

$$H = \omega_0 \sigma_3^A(\tau) + \omega_0 \sigma_3^B(\tau) + \sum_{\mathbf{k},\lambda} \omega_k a_{\mathbf{k}\lambda}^\dagger a_{\mathbf{k}\lambda} \frac{dt}{d\tau} - \boldsymbol{\mu}_A(\tau) \cdot \mathbf{E}(x_A(\tau)) - \boldsymbol{\mu}_B(\tau) \cdot \mathbf{E}(x_B(\tau)). \tag{28}$$

$\lambda = 1, 2$ indicates the polarization, $\boldsymbol{\mu} = e\mathbf{r}$ is the dipole moment operator of the atoms (restricted to the subspace of the two atomic levels considered), and $\mathbf{E}(x(\tau))$ is the electric field operator, with the appropriate boundary conditions on the reflecting plate.

As shown in the previous section, the resonance interaction energy is due only to the radiation-reaction term and can be obtained through the effective Hamiltonian $(H_A^{eff})_{sr} + (H_B^{eff})_{sr}$ (terms referring to atoms A and B, respectively) on the correlated state $|\psi_\pm\rangle$ (see Equations (4)–(6)), taking only terms depending on the interatomic distance:

$$\delta E = -i \int_{\tau_0}^{\tau} d\tau' \chi_{ij}^F(x_A(\tau), x_B(\tau')) C_{ij}^{AB}(\tau, \tau') + (A \rightleftharpoons B), \tag{29}$$

where $i, j = x, y, z$. We first evaluate the electromagnetic field susceptibility $\chi_{ij}^F(x_A(\tau), x_B(\tau')) = \frac{1}{2}\langle 0 | [E_i(x_A(\tau)), E_j(x_B(\tau'))] | 0 \rangle$ and the atomic symmetric correlation function $C_{ij}^{AB}(\tau, \tau')$.

The field susceptibility in the comoving frame can be obtained from the two-point correlation function of the field [68]. The two-point correlation function of the electric field operator in the presence of the reflecting boundary, is the following (for brevity, we omit the time-dependence in the following expressions):

$$g_{ij}(x_A, x_B) = \langle 0|E_i(x_A)E_j(x_B)|0\rangle. \tag{30}$$

It can be written as the sum of a free part, $g_{ij}^{(0)}(x_A, x_B)$, and a boundary-dependent term, $g_{ij}^{(b)}(x_A, x_B)$:

$$g_{ij}(x_A, x_B) = g_{ij}^{(0)}(x_A, x_B) + g_{ij}^{(b)}(x_A, x_B), \tag{31}$$

where

$$g_{ij}^{(0)}(x_A, x_B) = -\frac{1}{4\pi^2}(\delta_{ij}\partial_0\partial_{0'} - \partial_i\partial_{j'})\frac{1}{(\Delta t - i\epsilon)^2 - |\Delta \mathbf{x}_-|^2}, \tag{32}$$

$$g_{ij}^{(b)}(x_A, x_B) = \frac{1}{4\pi^2}[(\delta_{ij} - 2n_i n_j)\partial_0\partial_{0'} - \partial_i\partial_{j'}]\frac{1}{(\Delta t - i\epsilon)^2 - |\Delta \mathbf{x}_+|^2}, \tag{33}$$

and n is the unit vector along the line joining the two atoms.

We now specialize our considerations to the two specific configurations considered for the scalar-field case in Section 2 and illustrated in Figures 1 and 2, that is, atoms aligned in a direction perpendicular or parallel to the plate, respectively.

3.1. Atoms Aligned Perpendicularly to the Plate

We first consider two atoms aligned along the z-direction, perpendicular to the boundary, and uniformly accelerating along the x-direction, as shown in Figure 1. Thus they move on the trajectory given by Equation (18). Because of the vector structure of the electromagnetic field, the calculation of the field susceptibility turns out to be more complicated than for the scalar field [68]. After lengthy algebraic calculations, involving a Lorentz transformation of the fields to the comoving frame, we obtain the following (in the locally inertial frame):

$$g_{\perp_{ij}}(x_A, x_B) = g_{\perp_{ij}}^{(0)}(x_A, x_B) + g_{\perp_{ij}}^{(b)}(x_A, x_B), \tag{34}$$

where

$$g_{\perp_{ij}}^{(0)}(x_A, x_B) = \frac{a^4}{16\pi^2}\frac{1}{(\sinh^2(\frac{a}{2}(\Delta\tau - i\epsilon)) - \frac{1}{4}a^2 L^2)^3} \times \left\{\frac{1}{4}a^2 L^2(\delta_{ij} - 2n_i n_j) + \left[\delta_{ij} + \frac{1}{2}a^2 L^2(\delta_{ij} - k_i k_j - 2n_i n_j) + aL(k_i n_j - k_j n_i)\right]\sinh^2\left(\frac{a}{2}\Delta\tau\right)\right\} \tag{35}$$

is the two-point correlation function of two atoms uniformly accelerated in vacuum [25], and

$$g_{\perp_{ij}}^{(b)}(x_A, x_B) = -\frac{a^4}{16\pi^2}\frac{(1 - 2n_i n_j)}{(\sinh^2(\frac{a}{2}(\Delta\tau - i\epsilon)) - \frac{1}{4}a^2 \mathcal{R}^2)^3} \times \left\{\frac{1}{4}a^2 \mathcal{R}^2(\delta_{ij} - 2n_i n_j) + \left[\delta_{ij} + \frac{1}{2}a^2 \mathcal{R}^2(\delta_{ij} - k_i k_j - 2n_i n_j) + a\mathcal{R}(k_i n_j + k_j n_i)\right]\sinh^2\left(\frac{a}{2}\Delta\tau\right)\right\} \tag{36}$$

is the contribution due to the presence of the boundary. In the equations above, $k = (1, 0, 0)$ is a unit vector along the acceleration. As discussed in [25], the function $g_{\perp_{ij}}(x_A, x_B)$ is not isotropic, displaying a non-diagonal component. In fact, in the present case, we have two specific directions in space: the direction perpendicular to the plate and that of the acceleration. Similar anisotropies were already found for a single uniformly accelerated atom near a boundary [31] or for two accelerated atoms in the free-space [25]. They arise from the spatially extended structure of the two-atom-plate system here considered, as well as from the vector character of the electromagnetic field. This peculiarity, as we now show, has deep consequences for the interaction energy between the two atoms.

In order to evaluate the resonance energy, we first focus our attention on the boundary-dependent term and calculate the linear susceptibility of the electric field. Using Equation (36), after lengthly algebraic calculations, involving a Fourier transform of the statistical function of the field, we finally obtain

$$\chi_{\perp_{ij}}^{F(b)}(x_A(\tau), x_B(\tau')) = \frac{1}{8\pi^2} \int_0^\infty d\omega (e^{i\omega\Delta\tau} - e^{-i\omega\Delta\tau}) \left(f_{ij}^{\perp(b)}(a, \mathcal{R}, \omega) \cos\left(\frac{2\omega}{a} \sinh^{-1}\left(\frac{a\mathcal{R}}{2}\right)\right) \right.$$
$$\left. + h_{ij}^{\perp(b)}(a, \mathcal{R}, \omega) \sin\left(\frac{2\omega}{a} \sinh^{-1}\left(\frac{a\mathcal{R}}{2}\right)\right) \right), \quad (37)$$

where we have introduced the functions $f_{ij}^{\perp(b)}(a, \mathcal{R}, \omega)$ and $h_{ij}^{\perp(b)}(a, \mathcal{R}, \omega)$ given in Appendix A (see Equations (A1) and (A2)).

Substituting Equation (37) and the atomic symmetric statistical function:

$$C_{ij}^{AB}(\tau, \tau') = \frac{1}{2} (\mu_{ge}^A)_i (\mu_{ge}^B)_j (e^{i\omega_0 \Delta\tau} + e^{-i\omega_0 \Delta\tau}), \quad (38)$$

into Equation (29), we finally obtain the boundary-dependent contribution to the resonant energy shift of the two accelerating atoms:

$$\delta E_\perp^{(b)} = \mp \frac{1}{4\pi} [\delta_{ij} (\mu_{ge}^A)_i (\mu_{eg}^B)_j P_{ij}^{\perp(b)}(a, \mathcal{R}, \omega_0) \pm ((\mu_{ge}^A)_x (\mu_{eg}^B)_z + (\mu_{ge}^A)_z (\mu_{eg}^B)_x) P_{xz}^{\perp(b)}(a, \mathcal{R}, \omega_0)], \quad (39)$$

where we have introduced the function $P_{ij}^{\perp(b)}(a, \mathcal{R}, \omega_0)$:

$$P_{ij}^{\perp(b)}(a, \mathcal{R}, \omega_0) = f_{ij}^{\perp(b)}(a, \mathcal{R}, \omega_0) \sin\left(\frac{2\omega_0}{a} \sinh^{-1}\left(\frac{a\mathcal{R}}{2}\right)\right) - h_{ij}^{\perp(b)}(a, \mathcal{R}, \omega_0) \cos\left(\frac{2\omega_0}{a} \sinh^{-1}\left(\frac{a\mathcal{R}}{2}\right)\right), \quad (40)$$

modulating the interaction as a function of $\mathcal{R}$ and of the atomic acceleration.

With a similar procedure, evaluation of the boundary-independent contribution, $\delta E_\perp^{(0)}$, to the resonance interaction energy yields the following [25]:

$$\delta E_\perp^{(0)} = \pm \frac{1}{4\pi} [\delta_{ij} (\mu_{ge}^A)_i (\mu_{eg}^B)_j P_{ij}^{\perp(0)}(a, L, \omega_0) \pm ((\mu_{ge}^A)_x (\mu_{eg}^B)_z - (\mu_{ge}^A)_z (\mu_{eg}^B)_x) P_{xz}^{\perp(0)}(a, L, \omega_0)], \quad (41)$$

where

$$P_{ij}^{\perp(0)}(a, L, \omega_0) = f_{ij}^{\perp(0)}(a, L, \omega_0) \sin\left(\frac{2\omega_0}{a} \sinh^{-1}\left(\frac{aL}{2}\right)\right) - h_{ij}^{\perp(0)}(a, L, \omega_0) \cos\left(\frac{2\omega_0}{a} \sinh^{-1}\left(\frac{aL}{2}\right)\right), \quad (42)$$

and the functions $f_{ij}^{\perp(0)}(a, L, \omega)$ and $h_{ij}^{\perp(0)}(a, L, \omega)$ are given by Equations (A3) and (A4) of Appendix A.

The complete resonance interaction energy of the accelerated two-atom system is then obtained by summing Equations (39) and (41):

$$\delta E_\perp = \delta E_\perp^{(0)} + \delta E_\perp^{(b)}. \quad (43)$$

The result (Equation (43)) is valid for any value of the parameters a, L, and $\mathcal{R}$. It is easy to show that in the *near-zone* limit, $L \ll a^{-1}$ and $\mathcal{R} \ll a^{-1}$, the linear susceptibility is well described by its stationary counterpart, and we recover the expression of the resonance interaction for two atoms at rest [25,44]. However, at higher orders in $a\mathcal{R}$ (and/or aL), corrections related to the accelerated motion of the two atoms become relevant, yielding a different scaling of the interaction energy with the distance, in analogy to the scalar-field case discussed in the previous section. Interestingly, a comparison with the scalar-field case shows the emergence of new features in the resonance interaction, due to the boundary, and related to the anisotropic structure of the electromagnetic field susceptibility. Indeed, from Equation (39), it follows that the effect of the acceleration on the interaction can be controlled by an appropriate choice of the dipoles' orientations and of the distance of the two atoms from the plate. For example, when the dipole moments are orthogonal to each other, with one along x and the other along z, the diagonal term in Equation (39) vanishes, and only the second (non-diagonal) term survives. The non-diagonal term is present only for $a \neq 0$, and its contribution is a peculiar characteristic of the non inertial atomic motion, giving a non-vanishing interaction energy, in a configuration where that for static atoms is zero. This term is thus a sharp signature of an accelerated motion. To numerically estimate this energy shift, we can assume $a = 10^{18}$ m/s^2 (2.2 × 10^{-6} eV, in our units), $z = 10^{-8}$ m ($\sim 5 \times 10^{-2}$ eV^{-1}), $L = 1.5 \times 10^{-8}$ m ($\sim 7.5 \times 10^{-2}$ eV^{-1}), and $\hbar \omega_0 = 4.17$ eV, obtaining $\delta E \simeq 4.4 \times 10^{-10}$ eV. This energy shift is about 4 orders of magnitude smaller than the Lamb shift for the $n = 2$ level of the hydrogen atom. Although quite small, we expect that such an energy shift should be measurable using high-resolution spectroscopy, provided the assumed constant acceleration could be reached.

The results above suggest investigation of whether analogous effects of acceleration manifest also for other geometric configurations of the two atoms system, for example, when both atoms are aligned parallel to the reflecting plane boundary. This configuration is considered in the next subsection.

3.2. Atoms Aligned Parallel to the Plate

We now consider the configuration of two atoms aligned along the y-direction, parallel to the boundary, which move with uniform proper acceleration along the x-direction, such that their trajectories are those given by Equation (24). As before, the distance between the two atoms remains constant during their motion. This configuration is illustrated in Figure 2.

The two-point correlation function of the field in the locally inertial frame of both atoms is

$$g_{\|ij}(x_A, x_B) = g_{\|ij}^{(0)}(x_A, x_B) + g_{\|ij}^{(b)}(x_A, x_B), \tag{44}$$

where $g_{\|ij}^{(0)}(x_A, x_B)$ is the two-point correlation function in free-space [25] and $g_{\|ij}^{(b)}(x_A, x_B)$ is the boundary-dependent contribution, which consists of a diagonal term:

$$\begin{aligned} g_{\|ij}^{(b)}(x_A, x_B) &= -\frac{a^4}{16\pi^2} \frac{(\delta_{ij} - 2n_i n_j)}{(\sinh^2(\frac{a}{2}(\Delta\tau - i\epsilon)) - \frac{1}{4}a^2 R^2)^3} \left\{ \frac{1}{4} a^2 \tilde{R}^2 (n_i n_j - p_i p_j) \right. \\ &\left. + \frac{1}{4} a^2 R^2 k_i k_j + \left[1 + \frac{1}{2} a^2 \tilde{R}^2 (1 - k_i k_j - 2 p_i p_j) \right] \sinh^2\left(\frac{a}{2}\Delta\tau\right) \right\} \quad (i = j) \end{aligned} \tag{45}$$

that is non-vanishing only for $i = j$, and a non-diagonal term:

$$\begin{aligned} g_{\|ij}^{(b)}(x_A, x_B) &= -\frac{a^4}{16\pi^2} \frac{1}{(\sinh^2(\frac{a}{2}(\Delta\tau - i\epsilon)) - \frac{1}{4}a^2 R^2)^3} \left\{ -a^2 z D(p_i n_j - p_j n_i) \right. \\ &\left. + [aD(k_i p_j - k_j p_i) + 2az(k_i n_j + k_j n_i) - 2a^2 z D(p_i n_j - p_j n_i)] \sinh^2\left(\frac{a}{2}\Delta\tau\right) \right\} \quad (i \neq j) \end{aligned} \tag{46}$$

that is different from zero only for $i \neq j$. We have here introduced the unit vector $\mathbf{p} = (0, 1, 0)$ and the distances $R = \sqrt{D^2 + 4z^2}$ and $\tilde{R} = \sqrt{D^2 - 4z^2}$). The boundary-dependent contribution to the linear susceptibility of the field is then obtained as

$$\begin{aligned} \chi_{\|ij}^{F(b)}(x_A(\tau), x_B(\tau')) &= \frac{1}{8\pi^2} \int_0^\infty d\omega (e^{i\omega\Delta\tau} - e^{-i\omega\Delta\tau}) \left(f_{ij}^{\|(b)}(a, D, z, \omega) \cos\left(\frac{2\omega}{a} \sinh^{-1}\left(\frac{aR}{2}\right)\right) \right. \\ &\left. + h_{ij}^{\|(b)}(a, D, z, \omega) \sin\left(\frac{2\omega}{a} \sinh^{-1}\left(\frac{aR}{2}\right)\right) \right), \end{aligned} \tag{47}$$

where the functions $f_{ij}^{\|(b)}(a,D,z,\omega)$ and $h_{ij}^{\|(b)}(a,D,z,\omega)$, given in Equations (A5) and (A6) of Appendix A, modulate the resonance interaction energy with the distance D and the atomic acceleration a.

Substituting Equations (47) and (38) into Equation (29), we find the boundary-dependent contribution to the resonant energy shift:

$$\delta E_{\|}^{(b)} = -\frac{1}{4\pi}\left[\delta_{ij}(\mu_{ge}^A)_i(\mu_{eg}^B)_j P_{ij}^{\|(b)}(a,D,z,\omega_0) + \left((\mu_{ge}^A)_x(\mu_{eg}^B)_y - (\mu_{ge}^A)_y(\mu_{eg}^B)_x\right) P_{xy}^{\|(b)}(a,D,z,\omega_0) \right.$$
$$\left. + \left((\mu_{ge}^A)_x(\mu_{eg}^B)_z + (\mu_{ge}^A)_z(\mu_{eg}^B)_x\right) P_{xz}^{\|(b)}(a,D,z,\omega_0) + \left((\mu_{ge}^A)_y(\mu_{eg}^B)_z - (\mu_{ge}^A)_z(\mu_{eg}^B)_y\right) P_{yz}^{\|(b)}(a,D,z,\omega_0)\right], \quad (48)$$

where

$$\begin{aligned} P_{ij}^{\|(b)}(a,D,z,\omega_0) &= f_{ij}^{\|(b)}(a,D,z,\omega_0)\sin\left(\frac{2\omega_0}{a}\sinh^{-1}\left(\frac{aR}{2}\right)\right) \\ &- h_{ij}^{\|(b)}(a,D,z,\omega_0)\cos\left(\frac{2\omega_0}{a}\sinh^{-1}\left(\frac{aR}{2}\right)\right). \end{aligned} \quad (49)$$

The resonance interaction energy between the accelerating atoms is finally obtained by adding Equation (48) to the free-space interaction energy $\delta E_{\|}^{(0)}$, given by the following [25]:

$$\delta E_{\|}^{(0)} = \frac{1}{4\pi}\left[\delta_{ij}(\mu_{ge}^A)_i(\mu_{eg}^B)_j P_{ij}^{\|(0)}(a,D,\omega_0) + \left((\mu_{ge}^A)_x(\mu_{eg}^B)_y - (\mu_{ge}^A)_y(\mu_{eg}^B)_x\right) P_{xy}^{\|(0)}(a,D,\omega_0)\right], \quad (50)$$

with

$$P_{ij}^{\|(0)}(a,D,\omega_0) = f_{ij}^{\|(0)}(a,D,\omega_0)\sin\left(\frac{2\omega_0}{a}\sinh^{-1}\left(\frac{aR}{2}\right)\right)$$
$$-h_{ij}^{\|(0)}(a,D,\omega_0)\cos\left(\frac{2\omega_0}{a}\sinh^{-1}\left(\frac{aR}{2}\right)\right) \quad (51)$$

(the functions $f_{ij}^{\|(0)}(a,D,\omega)$ and $h_{ij}^{\|(0)}(a,D,\omega)$ can be obtained from Equations (A3) and (A4) in Appendix A by exchanging subscripts z and y).

A comparison with the case of accelerated atoms aligned along the z-axis, considered in the previous subsection, shows the emergence of a new effect, related to the specific geometric configuration of the two-atom system with respect to the plane boundary. In fact, from the equations above, it follows that when the dipole moments are orthogonal to each other, one of them along y and the other in the plane xz, a new non-vanishing contribution to the interaction energy (not present for atoms located perpendicular to the boundary) arises. This contribution exists only when $a \neq 0$, and thus it is a peculiarity of an accelerated motion. This gives new additional possibilities to exploit the resonance interaction between accelerated atoms for detecting (non-thermal) effects of acceleration and, in general, physical effects of the accelerated motion on radiation-mediated interactions between atoms.

4. Summary

We have discussed the resonance energy shift of two identical atoms, one excited and the other in the ground state, prepared in a correlated (superradiant or subradiant) state, and moving with uniform acceleration near a perfectly reflecting plate. The atoms interact with the massless scalar field or the electromagnetic field in the vacuum state. Following the approach in Refs. [62,63], we have identified the contributions of source field and vacuum fluctuations to the resonance interaction. We have shown that Unruh thermal fluctuations do not influence the resonance interatomic interaction, which is obtained from the source-field term only. We show that, in cases of both the scalar and electromagnetic field, the presence of the plane boundary significantly affects the resonance interaction between the accelerated atoms. Non-thermal effects of acceleration appear, yielding a change in the distance dependence of the interaction. Finally, in the case of the electromagnetic field, we show, for different configurations of the two-atom-plate system, the emergence of new and different effects in the resonance interaction energy, for example, a non-vanishing interaction energy in configurations/dipole orientations for which the interaction is zero for inertial atoms. These effects, not present for atoms at rest, therefore provide a sharp signature of the non-inertial motion of the atoms. These findings could be exploited for the detection of the non-thermal effects of atomic acceleration in radiation-mediated interactions between non-inertial atoms.

Author Contributions: All authors equally contributed to the research reported in this paper and to its writing.

Acknowledgments: W.Z. acknowledges financial support from the National Natural Science Foundation of China (NSFC) under Grant Nos. 11405091, 11690034, 11375092, and 11435006; the Key Laboratory of Low-Dimensional Quantum Structures and Quantum Control of Ministry of Education under Grant No. QSQC1801; the China Scholarship Council (CSC); the Research program of Ningbo University under Grant No. XYL18027; and the K. C. Wong Magna Fund of Ningbo University. R.P. and L.R. gratefully acknowledge financial support from the Julian Schwinger Foundation.

Conflicts of Interest: The authors declare no conflict of interest. The funding sponsors had no role in the design of the study; in the collection, analyses, or interpretation of data; in the writing of the manuscript; or in the decision to publish the results.

Appendix A

In this Appendix, we give the expressions of the functions $f_{ij}^{\perp(\|)}$ and $h_{ij}^{\perp(\|)}$ used in Section 3.

The explicit expressions of the functions $f_{ij}^{\perp(b)}(a, \mathcal{R}, \omega)$ and $h_{ij}^{\perp(b)}(a, \mathcal{R}, \omega)$ are

$$\begin{cases} f_{xx}^{\perp(b)} = \frac{\omega(1+a^2\mathcal{R}^2)}{\mathcal{N}^4 \mathcal{R}^2}, \\ f_{yy}^{\perp(b)} = \frac{\omega(1+\frac{1}{2}a^2\mathcal{R}^2)}{\mathcal{N}^2 \mathcal{R}^2}, \\ f_{zz}^{\perp(b)} = \frac{\omega(2+\frac{1}{4}a^2\mathcal{R}^2+\frac{1}{8}a^4\mathcal{R}^4)}{\mathcal{N}^4 \mathcal{R}^2}, \\ f_{xz}^{\perp(b)} = f_{zx}^{\perp(b)} = -\frac{a\omega(1-\frac{1}{2}a^2\mathcal{R}^2)}{2\mathcal{N}^4 \mathcal{R}}, \end{cases} \quad (A1)$$

$$\begin{cases} h_{xx}^{\perp(b)} = -\frac{1+\frac{1}{2}a^2\mathcal{R}^2+\frac{1}{4}a^4\mathcal{R}^4}{\mathcal{N}^5 \mathcal{R}^3} + \frac{\omega^2}{\mathcal{N}^3 \mathcal{R}}, \\ h_{yy}^{\perp(b)} = -\frac{1}{\mathcal{N}^3 \mathcal{R}^3} + \frac{\omega^2}{\mathcal{N}\mathcal{R}}, \\ h_{zz}^{\perp(b)} = -\frac{2(1+\frac{5}{8}a^2\mathcal{R}^2)}{\mathcal{N}^5 \mathcal{R}^3} + \frac{a^2 \mathcal{R}\omega^2}{4\mathcal{N}^3}, \\ h_{xz}^{\perp(b)} = h_{zx}^{\perp(b)} = \frac{a(1+a^2\mathcal{R}^2)}{2\mathcal{N}^5 \mathcal{R}^2} + \frac{a\omega^2}{2\mathcal{N}^3}, \end{cases} \quad (A2)$$

with $\mathcal{N} = \mathcal{N}(a, \mathcal{R}) = \sqrt{1+\frac{1}{4}a^2 \mathcal{R}^2}$.

Explicit expressions of $f_{ij}^{\perp(0)}(a, L, \omega)$ and $h_{ij}^{\perp(0)}(a, L, \omega)$ are

$$\begin{cases} f_{xx}^{\perp(0)} = \frac{\omega(1+a^2 L^2)}{N^4 L^2}, \\ f_{yy}^{\perp(0)} = \frac{\omega(1+\frac{1}{2}a^2 L^2)}{N^2 L^2}, \\ f_{zz}^{\perp(0)} = -\frac{\omega(2+\frac{1}{4}a^2 L^2+\frac{1}{8}a^4 L^4)}{N^4 L^2}, \\ f_{xz}^{\perp(0)} = -f_{zx}^{\perp(0)} = \frac{a\omega(1-\frac{1}{2}a^2 L^2)}{2N^4 L}, \end{cases} \quad (A3)$$

$$\begin{cases} h_{xx}^{\perp(0)} = -\frac{1+\frac{1}{2}a^2L^2+\frac{1}{4}a^4L^4}{N^5L^3} + \frac{\omega^2}{N^3L}, \\ h_{yy}^{\perp(0)} = -\frac{1}{N^3L^3} + \frac{\omega^2}{N^{1/2}L}, \\ h_{zz}^{\perp(0)} = \frac{2(1+\frac{5}{8}a^2L^2)}{N^5L^3} - \frac{a^2L\omega^2}{4N^3}, \\ h_{xz}^{\perp(0)} = -h_{zx}^{\perp(0)} = -\frac{a(1+a^2L^2)}{2N^5L^2} - \frac{a\omega^2}{2N^3}, \end{cases} \qquad (A4)$$

with $N = N(a,L) = \sqrt{1+\frac{1}{4}a^2L^2}$.

Explicit expressions of $f_{ij}^{\|(b)}(a,D,z,\omega)$ and $h_{ij}^{\|(b)}(a,D,z,\omega)$ are

$$\begin{cases} f_{xx}^{\|(b)} = \frac{\omega(1+a^2R^2)}{\tilde{N}^4R^2}, \\ f_{yy}^{\|(b)} = \frac{\omega[4z^2-2D^2-\frac{1}{4}a^2R^2(D^2-12z^2)-\frac{1}{8}a^4R^4(D^2-4z^2)]}{\tilde{N}^4R^4}, \\ f_{zz}^{\|(b)} = \frac{\omega[z^2(16+2a^2R^2+a^4R^4)-D^2(2+\frac{3}{2}a^2R^2+\frac{1}{4}a^4R^4)]}{2\tilde{N}^4R^4}, \\ f_{xy}^{\|(b)} = -f_{yx}^{\|(b)} = -\frac{\omega a D(1-\frac{1}{2}a^2R^2)}{2\tilde{N}^4R^2}, \\ f_{xz}^{\|(b)} = f_{zx}^{\|(b)} = -\frac{\omega a z(1-\frac{1}{2}a^2R^2)}{\tilde{N}^4R^2}, \\ f_{yz}^{\|(b)} = -f_{zy}^{\|(b)} = -\frac{2\omega z D(3+a^2R^2+\frac{1}{4}a^4R^4)}{\tilde{N}^4R^4}, \end{cases} \qquad (A5)$$

$$\begin{cases} h_{xx}^{\|(b)} = -\frac{1+\frac{1}{2}a^2R^2+\frac{1}{4}a^4R^4}{\tilde{N}^5R^3} + \frac{\omega^2}{\tilde{N}^3R}, \\ h_{yy}^{\|(b)} = \frac{2D^2-4z^2+\frac{1}{4}a^2R^2(5D^2-4z^2)}{\tilde{N}^5R^5} + \frac{\omega^2[4z^2-\frac{1}{4}a^2R^2(D^2-4z^2)]}{\tilde{N}^3R^3}, \\ h_{zz}^{\|(b)} = \frac{D^2(1+\frac{1}{4}a^2R^2)-8z^2(1+\frac{5}{8}a^2R^2)}{\tilde{N}^5R^5} + \frac{\omega^2[a^2z^2R^2-D^2(1+\frac{1}{4}a^2R^2)]}{\tilde{N}^3R^3}, \\ h_{xy}^{\|(b)} = -h_{yx}^{\|(b)} = \frac{aD(1+a^7R^7)}{2\tilde{N}^5R^3} + \frac{\omega^2 aD}{2\tilde{N}^3R}, \\ h_{xz}^{\|(b)} = h_{zx}^{\|(b)} = \frac{az(1+a^2R^2)}{\tilde{N}^5R^3} + \frac{\omega^2 az}{\tilde{N}^3R}, \\ h_{yz}^{\|(b)} = -h_{zy}^{\|(b)} = \frac{6zD(1+\frac{1}{2}a^2R^2)}{\tilde{N}^5R^5} - \frac{2\omega^2 zD(1+\frac{1}{2}a^2R^2)}{\tilde{N}^3R^3}, \end{cases} \qquad (A6)$$

with $\tilde{N} = \tilde{N}(a,R) = \sqrt{1+\frac{1}{4}a^2R^2}$.

References

1. Unruh, W.G. Notes on black-hole evaporation. *Phys. Rev. D* **1976**, *14*, 870, doi:10.1103/PhysRevD.14.870 [CrossRef]
2. Hawking, S.W. Particle creation by black holes. *Commun. Math. Phys.* **1975**, *43*, 199, doi:10.1007/BF02345020. [CrossRef]
3. Fulling, S.A. Non-uniqueness of Canonical Field Quantization in Riemanian Space-Time. *Phys. Rev. D* **1973**, *7*, 2850, doi:10.1103/PhysRevD.7.2850 [CrossRef]
4. Davies, P.C.W. Scalar particle production in Schwarzschild and Rindler metrics. *J. Phys. A* **1975**, *8*, 609, doi:10.1088/0305-4470/8/4/022. [CrossRef]
5. Unruh, W.G.; Wald, R.M. What happens when an accelerating observer detects a Rindler particle. *Phys. Rev. D* **1984**, *29*, 1047, doi:10.1103/PhysRevD.29.1047. [CrossRef]
6. Crispino, L.C.B.; Higuchi, A.; Matsas, G.E.A. The Unruh effect and its applications. *Rev. Mod. Phys.* **2008**, *80*, 787, doi:10.1103/RevModPhys.80.787. [CrossRef]
7. Buchholz, D.; Solveen, C. Unruh effect and the concept of temperature. *Class. Quantum Gravity* **2013**, *30*, 085011,doi:10.1088/0264-9381/30/8/085011. [CrossRef]

8. Rosu, H.C. Hawking like effects and Unruh like effects: Towards experiments? *Grav. Cosmol.* **2001**, *7*, 1–17.
9. Raine, D.J.; Sciama, D.W.; Grove, P.G. Does a uniformly accelerated quantum oscillator radiate? *Proc. R. Soc. Lond. A* **1991**, *435*, 205, doi:10.1098/rspa.1991.0139. [CrossRef]
10. Padmanabhan, T. Physical interpretation of quantum field theory in noninertial coordinate systems. *Phys. Rev. Lett.* **1990**, *64*, 2471, doi:10.1103/PhysRevLett.64.2471. [CrossRef] [PubMed]
11. Narozhny, N.B.; Fedotov, A.M.; Karnakov, B.M.; Mur, V.D.; Belinskii, V.A. Boundary conditions in the Unruh problem. *Phys. Rev. D* **2001**, *65*, 025004, doi:10.1103/PhysRevD.65.025004. [CrossRef]
12. Ford, G.W.; O'Connell, R.F. Is there Unruh radiation? *Phys. Lett. A* **2006**, *350*, 17, doi:10.1016/j.physleta.2005.09.068. [CrossRef]
13. Schützhold, R.; Schaller, G.; Habs, D. Signatures of the Unruh Effect from Electrons Accelerated by Ultrastrong Laser Fields. *Phys. Rev. Lett.* **2006**, *97*, 121302, doi:10.1103/PhysRevLett.97.121302. [CrossRef] [PubMed]
14. Retzker, A.; Cirac, J.I.; Plenio, M.B.; Reznik, B. Methods for Detecting Acceleration Radiation in a Bose-Einstein Condensate. *Phys. Rev. Lett.* **2008**, *101*, 110402, doi:10.1103/PhysRevLett.101.110402. [CrossRef] [PubMed]
15. Vanzella, D.A.T.; Matsas, G.E.A. Decay of Accelerated Protons and the Existence of the Fulling-Davies-Unruh Effect. *Phys. Rev. Lett.* **2001**, *87*, 151301, doi:10.1103/PhysRevLett.87.151301. [CrossRef] [PubMed]
16. Martin-Martinez, E.; Fuentes, I.; Mann, R.B. Using Berry's phase to detect the Unruh effect at lower accelerations. *Phys. Rev. Lett.* **2011**, *107*, 131301, doi:10.1103/PhysRevLett.107.131301. [CrossRef] [PubMed]
17. Peña, I.; Sudarsky, D. On the Possibility of Measuring the Unruh Effect. *Found. Phys.* **2014**, *44*, 689, doi:10.1007/s10701-014-9806-0. [CrossRef]
18. Cozzella, G.; Landulfo, A.G.S.; Matsas, G.E.A.; Vanzella, D.A.T. Proposal for Observing the Unruh Effect using Classical Electrodynamics. *Phys. Rev. Lett.* **2017**, *118*, 161102, doi:10.1103/PhysRevLett.118.161102. [CrossRef] [PubMed]
19. Matsas, G.E.A.; Vanzella, D.A.T. The Fulling-Davies-Unruh effect is mandatory: The proton's testimony. *Int. J. Mod. Phys. D* **2002**, *11*, 1573, doi:10.1142/S0218271802002918. [CrossRef]
20. Audretsch, J.; Müller, R. Spontaneous excitation of an accelerated atom: The contributions of vacuum fluctuations and radiation reaction. *Phys. Rev. A* **1994**, *50*, 1755, doi:10.1103/PhysRevA.50.1755. [CrossRef] [PubMed]
21. Audretsch, J.; Müller, R. Radiative energy shifts of an accelerated two-level system. *Phys. Rev. A* **1995**, *52*, 629, doi:10.1103/PhysRevA.52.629. [CrossRef] [PubMed]
22. Passante, R. Radiative level shifts of an accelerated hydrogen atom and the Unruh effect in quantum electrodynamics. *Phys. Rev. A* **1998**, *57*, 1590, doi:10.1103/PhysRevA.57.1590. [CrossRef]
23. Zhu, Z.; Yu, H.; Lu, S. Spontaneous excitation of an accelerated hydrogen atom coupled with electromagnetic vacuum fluctuations. *Phys. Rev. D* **2006**, *73*, 107501, doi:10.1103/PhysRevD.73.107501. [CrossRef]
24. Rizzuto, L.; Spagnolo, S. Energy-level shifts of a uniformly accelerated atom between two reflecting plates. *Phys. Scr.* **2011**, *T143*, 014021, doi:10.1088/0031-8949/2011/T143/014021. [CrossRef]
25. Rizzuto, L.; Lattuca, M.; Marino, J.; Noto, A.; Spagnolo, S.; Zhou, W.; Passante, R. Nonthermal effects of acceleration in the resonance interaction between two uniformly accelerated atoms. *Phys. Rev. A* **2016**, *94*, 012121, doi:10.1103/PhysRevA.94.012121. [CrossRef]
26. Lattuca, M.; Marino, J.; Noto, A.; Passante, R.; Rizzuto, L.; Spagnolo, S.; Zhou, W. Van der Waals and resonance interactions between accelerated atoms in vacuum and the Unruh effect. *J. Phys. Conf. Ser.* **2017**, *880*, 012042, doi:10.1088/1742-6596/880/1/012042. [CrossRef]
27. Yu, H.; Zhu, Z. Spontaneous absorption of an accelerated hydrogen atom near a conducting plane in vacuum. *Phys. Rev. D* **2006**, *74*, 044032, doi:10.1103/PhysRevD.74.044032. [CrossRef]
28. Zhu, Z.; Yu, H. Fulling-Davies-Unruh effect and spontaneous excitation of an accelerated atom interacting with a quantum scalar field. *Phys. Lett. B* **2007**, *645*, 459, doi:10.1016/j.physletb.2006.12.068. [CrossRef]
29. Zhou, W.; Yu, H. Spontaneous excitation of a uniformly accelerated atom coupled to vacuum Dirac field fluctuations. *Phys. Rev. A* **2012**, *86*, 033841, doi:10.1103/PhysRevA.86.033841. [CrossRef]
30. Rizzuto, L. Casimir-Polder interaction between an accelerated two-level system and an infinite plate. *Phys. Rev. A* **2007**, *76*, 062114, doi:10.1103/PhysRevA.76.062114. [CrossRef]
31. Rizzuto, L.; Spagnolo, S. Lamb shift of a uniformly accelerated hydrogen atom in the presence of a conducting plate. *Phys. Rev. A* **2009**, *79*, 062110, doi:10.1103/PhysRevA.79.062110. [CrossRef]

32. Rizzuto, L.; Spagnolo, S. Energy level shifts of a uniformly accelerated atom in the presence of boundary conditions. *J. Phys. Conf. Ser.* **2009**, *161*, 012031, doi:10.1088/1742-6596/161/1/012031. [CrossRef]
33. Zhu, Z.; Yu, H. Position-dependent energy-level shifts of an accelerated atom in the presence of a boundary. *Phys. Rev. A* **2010**, *82*, 042108, doi:10.1103/PhysRevA.82.042108. [CrossRef]
34. She, W.; Yu, H.; Zhu, Z. Casimir-Polder interaction between an atom and an infinite boundary in a thermal bath. *Phys. Rev. A* **2010**, *81*, 012108, doi:10.1103/PhysRevA.81.012108. [CrossRef]
35. Noto, A.; Passante, R. Van der Waals interaction energy between two atoms moving with uniform acceleration. *Phys. Rev. D* **2013**, *88*, 025041, doi:10.1103/PhysRevD.88.025041. [CrossRef]
36. Marino, J.; Noto, A.; Passante, R. Thermal and Nonthermal Signatures of the Unruh Effect in Casimir-Polder Forces. *Phys. Rev. Lett.* **2014**, *113*, 020403, doi:10.1103/PhysRevLett.113.020403. [CrossRef] [PubMed]
37. Antezza, M.; Braggio, C.; Carugno, G.; Noto, A.; Passante, R.; Rizzuto, L.; Ruoso, G.; Spagnolo, S. Optomechanical Rydberg-atom excitation via dynamic Casimir-Polder coupling. *Phys. Rev. Lett.* **2014**, *113*, 023601, doi:10.1103/PhysRevLett.113.023601. [CrossRef] [PubMed]
38. Bagarello, F.; Lattuca, M.; Passante, R.; Rizzuto, L.; Spagnolo, S. Non-Hermitian Hamiltonian for a modulated Jaynes-Cummings model with $\mathcal{PT}$ symmetry. *Phys. Rev. A* **2015**, *91*, 042134, doi:10.1103/PhysRevA.91.042134. [CrossRef]
39. Zhou, W.; Passante, R.; Rizzuto, L. Resonance interaction energy between two accelerated identical atoms in a coaccelerated frame and the Unruh effect. *Phys. Rev. D* **2016**, *94*, 105025, doi:10.1103/PhysRevD.94.105025. [CrossRef]
40. Power, E.A.; Thirunamachandran, T. Quantum electrodynamics in a cavity. *Phys. Rev. A* **1982**, *25*, 2473, doi:10.1103/PhysRevA.25.2473. [CrossRef]
41. Meschede, D.; Jhe, W.; Hinds, E.A. Radiative properties of atoms near a conducting plane: An old problem in a new light. *Phys. Rev. A* **1990**, *41*, 1587, doi:10.1103/PhysRevA.41.1587. [CrossRef] [PubMed]
42. Spagnolo, S.; Passante, R.; Rizzuto, L. Field fluctuations near a conducting plate and Casimir-Polder forces in the presence of boundary conditions. *Phys. Rev. A* **2006**, *73*, 062117, doi:10.1103/PhysRevA.73.062117. [CrossRef]
43. Passante, R.; Spagnolo, S. Casimir-Polder interatomic potential between two atoms at finite temperature and in the presence of boundary conditions. *Phys. Rev. A* **2007**, *76*, 042112, doi:10.1103/PhysRevA.76.042112. [CrossRef]
44. Zhou, W.; Rizzuto, L.; Passante, R. Vacuum fluctuations and radiation reaction contributions to the resonance dipole-dipole interaction between two atoms near a reflecting boundary. *Phys. Rev. A* **2018**, *97*, 042503, doi:10.1103/PhysRevA.97.042503. [CrossRef]
45. Palacino, R.; Passante, R.; Rizzuto, R.; Barcellona, P.; Buhmann, S.Y. Tuning the collective decay of two entangled emitters by means of a nearby surface. *J. Phys. B At. Mol. Opt. Phys.* **2017**, *50*, 154001, doi:10.1088/1361-6455/aa75f4. [CrossRef]
46. Casimir, H.B.G.; Polder, D. The Influence of Retardation on the London-van der Waals Forces. *Phys. Rev.* **1948**, *73*, 360, doi:10.1103/PhysRev.73.360. [CrossRef]
47. Salam, A. *Molecular Quantum Electrodynamics: Long-Range Intermolecular Interactions*; Wiley: Hoboken, NJ, USA, 2010.
48. Compagno, G.; Passante, R.; Persico, F. *Atom-Field Interactions and Dressed Atoms*; Cambridge University Press: Cambridge, UK, 1995.
49. Rizzuto, L.; Passante, R.; Persico, F. Nonlocal Properties of Dynamical Three-Body Casimir-Polder Forces. *Phys. Rev. Lett.* **2007**, *98*, 240404, doi:10.1103/PhysRevLett.98.240404. [CrossRef] [PubMed]
50. Rizzuto, L.; Passante, R.; Persico, F. Dynamical Casimir-Polder energy between an excited- and a ground-state atom. *Phys. Rev. A* **2004**, *70*, 012107, doi:10.1103/PhysRevA.70.012107. [CrossRef]
51. Berman, P.R. Interaction energy of nonidentical atoms. *Phys. Rev. A* **2015**, *91*, 042127, doi:10.1103/PhysRevA.91.042127. [CrossRef]
52. Donaire, M.; Guerout, R.; Lambrecht, A. Quasiresonant van der Waals Interaction between Nonidentical Atoms. *Phys. Rev. Lett.* **2015**, *115*, 033201, doi:10.1103/PhysRevLett.115.033201. [CrossRef] [PubMed]
53. Barcellona, P.; Passante, R.; Rizzuto, L.; Buhmann, S.Y. Van der Waals interactions between excited atoms in generic environments. *Phys. Rev. A* **2016**, *94*, 012705, doi:10.1103/PhysRevA.94.012705. [CrossRef]
54. Milonni, P.W.; Rafsanjani, S.M.H. Distance dependence of two-atom dipole interactions with one atom in an excited state. *Phys. Rev. A* **2015**, *92*, 062711, doi:10.1103/PhysRevA.92.062711. [CrossRef]

55. Förster, T. *Modern Quantum Chemistry*; Academic: New York, NY, USA, 1965.
56. Juzeliūnas, G.; Andrews, D.L. Quantum Electrodynamics of Resonance Energy Transfer. *Adv. Chem. Phys.* **2000**, *112*, 357, doi:10.1002/9780470141717.ch4. [CrossRef]
57. Kurizki, G.; Kofman, A.G.; Yudson, V. Resonant photon exchange by atom pairs in high-Q cavities. *Phys. Rev. A* **1996**, *53*, R35, doi:10.1103/PhysRevA.53.R35. [CrossRef] [PubMed]
58. Agarwal, G.S.; Gupta, S.D. Microcavity-induced modification of the dipole-dipole interaction. *Phys. Rev. A* **1998**, *57*, 667, doi:10.1103/PhysRevA.57.667. [CrossRef]
59. Shahmoon, E.; Kurizki, G. Nonradiative interaction and entanglement between distant atoms. *Phys. Rev. A* **2013**, *87*, 033831, doi:10.1103/PhysRevA.87.033831. [CrossRef]
60. Incardone, R.; Fukuta, T.; Tanaka, S.; Petrosky, T.; Rizzuto, L.; Passante, R. Enhanced resonant force between two entangled identical atoms in a photonic crystal. *Phys. Rev. A* **2014**, *89*, 062117, doi:10.1103/PhysRevA.89.062117. [CrossRef]
61. Notararigo, V.; Passante, R.; Rizzuto, L. Resonance interaction energy between two entangled atoms in a photonic bandgap environment. *Sci. Rep.* **2018**, *8*, 5193, doi:10.1038/s41598-018-23416-0. [CrossRef] [PubMed]
62. Dalibard, J.; Dupont-Roc, J.; Cohen-Tannoudji, C. Vacuum fluctuations and radiation reaction: Identification of their respective contributions. *J. Phys.* **1982**, *43*, 1617, doi:10.1051/jphys:0198200430110161700. [CrossRef]
63. Dalibard, J.; Dupont-Roc, J.; Cohen-Tannoudji, C. Dynamics of a small system coupled to a reservoir: Reservoir fluctuations and self-reaction. *J. Phys.* **1984**, *45*, 637, doi:10.1051/jphys:01984004504063700. [CrossRef]
64. Menezes, G.; Svaiter, N.F. Radiative processes of uniformly accelerated entangled atoms. *Phys. Rev. A* **2016**, *93*, 052117, doi:10.1103/PhysRevA.93.052117: [CrossRef]
65. Menezes, G.; Svaiter, N.F. Vacuum fluctuations and radiation reaction in radiative processes of entangled states. *Phys. Rev. A* **2015**, *92*, 062131, doi:10.1103/PhysRevA.92.062131. [CrossRef]
66. Zhou, W.; Yu, H. Boundarylike behaviors of the resonance interatomic energy in a cosmic string spacetime. *Phys. Rev. D* **2018**, *97*, 045007, doi:10.1103/PhysRevD.97.045007. [CrossRef]
67. Craig, D.P.; Thirunamachandran, T. *Molecular Quantum Electrodynamics*; Dover Publ.: Mineola, NY, USA, 1998.
68. Takagi, S. Vacuum Noise and Stress Induced by Uniform Acceleration: Hawking-Unruh Effect in Rindler Manifold of Arbitrary Dimension. *Prog. Theor. Phys. Suppl.* **1988**, *88*, 1, doi:10.1143/PTPS.88.1. [CrossRef]

© 2018 by the authors. Licensee MDPI, Basel, Switzerland. This article is an open access article distributed under the terms and conditions of the Creative Commons Attribution (CC BY) license (http://creativecommons.org/licenses/by/4.0/).

MDPI
St. Alban-Anlage 66
4052 Basel
Switzerland
Tel. +41 61 683 77 34
Fax +41 61 302 89 18
www.mdpi.com

Symmetry Editorial Office
E-mail: symmetry@mdpi.com
www.mdpi.com/journal/symmetry

www.ingramcontent.com/pod-product-compliance
Lightning Source LLC
LaVergne TN
LVHW071953080526
838202LV00064B/6740